Michael F. Roberts
Anne E. Kruchten

Receptor Biology

Michael F. Roberts and Anne E. Kruchten

Receptor Biology

WILEY-VCH
Verlag GmbH & Co. KGaA

Authors

Michael F. Roberts
Linfield College
Biology Department
McMinnville
97128 Murdock 216 OR
United States

Anne E. Kruchten
Linfield College
Biology Department
900 SE Baker Street
97128 McMinnville OR
United States

Cover
Frontcover picture: © Getty Images, ID 470751895

Library of Congress Card No.: applied for

British Library Cataloguing-in-Publication Data
A catalogue record for this book is available from the British Library.

Bibliographic information published by the Deutsche Nationalbibliothek
The Deutsche Nationalbibliothek lists this publication in the Deutsche Nationalbibliografie; detailed bibliographic data are available on the Internet at <http://dnb.d-nb.de>.

Print ISBN: 978-3-527-33726-2
ePDF ISBN: 978-3-527-80015-5
ePub ISBN: 978-3-527-80017-9
Mobi ISBN: 978-3-527-80016-2

Typesetting SPi Global, Chennai, India
Printing and Binding Markono Print Media Pte Ltd, Singapore

Printed on acid-free paper

To our mentors: Warren Porter, University of Wisconsin – Madison and David Bernlohr, University of Minnesota.

To our families:

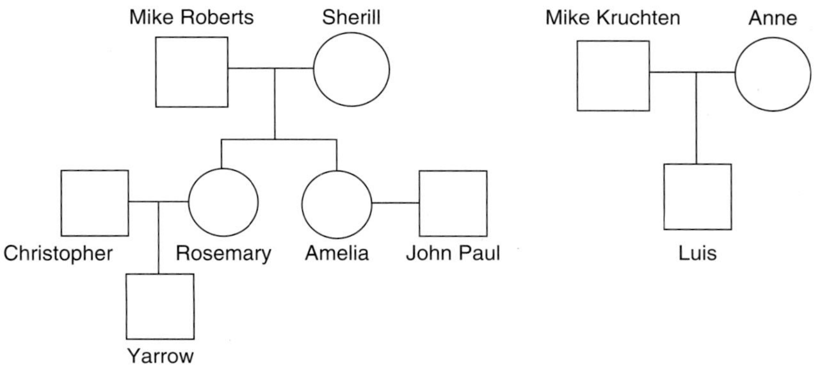

Contents

Acknowledgments

Drs Kent Thornburg and George Olsen (Oregon Health and Sciences University) devoted much time and thought to an early version of the manuscript, and made valuable comments at all levels. Dr Paul Kolenbrander (National Institutes of Health) provided valuable insights regarding bacterial signaling. Drs Christian Burvenich and Eddy Roets (University of Ghent) and Erik Raman (University of Antwerp) were valued colleagues and mentors in receptor pharmacology during MR's sabbatical research in Belgium. Numerous colleagues at the Mayo Clinic were helpful mentors in cancer biology and receptor signaling during AK's postdoctoral fellowship.

We also thank our Linfield students and colleagues as sources of assistance and stimulation. John Syring gave a valuable critical reading of the chapter on receptor evolution. Linfield students Chelsey Nieman, John Frank, Bonnie Hastings, Eric Lemieux, Chelan Guischer, Jacob Priester, Christine Lewis, and Henry Simons gave valuable suggestions and editorial assistance. Lige Armstrong of the Linfield Library Faculty Development Laboratory, provided assistance with illustrations. In addition, Dr Miranda Byse (Linfield graduate) read parts of the manuscript and worked with MR on signaling experiments.

We are also pleased to acknowledge the scientific and editorial assistance of the editors at Wiley-Blackwell, especially Dr Gregor Cicchetti, Ms Anne DuGuerny, and Ms Stefanie Volk.

Finally, other members of the Biology Department at Linfield College made the thinking and writing process especially enjoyable, and we thank them for their collegiality and conversations concerning the book.

"Drinking coffee with people cleverer than oneself is not a waste of time, but one of the best ways of expanding horizons."
David Colquhoun, 2006

Part I
Introduction

1
Introduction

> The beauty of reductionism is that it gives you something to do next.
>
> *Steve Jones [1]*

Biological processes require communication between cells and between individuals. In all kinds of living organisms, this communication begins at the molecular level. Small **signaling molecules** (proteins, amino acids, steroids, and other substances) are the messages that pass from one cell to the next; large protein **receptors** are the receivers of the message. Receptors bind the smaller molecules much as a lock receives a key or a glove receives a hand [2]. Other proteins in the cell membrane associated with the receptors convey the message to the interior of the cell.

Very few biochemical or physiological functions in our bodies are not somehow touched by these molecules or by the process of cellular communication. Here are some examples of how receptors are involved in a variety of biological processes:

- Sperm and egg meet, recognize each other, and bind by a receptor mechanism.
- Embryos develop by cell communication: one cell releases a **hormone** that binds to a receptor on another cell, and the second cell changes its shape and function, initiating the process of **differentiation.**
- Hormone-like **neurotransmitters** are released from one cell (a nerve) and bind to receptors on the surface of a nearby cell (another nerve or a muscle) to cause thought or movement.
- The digestive system propels food and releases enzymes according to the binding of hormones to cells lining the digestive tract.
- Immune system cells contain on their surfaces receptors that are able to recognize foreign proteins and attack invading cells.
- Diseases often act by subverting normal receptor function.

This introductory chapter covers general concepts of communication and how chemical communication compares with human communication; how evolution applies to receptor molecules; and how a pure chemical entity such as a receptor can initiate such large-scale functions as thought.

1.1
Receptors and Signaling

1.1.1
General Aspects of Signaling

Signaling is the means by which a cell knows what is happening in its surroundings, and is also the method by which one cell instructs nearby cells to alter their behavior. Organismal cell signaling involves molecular interactions, but the biological mechanisms of signaling are analogous to the ones humans use for verbal communication.

1.1.2
Verbal and Physiological Signals

Any sort of signaling requires that the sender and receiver are capable of interpreting the signals in the same way [3]:

- The sender must relay a characteristic signal, and it must be received by a characteristic device;
- The signal is *arbitrary:* it bears no real relation to the process it starts but is simply a way of obtaining a response in the receiver;
- The signal is *simpler* than the process it sets in motion.

These rules are easily understood in terms of human communication:

- The signals are the words of the language, and the receiver is the hearing/thinking/acting apparatus of another person;

Receptor Biology, First Edition. Michael F. Roberts and Anne E. Kruchten.

- Each language uses different words, yet all people can express the same thoughts.
- Any word (e.g., HELP) evokes in its hearer a set of thoughts or behaviors that are much more complex than the word itself.

The units of cellular communication abide by these same rules:

- The correct signal is the **drug** or hormone, the correct receiver is the **cell surface receptor** or **nuclear receptor**.
- It is arbitrary that one amino acid (e.g., glutamic acid) is an excitatory **transmitter** in the nervous system, whereas another amino acid (e.g., glycine) is an inhibitory transmitter.
- The binding of a single transmitter molecule to its receptor is adequate to start a cascade of intracellular events that amplifies the signal into a complex biochemical response.

In addition to these constraints, three more generally apply to biological communication:

- The receptor must be present on the correct **tissue**, it must be **selective** or specific to the hormone, and the receptor must *not* be present in tissues where the response is not desired; thus, the timing of the message must be coordinated with the presence of the receptor for that message.
- The signal must always mean the same thing to a particular receptor–**effector** mechanism.
- Some transmitters act on more than one type of receptor, often activating antagonistic cellular processes.

The analogies drawn between human communication and chemical communication are symbolic, yet the correspondence between the two systems is being strengthened as we find more instances where human interactions are being found to be at least partly chemical (e.g., the importance of **pheromone**-like substances in human behavior [4]).

1.1.3 Criteria for Recognizing Transmitters and Receptors

This book refers to signaling molecules in several ways. The most general term is **ligand**, which means any molecule that binds to a receptor. A ligand that activates its receptor is called an **agonist**. Hormones, transmitters, and pheromones are all agonists, and are naturally produced by organisms for signaling.

1.1.4 Agonists

The substances that serve as agonists are often also important as metabolic molecules within the cell. Thus, simply showing that a cell produces acetylcholine, for example, does not demonstrate its role as a transmitter. For a substance to be accepted as a specific transmitter or hormone, it must be shown to: [5]

- be synthesized, stored, and released from the proper type of cell (e.g., neuron or **endocrine** cell);
- have a specific mechanism for removal from the extracellular space near the target cell;
- be effective as an agonist if added to the target cell by experimenters.

1.1.5 Receptors

Cells can be activated by processes other than receptor mechanisms. To be accepted as a receptor mechanism, a process must be shown to [6]

- be activated by one or only a few substances;
- bind these substances with high **affinity**;
- be able to transmit the binding event to the cell interior.

These criteria for identifying receptors are not just for convenience; each has its basis in receptor structure, and later chapters show how these criteria are derived from, and actually define, the molecular mechanisms by which receptors operate.

1.1.6 Receptor–Enzyme Similarities

Enzymes are familiar proteins: they have **active sites** at which small **substrate** molecules bind and are converted to **products**. The relation between a receptor and its agonist is quite similar, at least at the binding step, to the action of enzymes: the receptor binds the agonist with high **affinity** because of the match between the shape and electric charge distribution of both molecules. The act of binding alters the shape of the receptor at another location; this change in shape is transmitted to other cellular proteins, thus stimulating further cellular processes.

As useful as the enzyme analogy is, however, enzyme action is unlike the receptor mechanism in some ways:

Table 1.1 Locations and properties of the four receptor superfamilies.

	Ion channel receptors	G-protein-coupled receptors	Receptor tyrosine kinases	Nuclear receptors
Location	Plasma membrane	Plasma membrane	Plasma membrane	Nucleus
Effector	Ion channel	Enzyme or ion channel	Enzyme	Regulation of gene action
Time scale	Milliseconds–seconds	Seconds–minutes	Minutes–hours	Hours–days
Examples	Nicotinic receptors	Adrenoceptors	Insulin receptors	Steroid receptors

- A receptor-binding event has no "product" because the agonist is unaltered by its interaction with the receptor.
- The receptor–agonist complex has an additional role after binding: the conversion of the binding signal to an intracellular event, such as enzyme activation or gene **transcription**.

Enzymes are important intracellular biochemical regulators; receptors are important regulators at the interface of the cell. Because of this location, they have a crucial role as molecular guardians, controlling the initial encounters between cells and chemicals in their environments.

1.2 Types of Receptors and Hormones

1.2.1 Receptor Superfamilies

A protein superfamily is a group of proteins that share structure, sequence, and functional features suggesting they are derived from the same common ancestral protein [5]. At present, researchers recognize four large superfamilies[1] of receptors: three reside in the cell membrane and one remains within the cytoplasm of the cell. Almost a thousand types of cell surface receptors belong to a single superfamily, the G-protein-coupled receptors. Their DNA thus comprises about 5% of all human genes. Another large superfamily of receptors is the fast ion channels that mediate neurotransmission in the central nervous system and skeletal muscles. A small superfamily, the receptor kinases, mediates metabolic, developmental, and immunological processes (Chapter 8). Also present in small quantities in the cytoplasm of the cell are the nuclear receptors that control transcription of new proteins. Table 1.1 summarizes the properties of the four superfamilies or receptors [7].

These four types are easily distinguished by shape, they each have characteristic agonists, and each causes characteristic intracellular changes. Figure 1.1 shows general structures of the four superfamilies.

The first superfamily (Chapter 6) consists of **ion channels** such as the nerve Na^+ channel that is activated by acetylcholine. These receptors consist of several protein chains held together in a ring. Each protein has four transmembrane regions. Together the separate chains form the pore through which the Na^+ ion moves when the agonist binds. The inward flux of Na^+ **depolarizes** the cell, causing it to generate an action potential.

The second superfamily (Chapter 7) consists of receptors such as the one for the neurotransmitter **norepinephrine** (NE) on heart muscle cells. This receptor has seven transmembrane regions, meaning the single receptor molecule passes through the cell membrane seven times and has both intracellular and extracellular regions. When a transmitter such as NE binds, it causes the receptor to activate a multiprotein assemblage in the membrane that produces an intracellular **second messenger** (such as **cyclic adenosine monophosphate (cAMP)**) that activates the cell by altering the level of phosphorylation of cellular enzymes. In the heart muscle, one effect of NE is to increase the strength of the heartbeat.

The third superfamily (Chapter 8) consists of the **receptor kinases**, growth factor receptors for substances such as the proteins insulin and **epidermal growth factor**. These receptors have a single transmembrane region, and their cytoplasmic end is an enzyme – a kinase. The binding of the hormone to the outer portion activates the kinase to phosphorylate cellular enzymes that regulate nutrient transport and cell division.

The fourth superfamily (Chapter 9) consists of the **intracellular receptors**, the proteins that bind

1 Several terms are used in receptor literature to denote classes of receptors: *superfamily, family, class, group, and clan*, often inconsistently. We use the term *superfamily* to refer only to the four major groups of receptors, and use the other terms in order: within each superfamily are found *families* (e.g., the several types of ion channels); within each family are found *classes* (e.g., the types of Ca^{2+} channel). *Group* will be used informally and sparingly, *clan* not at all.

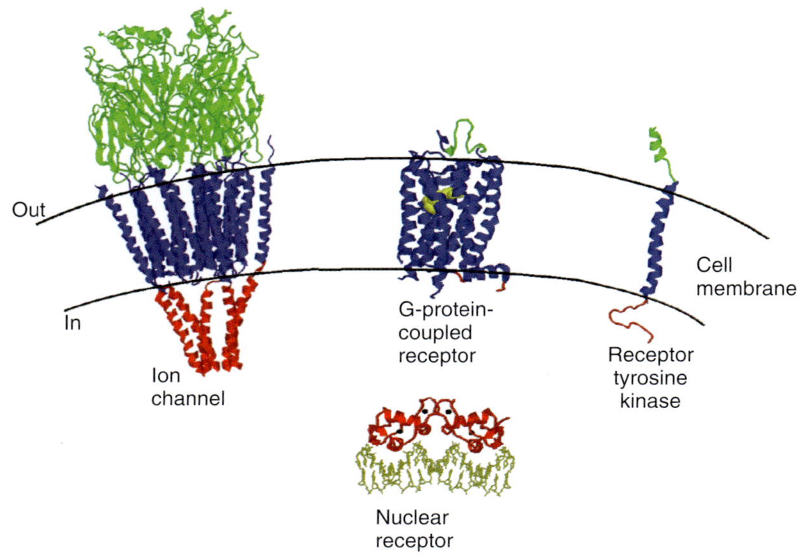

Figure 1.1 Overview of the four major receptor types. (A) Ion channel with extracellular **domain** labeled in red and transmembrane chains labeled in green; (B) G-protein-coupled receptor with extracellular domains labeled in red, seven transmembrane domains labeled in green; (C) receptor tyrosine kinase with transmembrane domain labeled in green, extracellular regions in red, and intracellular regions in blue. Black lines represent lipid bilayer; and (D) nucleus (dashed line) with nuclear receptors labeled in blue dimerizing on a DNA template labeled in black. Images were created using Rasmol [8] from PDB ID [9], PDB ID 1F88 [10], PDB ID [11], and PDB ID [12].

steroids, thyroid hormone, and certain vitamins. These fat-soluble ligands diffuse through the cell membrane to the interior of the cell, where they bind to and activate receptor proteins that enhance synthesis of new proteins within the cell.

Receptors are involved in cellular processes, such as metabolism and ionic changes, as well as cell division, growth, and protein synthesis. This book also covers receptor actions at a higher, organismal level. Embryonic development (Chapter 11), disease (Chapter 12), and the activities of the mind (Chapter 13) all involve integration of many physiological systems, all bound by the same receptor process as cell–cell communication.

All classes of receptors are encoded by genes within each cell. Genes for receptors are also subject to mutation and evolve by natural selection. As a consequence, receptors will change over time, allowing us to draw evolutionary inferences from the present phylogenetic distribution of genes for families of receptor molecules. The "fossil record" of proteins is thus found not in the rocks of the world but in the diversity of present-day organisms. The four superfamilies of receptors described are all widespread in eukaryotes. Some superfamilies are also present in prokaryotes, and the study of their distribution among all organisms (Chapter 14) gives researchers an understanding of their functions and role in organismal adaptations.

The relationships among protein families suggest that their genes have mutated, changed location, and duplicated many times, each time allowing the production of new protein molecules with similar functions. These similarities indicate further that protein function can change over time, and that new proteins with completely different functions can arise from gene mutations. This seems to be how some receptors arose, and how the families of receptors have changed.

1.3 Receptors Are the Chemical Expression of Reality

Because receptors are at the interface between cells and their environments, they are the first cellular units to receive environmental information and provide crucial information about the surroundings. For example, animals know that nighttime is the time to sleep, even though their brains have no way of directly sensing the light or dark. Visual information from the eyes goes to the pineal gland, which produces the hormone melatonin in inverse proportion to the amount of ambient light. Melatonin is therefore the *chemical expression of* darkness [13].

In an analogous manner, other hormone and receptor systems give information about the food taken in by organisms. Insulin is produced in the pancreas following a meal when blood glucose levels rise. Insulin is therefore the *chemical expression of plenty.* When food is scarce, the adrenal gland produces the steroid cortisol as a means of liberating glucose from storage forms in cells. Cortisol can be seen as the *chemical expression of starvation.* As melatonin, insulin, and cortisol all act on cellular receptors, we view receptor mechanisms as an important way that organisms have of knowing what reality is.

As the foregoing suggests, receptors are complex, as are their interactions with cellular processes. However, we hope that this complexity will be made comprehensible by the approach we are taking: the thousands of different receptors fall into only four fundamental superfamilies; each has a unique structure and a unique way of activating the cell, so it is possible to identify an unfamiliar receptor if one knows only a few things about it. Knowledge of receptor function illuminates the many interactions among proteins in the body and gives researchers important information on higher level functions of cell physiology (e.g., the normal workings of the mind or the aberrant interactions involved in disease states).

The book is divided into three parts: first is a general discussion of cell membranes, proteins, hormone types, and receptor theory.[2] Next follows one chapter on each of the four receptor types. Finally, several chapters outline receptor-mediated biological processes such as embryonic development, disease, mechanisms of the mind, and the evolution of these remarkable molecules.

Pharmacology texts generally focus on hormones and the kinetics of drug actions. We have written this book for students who wish to become more familiar with receptors themselves: the mechanisms by which they act, the sorts of processes they direct, and their evolution as molecules. It is meant for students at the advanced undergraduate and early graduate levels and requires an understanding of fundamental chemical and biological principles, a general knowledge of evolutionary thought, and a grasp of physiological interactions – all concepts that are part of any good general biology course. The text builds on these ideas to help students form a more complex understanding of pharmacology and cellular biochemistry.

Evolutionary inferences provide information that allows the study of receptors to be not only exciting and useful but also conceptually possible: despite the bewildering array of cell surface receptor types, they fall into just four major categories and interact with only a few dozen other membrane effector proteins that transmit the binding event into a biochemical process. Thus, genetic relationships among receptors are relatively simple, and their use of similar biochemical mechanisms shows that the important problems of cell-to-cell signaling have needed to be solved only a few times in evolution.

We wrote the book because in our teaching and research we see the importance of receptor mechanisms and intracellular signaling across all kingdoms of organisms and in many types of cellular processes. Even so, it is difficult to find a book that gives them complete coverage (structure, mechanism of action, evolutionary history) without being written specifically for professionals. The two unifying themes of the book are (i) the receptor concept itself: the idea that biological communication is involved in nearly all the activities of living things, and that receptor function is the mechanism of that communication and (ii) the role of natural selection and evolution in shaping receptor structure and function. We hope that this book will give a clear idea of the roles that hormones and their receptors play in our lives, from the reactions of individual cells to the behavior of whole organisms.

2 The term *theory* is often used by mistake in place of *hypothesis.* In proper usage, a theory is a hypothesis that has been tested and promoted to the level of widespread acceptance as a major concept in science. It is unfortunate that scientists themselves often misuse *theory* to mean *hypothesis,* as in "I have a theory about that" and non-scientists often pounce on this misuse to denigrate science, as in "evolution is only a theory." In this book, we restrict the use of the term *theory* to major scientific concepts, such as the *theory of evolution,* or *cell theory,* or *receptor theory.* All three of these ideas have been rigorously tested; they are no longer hypothetical, but have become key concepts in biological thinking. Other concepts, still provisional, are called *hypotheses.*

2
The Origins of Chemical Thinking

> A mystery is a phenomenon that people don't know how to think about – yet.
>
> *Daniel Dennett [14]*

2.1
Overview of Early Pharmacological History

2.1.1
The Development of a Chemical Hypothesis

The earliest Greek thinkers such as Thales (sixth century BCE) and Democritus (fourth century BCE) taught that life was material and that physical components of the environment were responsible for the organization of matter into living things. Thales also initiated an experimental approach to studying natural phenomena [15]. However, these early thinkers were unusual – the scholars who followed them had a non-material, non-experimental, non-molecular concept of the world. The *non-material* worldview promoted the idea that life processes were fundamentally different from processes in non-living systems. The *non-experimental* worldview encouraged the use of logic rather than the use of experiment to test ideas about natural phenomena. The *non-molecular* worldview is best seen by its two main hypotheses concerning the physical and biological spheres: the four "elements" (earth, air, fire, and water) and the "humoral" hypothesis (yellow bile, black bile, blood, and phlegm).

Plato (fourth century BCE) exemplified the non-material view, as he deemphasized observation and experiment, and claimed that our perceptions of matter were transitory and only what the mind perceived (via logic) was permanent [15]. Under his system, mind and body were considered to be separate entities – the senses give an inaccurate version of the world; only the mind provides "purity" of perception [16]. Aristotle's (fourth century BCE) and Galen's (second century CE) thinking opposed this attitude, as they appreciated the role of matter in life and advocated experimental approaches to the study of nature.

Descartes (seventeenth century CE), although a proponent of reason and experimentation, maintained that the body is a dual being, both mind and matter, and the workings of the mind are outside nature. He said that because *mechanism* describes non-human workings, then *other laws* must apply to human workings [17]. The concept of **mind–body dualism** of Descartes and others thus furthered non-material approaches to the study of life, and inhibited development of a systematic approach to the study of, among other things, the function of the brain.

However, chemical thinking did arise among some medieval thinkers: the physician Paracelsus in the sixteenth century was the first to take up the concepts of the earliest Greeks, teaching that the body was composed of chemicals and that illnesses were the result of chemical imbalances. He anticipated modern thinking in two of his teachings: that within a natural product the curative agent could be a single substance, and that the curative or poisonous functions of a drug were directly proportional to its concentration [18].

Felix Fontana in the late eighteenth century experimentally confirmed Paracelsus' view that a crude drug exerted its effect through a specific active principle that acted on a discrete tissue in the organism. His contemporary Peter Daries showed further that the effect was proportional to the concentration of drug applied. Setürner in the early nineteenth century was the first actually to isolate a pure drug when he obtained morphine from opium. This achievement initiated a period of rapid change: before many decades had passed, the chemical natures of many pharmacological substances were determined, and the new drug manuals, or **pharmacopoeias**, were based on pure substances rather than on crude plant extracts. Organic chemistry was emerging as

Receptor Biology, First Edition. Michael F. Roberts and Anne E. Kruchten.

a discipline in its own right, and the elucidation of chemical structure soon led to the development of drugs from synthetic sources rather than natural ones.

2.1.2 Chemical Structure and Drug Action

The three principal researchers in this area in the early-to-middle nineteenth century were François Magendie and his students Claude Bernard and James Blake. Magendie was the author of the first modern physiology textbook [19], initiated the study of experimental pharmacology, and was the first to show that drugs act on specific organs. Blake showed that related chemical compounds had differing effectiveness in activating tissues. From experiments with chemically related inorganic salts, Blake developed the concept that the biological effect of a substance depended on its structure and chemical nature. He found in general that the biological effects of different metallic ions fell into predictable patterns; these groupings based on biological activity actually predated Mendeleev's periodic chart of the elements, which was based on inorganic chemical combinations [20].

2.1.3 The Site of Drug Action

It is only in the last century or so that material, experimental, and chemical principles have been applied consistently to help us understand biological processes. The mechanisms of drug–tissue interactions were developed by Bernard [21], who made important discoveries with one particular tissue, the nerve–muscle junction in the skeletal muscle of the frog. He showed that curare, the South American Indian arrow poison, paralyzes muscles by acting on a structure in the junction (**synapse**) between the nerve and the muscle. We now know that curare binds to but does not activate the same receptor to which the neurotransmitter acetylcholine (ACh) binds. Curare thus antagonizes the transmitter action of ACh (Figure 2.1).

Bernard's experiments in the 1840s and 1850s were thus the earliest demonstrations of the existence of receptors. He also showed the existence of the nerve–muscle synapse and established that nerves and muscles were separate cells. It would be 50 years before any of these ideas would be developed theoretically and experimentally by others.

2.2 Modern Pharmacology

2.2.1 Langley and Ehrlich: the Origins of the Receptor Concept

John Langley was a student at Cambridge University in 1875 when he began studies of the actions of two **alkaloid**s, **atropine** and **pilocarpine**, on the heart. He found that pilocarpine slowed the heart and that atropine opposed the action of pilocarpine. Further, the way atropine worked suggested to him that it was acting at the same location as pilocarpine, and he proposed that both substances had affinity for the same site, for which both had binding affinity [22]. He did not at that time propose the molecular nature of the site, although it is clear that his thinking was going in that direction. Langley's research interests changed for a quarter century, and only after 1900 did he return to work on the receptor concept [23].

Paul Ehrlich also began his studies of specific binding processes in the 1870s. His initial work was with dyes used in histological staining of tissues for preparation of microscope slides. He proposed that the dyes used to stain cells did not simply adhere nonspecifically to the material but that they showed a chemical affinity for certain molecules in the cell. In succeeding years, he continued to lay the framework for the concept of a definite cell-surface molecule that acted in cell–cell communication [24]. He was by this time concerned with the genesis of the immune response; he discovered mast cells and proposed the currently used classification system for leukocytes based on cytoplasmic granulation [25]. Ehrlich also proposed that immune system cells have the ability to attract and bind foreign substances with specific chemical affinities to the immune cell surface [26].

Even though Ehrlich had developed the concept of specific binding for immune system activation, he borrowed the descriptive term "lock and key" from Fischer, who in 1894 had proposed a specific mechanism of binding for enzymes and substrates in biochemical reactions [27]. Even though the lock-in-key interaction is now better thought of as an "induced fit" interaction [2], Ehrlich's thinking about receptors was modern: he conceived of a receptor as having a special active site with a high affinity for the foreign chemical. He also anticipated the current immunological concept that when an immune system cell binds an **antigen**, receptors for the antigen are, as a result, manufactured in large numbers and act in the blood as binding agents for the antigen [26].

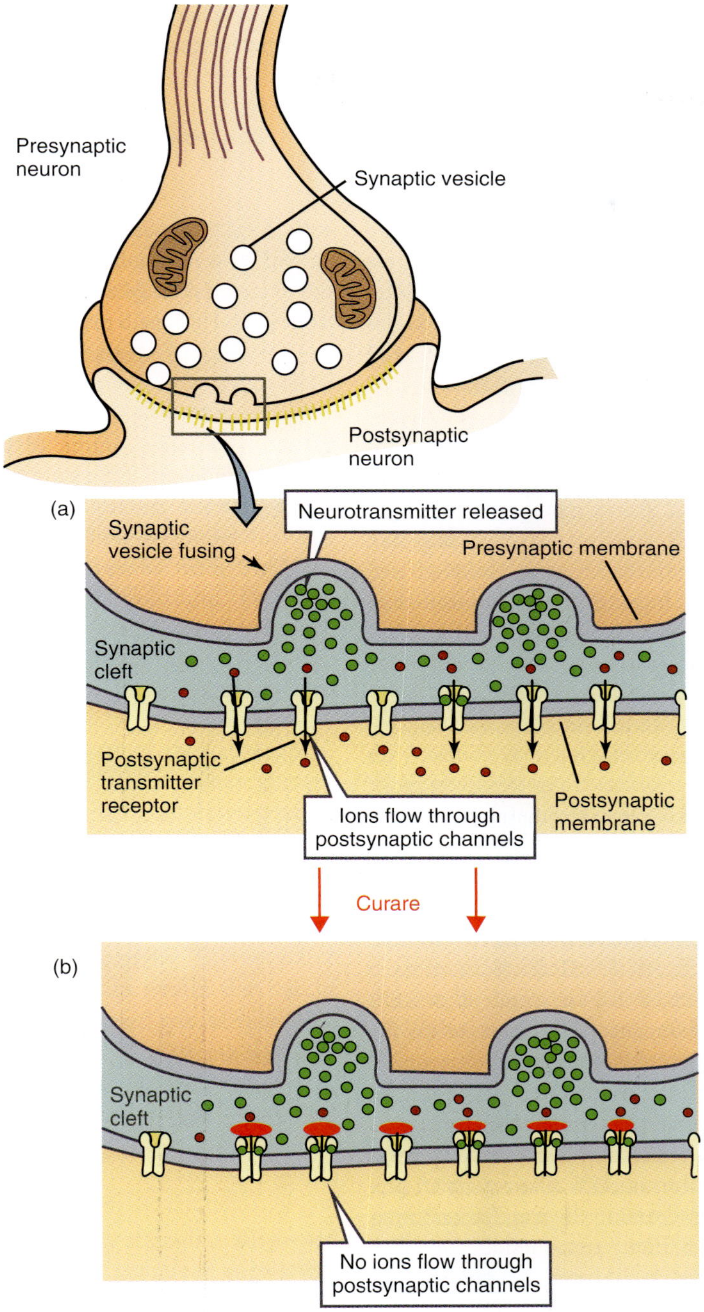

Figure 2.1 (a) Axonal ending of a neuron at a postsynaptic cell (nerve or skeletal muscle). The axon is not directly connected to the muscle cell, but is separated by a gap or synapse. The transmitter (green spheres) is synthesized in the ending within vesicles; it is released in response to an action potential. The transmitter diffuses to receptors on the postsynaptic cell. (b) An antagonist such as curare (red ovals) binds to the same receptor at the active site, preventing the transmitter from binding and thereby preventing the effects of nerve stimulation. http://kin450-neurophysiology.wikispaces.com/Synaptic+Transmission. Accessed 7 February 2015.

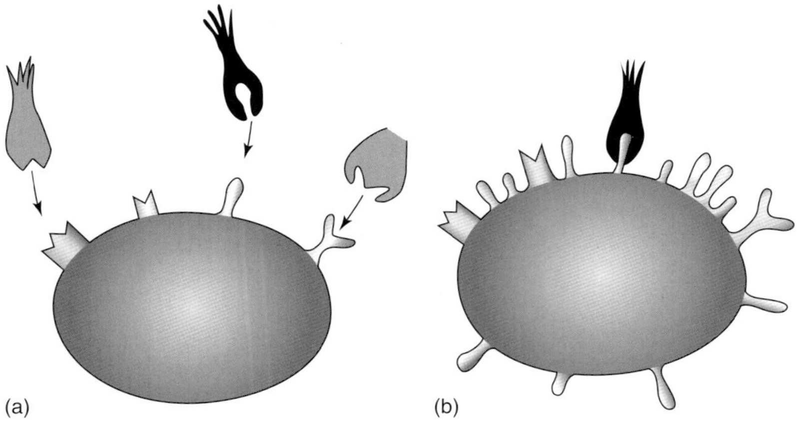

Figure 2.2 Ehrlich's representation of "side chains" representing his receptor concept. The fish-shaped objects binding to the chains are toxins or immunologic antigens. Here he shows a cell surface with a "side chain" projecting off the cell. This chain has a specific binding site for a toxin or antigen. Plate 6 in Ref. [23]. Used with permission.

Ehrlich had a great ability to see in his mind the structures of concepts he had developed as hypotheses, from chemical bonds and formulas to molecular receptors. In his early publications, he drew fanciful yet conceptually accurate pictures of his cell-surface receptors. Representative diagrams are shown in Figure 2.2.

However, Ehrlich did not immediately apply his side chain hypothesis, developed for the immune response, to the relation between drugs and their sites of action. Thus he initially saw drug–cell reactions as fundamentally different from immune reactions [23].

Langley finally returned to the concept of cell-surface receptive substances. Since the 1880s he had been studying the **autonomic nervous system**, which is the system that controls contraction of cardiac muscle in the heart and **smooth muscle** (e.g., in blood vessels) and glands. The system consists of **preganglionic** motor nerves originating in the brain or spinal cord and ending in ganglia (subunits of the nervous system that exist outside the brain and cord). In the ganglia, the preganglionic nerves synapse with **postganglionic** nerves that innervate the various internal organs. Langley was the first to demonstrate the presence of this synapse by showing that **nicotine**, which mimics the effects of ACh, activates postganglionic neurons *emerging from* the ganglia even after the nerves *to* the ganglia have been destroyed (Figure 2.3). He concluded that the effect of nicotine must therefore be on receptors of the **postsynaptic** nerve cells of the ganglia [28].

He also repeated Bernard's experiments on the effects of curare on the neuromuscular junction. He showed that nicotine added to a skeletal muscle can cause contraction even if the nerve has degenerated, and that curare antagonizes the effect of nicotine [29]. To explain the actions of these drugs on nerves and muscles, he had to formalize and make concrete his earlier (1870s) general proposal of a tissue site for drug action, and by 1905 he suggested that both nicotine and curare bind to the same structure on the muscle (Figure 2.1).

Because a muscle could be activated by direct electrical stimulation following a dose of curare that prevented **nicotinic** stimulation, Langley proposed that the cellular site for the actions of curare and nicotine was different from the site affected by direct electrical stimulation. He then called the binding site the "receptive substance" of the cell [29], and proposed that it somehow transmitted the binding information to the contractile mechanism. This is the first suggestion of a **signal transduction** mechanism for receptor events, an idea to which we return in Chapter 10.

In 1906, Langley proposed that nerves work by releasing a chemical substance that diffused from nerve to muscle or from nerve to nerve [30]. Also by 1906, Ehrlich became convinced by Langley's work that the receptive substances for drugs and for the immune system worked by the same mechanisms. In his 1907 work on the development of drug resistance by trypanosome parasites (single-celled organisms that cause sleeping sickness), he proposed that a receptor for the anti-trypanosome drug existed, and that resistance to the drug by the parasite involved a decrease in affinity of its receptor for the drug [23].

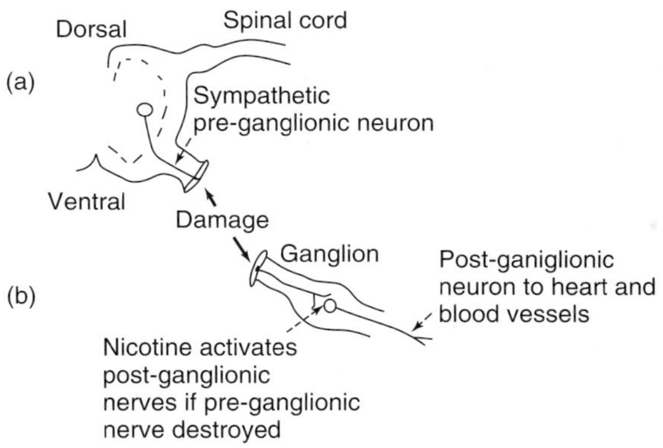

Figure 2.3 (a): Cross section of the spinal cord in the chest region; a sympathetic preganglionic nerve exits and synapses in the ganglion at (b) with a postganglionic neuron. Langley discovered the synapse at (b) by cutting the presynaptic nerve between (a) and (b), then adding nicotine to the ganglion.

2.2.2 Maturation of the Receptor Concept

Thus, by 1907, the receptor concept had matured in the minds of its creators Langley and Ehrlich to the point where it was applied to several biological actions: tissue staining; immune system responses; drug actions on nerves, muscles, and glands; and parasite responses to poisons. The origins of the modern concepts of binding affinity and signal transduction had also been proposed.

The relation between receptor binding and enzyme function also became clear through the work of Fischer, Michaelis, and Menten. Fischer [27] had earlier noted that binding of substrate to enzyme seemed to be similar to the binding of hormone to receptor, and Michaelis and Menten [31] formalized the mathematics of this idea into the current model of enzyme kinetics.

Elliott [32] was the first to suggest that a specific substance acted as a transmitter when he proposed epinephrine (EP) as the **sympathetic nervous system** activator. However, the first neurotransmitter to be identified chemically was ACh, isolated by Loewi [33] in 1921. In his 1936 Nobel Lecture, Loewi suggests that neurotransmitters such as EP and ACh act on "spots" that are "receptive" to the substances. Thus, even though the "receptor concept" was several decades old, in the 1920s receptors were not thought by all physiologists to be molecular structures; they were "virtual objects," postulated to exist as a means of making sense of pharmacological data, but with unknown biological nature, structure, and detailed function. Our understanding of the structure and nature of receptors had to await further developments in chemistry – knowledge of the forces involved in chemical binding, understanding of the molecules making up biological membranes, and development of techniques for sequencing and visualizing proteins.

2.3 Phylogenetics of Signaling

2.3.1 The First Communicators

As indicated earlier, transmitters and receptors were discovered in multicellular animals, and the concepts of communication were derived from the features of cell–cell signaling that are exhibited by these organisms. The idea that single-celled organisms could also communicate occurred to scientists only later; the earliest reports of bacterial and unicellular eukaryotic communication came in the mid-1900s [34].

Communication in single-celled organisms occurs at two levels: first, it is crucial for single-celled organisms to find food and to escape predation or noxious environments; second, some sort of interindividual communication is also adaptive between separate cells. Such communication acts as a "social behavior control mechanism," adaptive for managing activities of large numbers of cells in feeding, mating, defense, or aggregation. These responses are actually adaptive for the survival of the individual cells of the population.

Then, when multicellular organisms arose, they had in place a set of adaptations for environmental communication that also became adaptive for requirements of cellular organization. The individual

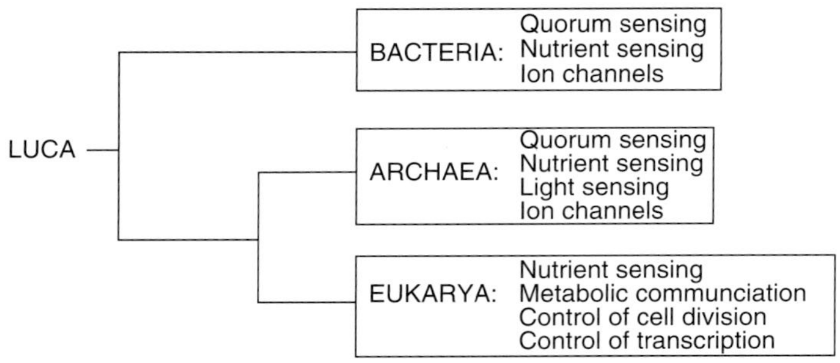

Figure 2.4 Phylogenetic diagram of the three domains of living organisms, with overall indication of the receptor types present in each and some functions of each. LUCA: "last universal common ancestor," the hypothesized cellular ancestor of living organisms. The Eukarya and Archaea are more closely related to each other than either is to the bacteria.

cells of an organism such as a plant or an animal are required to cooperate, or else the organism will not survive. Cooperation involves, among other things, giving up independent reproduction or predation on nearby cells. The evolution of multicellularity therefore uses the same mechanisms that function within communities of single-celled organisms.

These ideas suggest that chemoreception serves important biological functions in all organisms, and we can organize our treatment of receptor function both along conceptual lines and along molecular (specific receptor) lines. The original receptors were likely proteins of two sorts:

- Sensors for amino acids or sugars for transport into the cell. These proteins are widespread in bacteria and seem to be the phylogenetic basis for the transporter and ion channel group of receptors to be covered later in the book.
- Sensors for social aggregation signals; this process is called **quorum sensing**, and is a feature of all unicellular organisms, whether prokaryotic or eukaryotic. Although the actual receptor molecules are not always homologous between eukaryotes and prokaryotes, the functions of chemosensation are clearly adaptive and widespread. Transporters are covered in detail in Chapter 6, and quorum sensing is considered in Chapter 11.

Figure 2.4 illustrates a conceptual organization of the various processes and the receptors involved in each.

This figure gives an overview of the processes of reception in all domains of life. The Bacteria and Archaea use homologous mechanisms for quorum sensing, and also possess some receptor types that are homologs of eukaryotic receptors (ion channels). The Bacterial and Archaeal ion channels were thus likely passed on to the first eukaryotes when they arose, but the other receptor types are not common to all the Domains. Although the G-protein-coupled receptors are probably derived from Archaeal membrane proteins, true G-protein-coupled receptors do not exist in Archaea, and the receptor kinase and nuclear receptor group are also only found in eukaryotes.

The evolution of reception, of signaling molecules, and of receptors is therefore a complex and fascinating challenge, and the many receptors and signal molecules produce a confusing array of apparently unrelated molecules. Therefore, this book attempts to provide a framework for understanding these signaling processes by keeping a focus on the evolutionary process and on the evolutionary histories of organisms and their proteins.

Part II
Fundamentals

Introduction to the Cellular Processes of Receptor Action

The chemical and historical background of the previous chapters serves as a preparation for the following chapters that discuss the properties of cellular membranes and their proteins, the types of molecules that are used as signals, and a theoretical discussion of the ways these signaling molecules interact with receptors.

- Membranes and proteins – Chapter 3.
- Hormones and other first messengers – Chapter 4.
- Receptor theory – Chapter 5.

3
Membranes and Proteins

> The basic function of cell membranes is to separate the vital chemicals and structures needed to maintain the cell from the environment, ... membranes also contain structures that help regulate the types and amounts of molecules pulled from the environment.
>
> *Darrell Vodopich and Randy Moore [35]*

3.1
Membranes

3.1.1
The Cytoplasmic Membrane – the Importance of Cell Membranes

Membranes are the boundaries of cells. These boundaries are much like any other physical barriers in that they have two related and partially contradictory functions:

- Cellular processes can only occur if the components of biochemical pathways, enzymes, and substrates, are kept in close association. Membranes are barriers to the diffusion of biochemical substances out of the cell as well as to the entry of other materials that might disrupt cell function.
- Membranes cannot be absolute barriers otherwise the cell could not obtain substrates nor excrete waste products of metabolism. Thus, membranes must be selective, semipermeable barriers that maintain the dynamic integrity of cell structure and function (Figure 3.1).

Each group of membrane molecules has properties that promote the idea of dynamic integrity:

- **Hydrophobic** lipids (**phospholipids**) prevent entry of most foreign water-soluble substances and the loss of important water-soluble metabolites.
- **Integral membrane** proteins act as ion channels to regulate the flow of charged particles.
- **Glycoproteins** with carbohydrate side chains give identity to the cell.
- **Receptor** proteins bind to agonists, regulate cell-to-cell communication, and regulate cell behavior.
- Enzymes located in the membrane transmit hormonal signals to the interior of the cell.
- Structural proteins maintain the shape of the cell and maintain physical connections among the metabolic proteins of the membrane.
- **Cholesterol** controls fluidity of the membrane.

3.1.2
History of Membrane Models

Membranes are not visible under light microscopy, so until the advent of electron microscopy, our knowledge of their structure was based almost entirely on inferences from physiological studies. Once cells were suspected of being the fundamental units of life, the cell surface was thought to be a membrane of some type, but its structure was not known for nearly a century after that, and the detailed structures of cell membranes were only elucidated in the 1970s.

However, even in the mid-1800s, information accumulated to give an indication of many of the properties of cell membranes: Overton [36] found that cellular permeability of a substance was proportional to its lipid solubility. This suggested that the membrane was a lipid structure. However, the chemical nature of the lipids was not known at the time. Until the mid-1900s, membrane lipids were assumed to be triacylglycerides, or the normal fats found as components of fatty tissue. Phospholipids were known, but were not thought to be components of membranes.

Earliest models of the cell membrane therefore proposed that triacylglyceride lipids had enough of a **polar–nonpolar** structure to serve as membrane components. The fats were thought to be arrayed as a monolayer around the cell surface, because a polar lipid forms a monolayer when added to a water surface – the polar end bonds with the water and the nonpolar end projects into the air [37] (Figure 3.2).

In 1925, Gorter and Grendel [38] developed the **lipid bilayer** hypothesis to explain their observations

Receptor Biology, First Edition. Michael F. Roberts and Anne E. Kruchten.

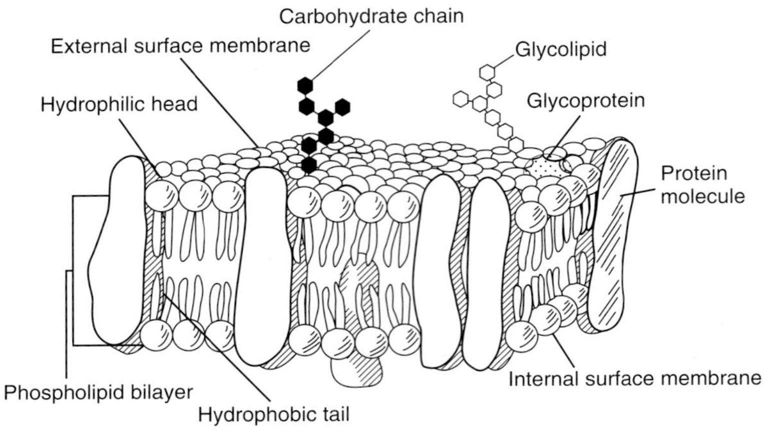

Figure 3.1 The fluid mosaic model of the cell membrane. The characteristics of this model are the phospholipid bilayer and the presence of globular proteins (large molecules) within and on the surface of the membrane. The portions of the proteins facing the extracellular and cytoplasmic solutions are hydrophilic; central parts are hydrophobic. Cholesterol molecules within the membrane are not shown here. https://s3.amazonaws.com/engrade-myfiles/4034298582543629/fluidmosaic.jpg. Accessed 7 February 2015.

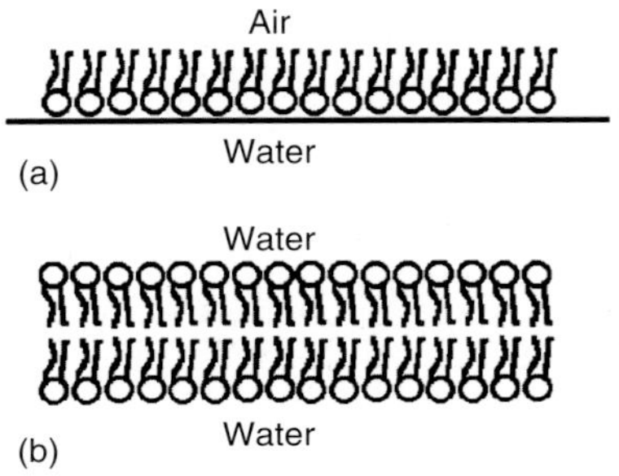

Figure 3.2 Spontaneous arrangements of phospholipids on water surface. (a) Monolayer of lipid on water surface in contact with air. (b) Lipid bilayer within the water environment.

on isolated lipids of erythrocytes. They removed the cellular contents of a known number of erythrocytes under hypo-osmotic conditions, and then arrayed the membrane components into a monolayer on an aqueous surface. They determined the surface area of the lipids on this surface and also computed the expected surface area of that number of cells. The flat area of the lipid was twice the area expected from that number of cells (Figure 3.2), suggesting that each cell contained two lipid layers rather than one. Later measurements showed that Gorter and Grendel made errors in their measurements and in their computations of membrane areas. However, their errors compensated for each other, leading to the correct conclusion. More recent experiments have confirmed their inference that polar lipids submerged in water form a bilayer, with the polar regions on the outside and the nonpolar tails facing each other [39].

3.1.2.1 The Roles of Proteins in Membranes

In the 1930s, Danielli and Davson [40] proposed in a series of papers that the properties of membranes were not simply explained by their lipids. They found that the surface tension of natural membranes was too low to be accounted for by a pure lipid bilayer, and proposed that proteins were also an important component of membranes. They found that adding solutions of proteins (e.g., hemoglobin and ovalbumin) to pure lipid membranes decreased surface tension because the proteins adsorbed to the surface of the lipid. They thus proposed that in natural membranes, proteins are superficial to the lipid, lying atop the outer surface on the extracellular side and beneath the inner surface on the cytoplasmic side.

This idea was consistent with evidence concerning cell permeability of different substances. In general, lipid solubility is a good predictor of rate of passage across a membrane. However, Danielli and Davson observed that some molecules were able to pass across a membrane at rates much higher than is predicted from their lipid solubilities: for example, water, urea, small alcohols, and gases such as CO, CO_2, and NO have permeabilities higher than their lipid solubilities suggest [41]. Danielli therefore proposed the concept of "active regions" on the cell membrane to explain the rapid movement of some substances across the membrane. He also suggested that **facilitated diffusion**, in which a substance may cross a membrane unexpectedly, is accomplished by a membrane-bound carrier protein. The Danielli–Davson model of cell membranes is shown in Figure 3.3.

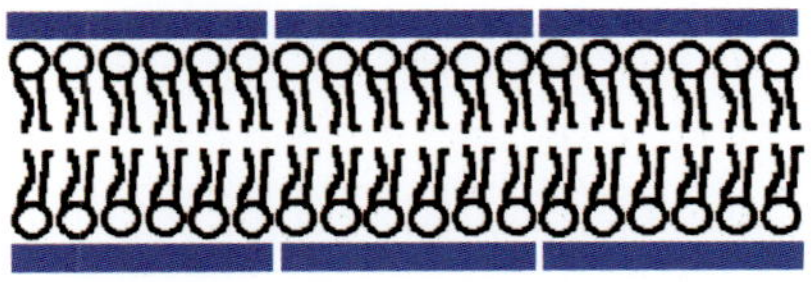

Figure 3.3 Danielli–Davson model of membrane lipids and proteins. The proteins (blue) were thought to be flat sheets lying on the lipid bilayer inside and outside the cell.

3.1.2.2 Challenges to the Danielli–Davson Model

However, Danielli and Davson's model did not explain several observations concerning the structure and behavior of membrane components: first, some membrane proteins can be easily removed in solutions of low salt content, while other proteins are much more tightly bound, and can be removed only following digestion of the membrane with detergents. These observations suggest the presence of at least two types of membrane proteins: one **superficial** (attached only loosely to the surface of the membrane) and another **integral** (located wholly or primarily within the plane of the membrane). Second, Frye and Edidin [39] showed that proteins can move laterally within the membrane, suggesting that it is unlikely that the outer proteins form rigid sheets. In addition, the Danielli model does not explain the observation that intact integral membrane proteins are not flat but are folded over a substantial part of their length into alpha helices (see section on proteins later in this chapter).

3.1.2.3 A New View of Membrane Proteins

Lenard and Singer [42] proposed that membrane proteins are also amphipathic, having a polar (**hydrophilic**) end and a nonpolar (hydrophobic) end, and are folded such that their hydrophobic portions are embedded in the membrane and their hydrophilic portions extend into the aqueous environments on both sides of the membrane (Figure 3.1). Studies of membrane proteins have indeed shown that cytoplasmic and extracellular domains are polar, while membrane domains are nonpolar.

This revised view of membrane proteins explained not only their relation to membranes but also the distinction between the water-soluble cytoplasmic proteins and the lipid-soluble membrane proteins. It also explained why some proteins, such as hemoglobin and albumin, could remain in solution *in vitro* or in the cytoplasm but others, such as the membrane proteins, were not soluble except in hydrophobic solvents.

3.1.2.4 The Modern Concept of Membranes – the Fluid Mosaic Model

The above observations were brought together into a comprehensive model of membrane structure by Singer and Nicolson [43] in 1972 (Figure 3.1). The name for this model reflects the authors' view that the membrane is a dynamic structure in which the components are free to move in the plane of the membrane, even while the structure as a whole is strong enough to give structural integrity to cells. This membrane model retains the lipid bilayer and the proteins, but places the proteins within the plane of the membrane rather than totally peripheral to it. Certain proteins are able to remain in the membrane if they have long stretches of non-polar amino acids that form hydrophobic bonds with the fatty acid chains of the lipid bilayer. X-ray diffraction studies and electron photomicrographs of membranes have confirmed the lipid bilayer structure and the integral peripheral protein model.

3.1.3 Membrane Components

A cell membrane is 9–12 nm thick, and consists of a bilayer of phospholipid molecules, proteins, **cholesterol**, glycoproteins, and glycolipids. Membranes also contain small quantities of water and metal ions. Their composition is about half lipid and half protein by weight.

3.1.3.1 Membrane Lipids

The lipids belong to three groups: phospholipids, cholesterol, and **sphingomyelin**. The main component of cell membranes is a type of polar lipid called a **phospholipid** , in which one of the nonpolar chains of a triacylglycerol molecule is replaced with a polar molecule such as a phosphate or a phosphate bound to another organic compound (Figure 3.4b).

Components of the Phospholipid Membrane: In the center of Figure 3.4b is the backbone of glycerol to which the lipid chains and the polar unit attach. In general, the phospholipids contain two different fatty acid groups, one **saturated** and the other **unsaturated**; the unsaturated chain is generally attached to the center carbon of the glycerol. The polar groups attached to the phosphate are most often structures based on choline: a **quaternary nitrogen** with one or more methyl groups. In other phospholipids, ethanolamine, serine, threonine, various forms of glycerol, or various forms of inositol make up the polar end of the molecule. The unsaturated

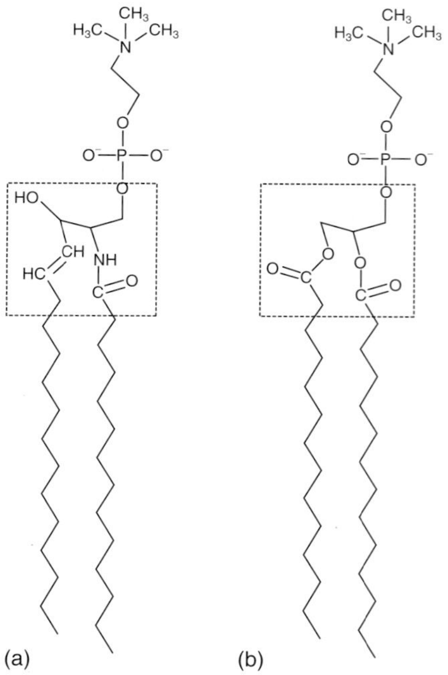

Figure 3.4 Molecular structures of (a) sphingomyelin, a **sphingolipid** and (b) phosphatidylcholine, a glycerol-based **phospholipid.** Both have phosphatidyl portions. The dotted squares enclose the head of the sphingosine group (a) and the glycerol group (b). Each vertex of the jagged lines represents a CH_2 unit of the fatty acid.

fatty acid forms a "kink" that prevents close packing of the fatty acid chains and keeps the membrane more fluid.

Sphingolipids: Some membrane lipids in eukaryotes are *sphingolipids*, based on a sphingosine structure rather than on glycerol [44] (Figure 3.4a). They have polar/nonpolar properties, but the fatty acid is not attached via an ester linkage; the polar portion may be a phosphate, hydroxyl, sugar, or polysaccharide. Sphingolipids are found primarily in the outer leaflet of the cell membrane, where they serve protective and regulatory roles [45].

Cholesterol: Cholesterol is not found in all membranes: it is common in vertebrates but absent in bacteria. It is important for maintaining fluidity of membranes, as it disrupts the tight hydrophobic bonds that form between fatty acid tails of adjacent lipids in the plane of the membrane. Its structure is shown in Figure 3.5.

3.1.3.2 Asymmetry and Heterogeneity in Membrane Lipids

The phospholipid molecules of the membrane are unevenly distributed. The extracellular surface of

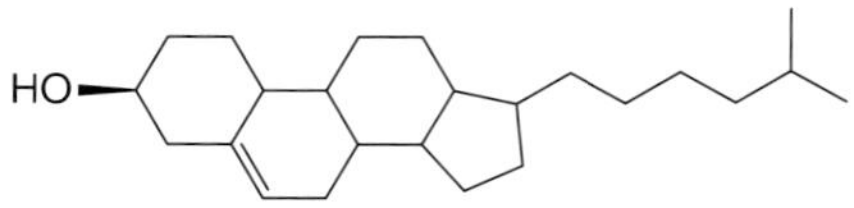

Figure 3.5 Molecular structure of cholesterol.

some membranes has primarily phosphatidyl choline and sphingomyelin; the inner surface has phosphatidyl ethanolamine, phosphatidyl serine, and **phosphatidyl inositol**. In at least one case, the function of this asymmetry is understood: phosphatidyl inositol on the inner membrane surface serves as a source for the production of some intracellular messengers, as is discussed in Chapter 7.

Even within one membrane surface, the composition of phospholipids differs from one region to another. In addition, the functions of membrane-bound enzymes and receptors are affected by changes in membrane fluidity that arise from changes in temperature. For example, the activation of the membrane enzyme adenylyl cyclase by the beta-**adrenoceptor** (Chapter 7) is markedly reduced below 24 °C, because at that temperature the inner portion of the lipid bilayer restricts mobility of inner membrane proteins, and receptor activation requires lateral movement in the plane of the membrane [46].

3.1.3.3 Membrane Construction and Insertion of Proteins

The cells of eukaryotes are more complex structurally than are those of bacteria, and contain several types of **organelles** that perform cellular functions. Most of these organelles are bounded by their own lipid bilayers; the basic structure of each of these bilayers is the same but each has a different protein composition. However, a functional relationship exists among some of the cell's membranes: the nuclear membrane is continuous with the endoplasmic reticulum (ER), the site of protein and fat synthesis. The Golgi body is a series of membranes that package proteins for release from the cytoplasm. These membranes form by budding off the ER; they then fuse with the plasma membrane of the cell surface during the processes of protein secretion and insertion of protein into the membrane. Thus, the nuclear membrane, ER, Golgi apparatus, and plasma membrane are continuous in a way that makes them parts of a functional and structural whole.

All proteins are synthesized on the ribosomes as amino acid sequences translated from RNA. In eukaryotes, most ribosomes are bound to the ER, and as the new peptide sequence is built, it threads its way across one ER membrane and into the space between the two layers of the ER. The placement of proteins

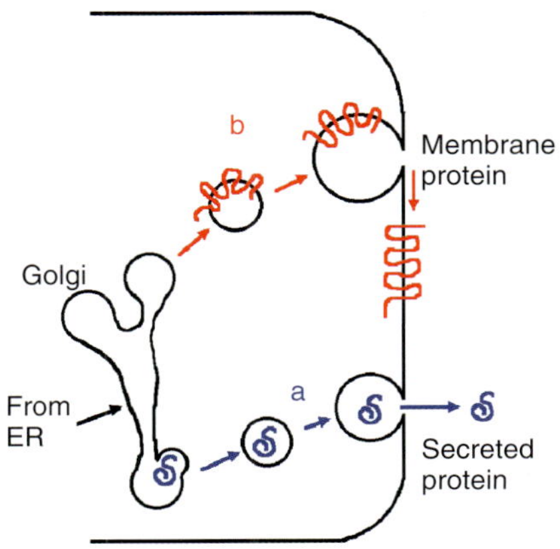

Figure 3.6 Post-translation processing of proteins. Golgi apparatus membranes shown at left. **a** (blue): Secreted protein, for example, a peptide hormone. Because most of the peptide is hydrophilic, it crosses the ER membrane and resides in the cisterna of the ER. The vesicle containing the peptide fuses with the cell membrane and the peptide finds itself outside the cell. **b** (red): Membrane protein, for example, a receptor. The signal peptide promotes uptake of the growing chain, but because of hydrophobic regions, it remains in the Golgi membrane. As the vesicle fuses with the cell membrane, the receptor remains part of the membrane.

into the ER occurs because a hydrophobic leader, the **signal peptide**, is part of the N-terminal end of the new polypeptide chain (Figure 3.6).

The signal peptides differ from one protein to the next, but all contain hydrophobic amino acids and bind selectively to specific receptors on the surface of the ER. For proteins that become hormones or secretory molecules, small vesicles bud off the ER membrane and are then transported to the Golgi body, where final processing takes place. For example, **glycosylated** proteins receive their carbohydrate chains in the Golgi body. The vesicles then continue on to the plasma membrane. Because ER, Golgi, and plasma membranes share a common structure, they can fuse. When a vesicle containing a protein arrives at the plasma membrane, the fusion of the two membranes places the *inner* portion of the membrane vesicle at the *outer* portion of the plasma membrane. In so turning inside out, the hormone is left floating in the extracellular space.

A similar process controls placement of proteins in the plasma membrane: they are synthesized by standard protein synthetic mechanisms, and their signal peptide begins to pull them across the ER membrane. However, they stop within the membrane because they contain several hydrophobic regions (Figure 3.6b). Thus, these proteins do not pass into the space between the layers of ER; they remain in the ER membrane as it coalesces with the cell membrane. Because their N-terminal region extends into the ER space, it is topologically on the outside of the cell membrane at coalescence. The receptor protein thus originally projects into the lumen of the vesicle and then projects outward on the cell membrane through a process called **vectorial discharge**. Hormones are concentrated in a solution within the vesicle for secretion; receptors are concentrated on the inner surface of the vesicle for carriage to the cell membrane [47].

3.2 The Nature and Function of Proteins

Proteins are polymers of amino acids that form into complex structures to accomplish their roles in the cell. These roles include acting as contractile elements, as enzymes, as transport molecules, and as receptors. Some receptor agonists are also proteins.

Proteins are generally large molecules with structural and electrical heterogeneity that derive from their folded three-dimensional structures. These structures are in turn determined by interactions among their component amino acids [48]. Each amino acid in the chain attracts or repels others nearby as the chain folds, and thus the linear arrangement of the amino acids determines the size and shape of the molecule: some proteins are globular (roughly spherical), others filamentous (linear), others a combination of these (Figure 3.7).

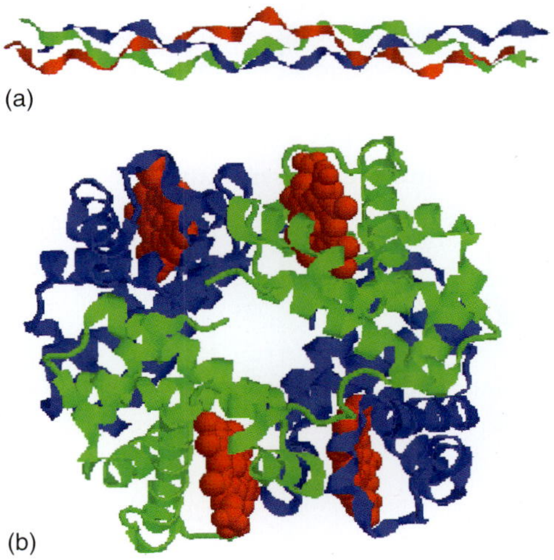

Figure 3.7 Linear and globular proteins. (a) Three linear peptide strands of collagen (red, blue, and green). (b) Four globular peptide strands of hemoglobin (blue and green), each with iron atom bound to heme (red). Images were created using Rasmol [8] from PDB ID 1CGD [49] and PDB ID 1C7D [50].

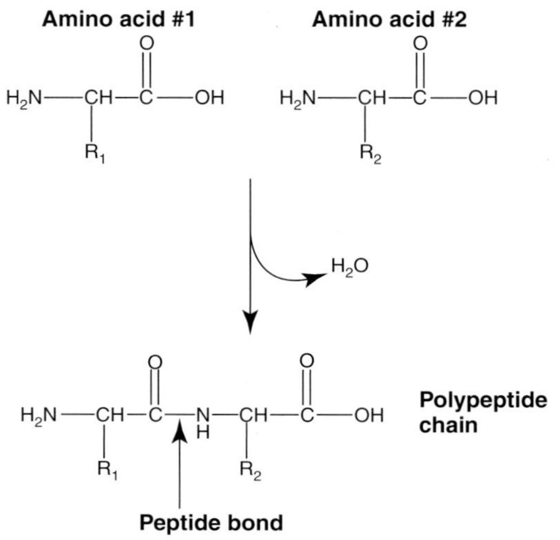

Figure 3.8 Formation of the peptide bond in proteins when the terminal hydroxyl from one amino acid combines with the initial hydrogen of another.

3.2.1 Linear and Three-Dimensional Structures

Proteins are infinitely variable in shape and function. For a small protein of 100 amino acids, for example, the amino acids can be rearranged in 20^{100} (about 10^{260}) different ways (to illustrate the size of this number, consider that the universe contains slightly fewer than 10^{80} atoms). The linear sequence of amino acids forms the **primary structure** of the protein; the folding of the primary structure into an alpha helix or beta pleated sheet is termed the **secondary structure**. Further large-scale foldings of the helices make a third level of structure, the **tertiary** structure, such as a globule or a filament. Finally, two or more polypeptide chains forming a functional complex represents the **quaternary structure**, as seen in hemoglobin (Figure 3.7).

3.2.2 Primary Structure

The 20 amino acids that compose all proteins are variants on a common theme: each consists of an axis with a pair of carbon atoms, one with a carboxyl group attached, and the other (the alpha carbon) with any 1 of 20 chains (the R groups) of different length and polarity that give structural and chemical identity to the amino acid. These R groups determine whether the amino acid as a whole is acidic or basic, hydrophilic or hydrophobic, large or small.

The peptide chain is built by combining adjacent amine and carboxyl groups with the release of a molecule of water. The bond thus formed is called the **peptide bond**. The amino group of the first amino acid in a protein and the carboxyl group of the last one therefore remain intact, and form the amino and carboxy termini, respectively, of the protein (Figure 3.8).

The CO–NH group that makes up the peptide bond shares electrons in such a way that the bond is planar and the atoms remain in a constrained relation to each other. The alpha carbon is the only place where bonds may rotate, and a peptide is best seen as an articulated set of flat plates.

The amino acid sequence of a protein is not random, and each functional region of a protein has a characteristic type of amino acid sequence. For example, the membrane-spanning portion of a receptor has a preponderance of hydrophobic amino acids. The existence of one or more membrane-spanning regions can be inferred from **hydropathy plots**, in which the amino acids are given numerical values reflecting their polarities.

Polarity has been computed by Kyte and Doolittle [51], and each amino acid given a numerical value proportional to its lipophobic nature. The most polar amino acid has the value −4.5 and the least polar has the value +4.5 (Figure 3.9, Box 3.1). Each amino acid in a protein sequence is scored and the values averaged over a small number of amino acids. Then the hydropathy values are plotted on a graph in which the amino end of the protein is at left and the carboxyl end at right (Figure 3.10). Positive values represent protein regions that are likely within membranes, and negative values represent regions that are either cytoplasmic or extracellular [51].

Amino acid hydropathy values

	ile	val	leu	phe	cys	met	ala	gly	thr	ser	trp	tyr	pro	his	gln	asn	asp	glu	lys	arg
Amino acid	I	V	L	F	C	M	A	G	T	S	W	Y	P	H	N	Q	D	E	K	R
Hydropathy score	4.5	4.2	3.8	2.8	2.5	1.9	1.8	−0.4	−0.7	−0.8	−0.9	−1.3	−1.6	−3.2	−3.5	−3.5	−3.5	−3.5	−3.9	−4.5

Figure 3.9 Table of polarities of amino acids, in order of hydropathy score (positive = nonpolar).

Box 3.1 Relation of Hydropathy Plot to Protein Domains

Domains are easily visualized by a hydropathy plot, which expresses the polarity of a sequence of amino acids.

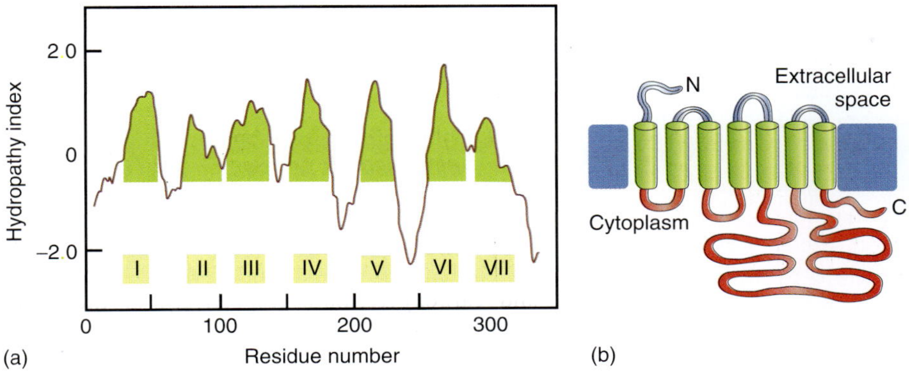

Figure 3.10 Relationship between receptor protein hydropathy plot and orientation within the membrane. (a) Plot of hydropathy values (+ represents hydrophobic amino acids) for rhodopsin (about 350 amino acids). I–VII: membrane domains. (b) Probable orientation of the seven transmembrane (TM) domains within the membrane. The entire molecule forms a roughly circular structure rather than a linear structure. Modified from Figures 14.9 and 14.10 in Ref. [52]. Used with permission.

The properties of the 20 amino acids are given in Table 3.1.

3.2.3 Secondary Structure

The identifying R groups extend away from the axis of the chain. Folding, whether in an aqueous solution or within a membrane, is determined by the interactions of the side groups, and by covalent modifications of the amino acids themselves.

Side Group Interactions: These interactions exist because of polarity differences between protein and environment: the polar side chains tend to seek water or other polar environments; the nonpolar ones tend to seek hydrophobic conditions. Because water is polar, the hydrophilic side chains of a soluble protein tend to be arrayed on the outer surface and the hydrophobic side chains tend to remain on the inner portions of the folded structure. In proteins that insert into cell membranes, the opposite holds: the nonpolar side chains are in contact with, and anchored by, the hydrophobic lipids of the membrane. The same types of forces govern protein folding as govern ligand-receptor binding.

Glycosylation: This is a sort of covalent modification that involves the enzymatic addition of carbohydrate polymers to specific sites on certain proteins. Two types of carbohydrate additions exist, the N-linked (sugar bound to the amino group of asparagine) and the O-linked (sugar bound to the hydroxyl of serine or threonine) **oligosaccharides**. The consequences of glycosylation are profound not only for cell localization of proteins but also for their function as binding sites for extracellular molecules. The main functions of glycoproteins are to serve as recognition sites for receptors, immune system molecules, and blood group antigens. Examples are shown in Figure 3.1.

Table 3.1 Properties of the amino acids.

Single-letter code	Name	Property
A, V, L, I P, W, F M, G	Alanine, valine, leucine, isoleucine, proline, tryptophan, phenylalanine, methionine, glycine	Nonpolar
S, T, Y C, Q, N	Serine, threonine, tyrosine, cysteine, glutamine, asparagine	Polar
K, R, H	Lysine, arginine, histidine	Basic
D, E	Aspartic acid, glutamic acid	Acidic

Disulfide Bridge: Bridge formation is a further form of covalent modification. Cysteine consists of the peptide axis plus a CH_2-SH group. Two cysteine residues in a peptide chain, either nearby or distant, can be covalently joined by formation of a C–S–S–C bond, thus linking the two portions of the peptide and altering the folding pattern that would occur in the absence of **disulfide bridge** formation. This type of linkage is found in some hormones (e.g., insulin) as well as in some types of receptors (e.g., the insulin and immunoglobulin receptor groups).

Higher Order Structures: Two main secondary structures (**alpha helix** and the **beta sheet**) form as the result of **hydrogen bonding** interactions between CO and NH groups on nearby amino acids.

The alpha helix forms when the linear sequence twists as the result of interactions between an amino acid and one that is four amino acids (a half-turn) ahead or behind it in the sequence (Figure 3.11).

The carboxyl group of one amino acid makes hydrogen bonds with the amino group of the other; the bonds are arrayed in the same direction as the helix runs, so the bonds tend to stabilize and strengthen the helix.

The **beta sheet** structure (Figure 3.12) forms between two adjacent non-helical protein sequences running in opposite directions.

Two adjacent strands may be parts of separate polypeptides, or they may be two regions of the same polypeptide. Most commonly, the structure occurs in regions of a peptide rich in small side groups such as glycine or alanine. In sheet structures, the CO groups of an amino acid on one polypeptide chain form hydrogen bonds with amino groups of an amino acid on the other chain, thus stabilizing the chains into a flat structure.

3.2.4 Tertiary Structure

Because of the variety of amino acid combinations, tertiary structures are enormously variable. Even so, among the thousands of proteins, two major functional shapes exist: those that form rigid structures, and those that form flexible binding structures such as enzymes or as receptors. Rigid structural proteins (collagen, laminin) are often linear with long stretches of identical alpha helices. Enzymes and receptor proteins are generally globular, with compact structures. They consist of a variety of regions, some in alpha helix **conformation** and others in beta sheets, though each region is usually quite short. The globular form

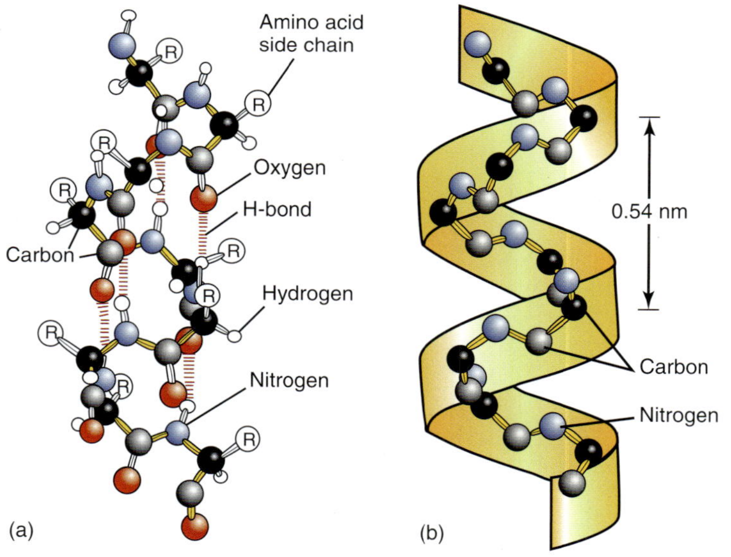

Figure 3.11 Alpha-helix structure of proteins. (a, b) The shape of the helix is stabilized by hydrogen bonds between adjacent turns of the helix. This structure is found in most proteins, but is especially important in membrane proteins as the transmembrane domain (Chapters 6–8). From Figure 3-7 in Ref. [53]. Used with permission.

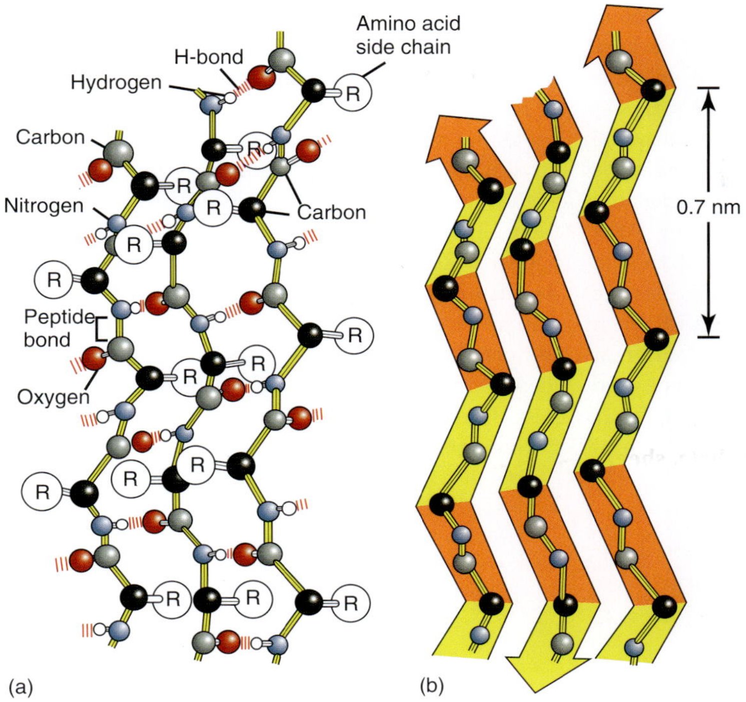

Figure 3.12 Pleated sheet structure of proteins. (a, b) Hydrogen bonds form between the CO of one strand to the NH of the parallel strand. This structure is seen in many proteins, but is often a part of the extracellular binding domain of ligand-gated ion channels (Chapter 6). From Figure 3–7 in Ref. [53]. Used with permission.

arises from multiple alterations of direction of the chain.

3.2.5 Protein Domains

At a level of organization above the tertiary structure lies the **domain** of a protein. This is a region of alpha helix or beta sheet, or a combination or these, that has a particular function. A large protein is made up of many domains strung together, each domain performing a different task for the protein. The domain concept of protein structure now makes it easier to understand the multiple-functioned proteins such as the receptors to be covered in later chapters. These proteins contain different sites for each of their functions: sites for binding hormone, sites for holding to the cell membrane, and sites for activating intracellular metabolic or genetic processes.

3.2.6 Proteomics

Organisms consist of a hierarchically arranged set of systems, organs, tissues, and cells. Function at any of the levels depends on interactions among the cells making up a lower level. Researchers have organized their molecular thinking to reflect a hierarchical scheme, and see cell biology as a functional relation between genes and their protein products, and between proteins and the processes they accomplish.

The complement of a body's genetic material has since 1920 been called its **genome**. In the mid-1990s Wilkins introduced the term **proteome** to highlight the variety of proteins that result from gene transcription and translation [54]. By 2001, over 30 "-ome" words had been coined [55] to illustrate the interactions among genes and proteins involved in the process of crafting organismal phenotypes. We will use only a few of these -omic concepts, those that are most relevant to receptor function:

Proteome [54]	The proteins produced by an individual, along with their functions
Interactome [55]	The set of molecular interactions within a cell. *Interactomics* studies the cellular effects of protein–protein contact and communication
Metabolome [56]	The set of metabolic substances in a cell. *Metabolomics* studies the metabolite profiles that remain as metabolic activities progress, and thus give an idea of overall cell function

The terms indicating the complexity of protein actions are not simply convenient rhetorical devices; they illustrate a fundamental difference in how proteins act as compared to how DNA (the genome) acts. Any organism's genome can be determined relatively easily because all DNA molecules have the same structure and differ only in base sequence; the differences in sequence do not alter the way DNA folds, replicates, or is transcribed. In contrast, though the amino acid sequences in protein molecules are defined by the nucleotide sequence in DNA, each DNA sequence can produce any of several different proteins and each protein folds in a different, and largely unpredictable, way once it is translated. The total number of proteins produced by an organism may thus be several times the number of genes it has, and understanding the cellular roles of proteins is far more complex than understanding the roles of DNA [57]. Thus, researchers actively probing the proteome realize that adequate understanding of cell physiology requires inter-laboratory initiatives such as the Human Proteome Project [58] and the Alliance for Cell Signaling [59]. Two promising approaches appear to be (i) the development of "protein chips," or **antibody**-based spots on small wafers that can trap and display proteins isolated from a cell and (ii) the concept of "medically relevant" tissues and the identification of the functional proteins within them [57]. These approaches are discussed in appropriate chapters that follow.

4
Hormones as First Messengers

> In animal societies, communication provides the ties that bind, and chemical communication is especially important for social insects and mammals.
>
> *G. Robinson [60]*

> Any organism may release neurotransmitters ... the presence of neurotransmitter compounds has been shown in organisms lacking a nervous system and even in unicellular organisms.
>
> *V. Roshchina [61]*

4.1
Hormones and Cellular Communication

Hormones, transmitters, and **local modulators** are small molecules released from one cell to affect the behavior of another cell [62]. They are called **first messengers** because their binding is the first event to occur in the chain of activation. The result of their binding may be the release within that cell of an intracellular signal, or **second messenger**. The second messenger then activates enzymes or genes that cause major changes in cell behavior.

4.1.1
Discovery of Hormones

By the nineteenth century, physiologists suspected that individual animal organs produced substances with important effects on the body as a whole; before the twentieth century, none of these substances had been identified. In 1901, Abel and Takamine independently discovered **epinephrine** as the hormone of the autonomic nervous system [63]; in 1902, Bayliss and Starling [64] were able to identify a specific substance, which they called *secretin*, as being responsible for one process – the secretion of bicarbonate by the pancreas. This work was thus the first to show that a chemical substance had definable physiological effects, and in 1905 Starling [65] proposed the term hormone, derived from the Greek word for "excite," or "arouse," to denote such active compounds. He used the term to refer to substances released by specialized endocrine cells and carried by the circulation to their target cells. He did not suggest they operated via receptors, though he clearly indicated that hormones had specific cellular effects. Though researchers still use the term hormone in Starling's sense, they have added the crucial concept that hormonal action requires the binding of the substance to a cellular receptor.

4.2
Types of Hormones

Starling only considered blood-borne substances as hormones, so his and Bayliss' work conflicted with the ideas of neurophysiologists of the time, who thought that only neural processes were involved in such chemical control. The resolution of that conflict only came after Loewi's 1921 discovery [33] that nerves communicated using neurotransmitters. His work led to the recognition that the nervous system works in the same way as the endocrine system: the release of a substance (called either a neurotransmitter or a hormone) from one cell (either a neuron or a gut cell) and its action on a different cell (either a muscle or a pancreas cell).

The 1956 discovery by Grobstein of locally acting substances [66] forced a further expansion of the concept of hormone action to substances that are produced in non-neural tissue and diffuse only a short distance from their origin to their site of action. The term **local modulators** is now used for such substances, though it is important to recognize that **neurotransmitters** are also a type of local modulator.

Some local modulators can also act as longer distance hormones. For example, the **catecholamine** norepinephrine is primarily a neurotransmitter in the brain and the autonomic nervous system, but it is also released into the blood from the **adrenal medulla,**

Receptor Biology, First Edition. Michael F. Roberts and Anne E. Kruchten.

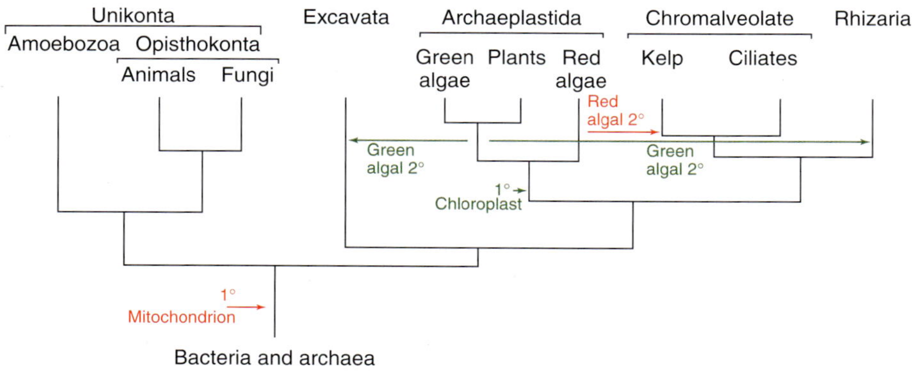

Figure 4.1 Simplified phylogeny of life, showing the five major eukaryotic groups – unikonta, excavata, archaeplastida, rhizaria, and chromalveolata. Organisms mentioned in this chapter are indicated. 1°: Primary endosymbiosis of mitochondrion (all eukaryotes) and chloroplast (line leading to archaeplastida. 2°: Two separate endosymbiotic events in which a eukaryotic green alga was incorporated into an early excavate and a rhizaria; also, a eukaryotic red alga was incorporated into an early brown alga (kelp). Modified from Figure 1 of Ref. [67]. Used with permission.

and acts on distant organs such as the heart. The feature common to all these substances is that their action follows specific receptor binding.

Hormone-like molecules are either water soluble or they are fat soluble. Solubility determines the place of action of hormones: water-soluble hormones (peptides, amino acids, catecholamines, and acetylcholine (ACh)) act on cell-surface receptors, and fat-soluble hormones (steroids, thyroid hormone, and vitamins A and D) act on intracellular receptors after first diffusing through the cell membrane.

4.2.1 Pheromones for Signaling between Individuals

So far we have been discussing chemical communication between different cells or organs within a single individual. However, abundant evidence exists for interindividual chemical communication as well. The term **pheromone** refers to such a signal if its mode of action is between individuals rather than within the same individual, though the mechanism is identical to that of hormones in the sense that a molecule is released from one cell and acts on a receptor on or in another cell. Pheromonal signals between single-celled organisms thus must have predated hormonal signals within multicellular organisms, and pheromones are best seen as the more fundamental signal type. Indeed, as we discuss in Chapter 9, the origin and continued existence of multicellular organisms depends on the type of intercellular communication first seen in unicellular organisms.

Figure 4.1 shows the diversity of organisms on Earth today, expressed as a phylogenetic diagram with the most likely relationships among the living groups.

Thus, Bacteria and Archaea represent the root, and eukaryotic organisms originated from an Archaeal-like ancestor by the endosymbiosis of a bacterial cell (now called the *mitochondrion*). Further diversity within the eukaryotes arose when another bacterial cell (the chloroplast) was incorporated at a later time (right side of Figure 4.1). This addition produced the lineage of photosynthetic organisms. All organisms known are capable of signaling; we thus begin this discussion with a brief overview of the variety of signaling processes, focusing on the signal molecules themselves; their receptors form the basis for later chapters.

4.2.2 Archaea and Bacteria

Both groups of prokaryotes, though largely single celled and solitary, also show multiple cell aggregations that can be thought of as "intentional communities" rather than random associations. Bacteria have sophisticated signaling processes called **quorum sensing** that initiate and sustain long-term associations. Each cell produces **autoinducer** (AI) molecules, pheromones that diffuse into the nearby environment and are picked up by other bacterial cells (Figure 4.2a).

Bacterial AIs belong to different chemical classes. AI-1 is *n*-acyl homoserine lactone (Figure 4.2b) and AI-2 is furanosyl borate diester (Figure 4.2c). AI-1 appears to induce aggregation in same-species

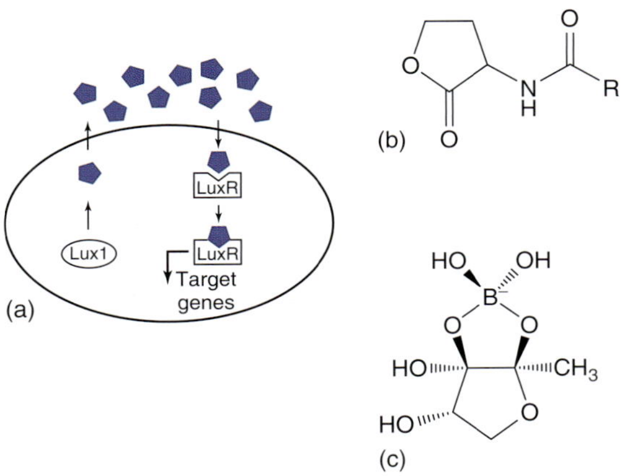

Figure 4.2 Quorum sensing in bacteria. (a) Gene Lux1 encodes an enzyme that synthesizes autoinducer (pentagon), which diffuses away from the cell. If enough cells are in the vicinity, large amounts of autoinducer are produced, and diffuse into nearby cells. There, they bind to transcription factor LuxR and activate target genes. (b) Structure of the autoinducer AI-1. (c) Structure of the autoinducer AI-2. Panel (a) from Ref. [68]. Used with permission.

bacterial complexes, while AI-2 is the signal for aggregation of different bacterial species [69]. AI-3 regulates inter-kingdom aggregation (i.e., between bacteria and eukaryotes). Its structure is not known, but as both epinephrine and norepinephrine activate the AI-3 receptor in bacteria, it may have a catecholamine-like structure [70].

Once inside the recipient cell, the inducer binds to a cellular transcription molecule and by a process reminiscent of nuclear receptor signaling (Chapter 9), activates transcription of bacterial genes. If enough AI-releasing cells are together, the signal each receives will be strong enough to indicate that a "quorum" of bacteria is present, and they will all express genes encouraging aggregation; in addition, some bacterial species use quorum sensing for bioluminescence, virulence factor expression, sporulation, or mating [68, 69]. It is thus clear that intercellular signaling is ancient and is of selective advantage to all organisms.

Signaling systems of eukaryotes have primarily been studied in insects and in vertebrates. However, though the literature is small, we describe a few examples of signaling in other eukaryotes to show the diversity and importance of this process.

4.2.3 Eukaryotes

4.2.3.1 Chromalveolates

This group includes the kelp, the giant seaweed also called *brown algae* that show features of multicellularity. It is thus not surprising that they have signal molecules and receptors that promote cell–cell communication. Though the receptors for these signals have not yet all been identified, the pheromones themselves are beginning to be known [71, 72] (Figure 4.3).

The planktonic chromalveolates (dinoflagellates) have receptors for carbohydrates (e.g., mannose – Figure 4.3a) that are carried on the surfaces of their prey. Contact causes receptor binding of mannose to the receptor, a lectin, followed by prey engulfment. Other marine chromalveolates, the diatoms, use large molecular weight aldehydes (Figure 4.3b) as signals to control population growth rates and competition with other species. Multicellular brown algae such as *Laminaria* and *Fucus* (Figure 4.3c,d) also use hydrocarbons as mating signals [71].

The substances shown in Figure 4.3b–d are typical of eukaryotic pheromones: a hydrocarbon backbone, frequent double bonds that produce kinks and bends in the molecule, and often a terminal alcohol group or oxygen atom. In addition, pheromones may be alcohols, small peptides, sterols, or steroids. Volatile compounds are often found in terrestrial organisms. Pheromones function in several ways: they are best known for their use as sexual attractants, but they also function as alarm calls, hatching signals, chemical trails, territorial markers, and signals to maintain social cohesion.

4.2.3.2 Unikonts – Amoebozoa, Fungi, Animals

The best-known Amoebozoan is the cellular slime mold *Dictyostelium discoideum.* Its signaling mechanisms have been well studied for decades because they are responsible for its transition from single-celled solitary amoebas to "multicellular" aggregates. Two types of signal are important – cyclic adenosine monophosphate (cAMP) and peptide-like substances [73] (Figure 4.4).

cAMP is the signal released from starving cells; it binds to a G-protein-coupled receptor (GPCR) on other cells to cause migration and aggregation into large multicellular "slugs" that migrate as a unit. Once the migrating slug reaches a suitable location for sporulation, the cells differentiate into spore cells and stalk cells. The stalks then grow into towers and the spores are released from tower tops. These spores travel in the air to new places with better food supply. The differentiation of stalk and spore cells is influenced by peptide-like substances such as glorin (Figure 4.4) that also act on GPCRs.

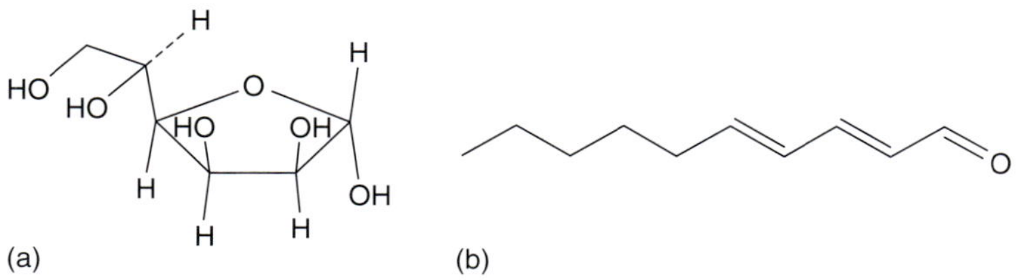

Figure 4.3 Molecular structures of chromalveolate signaling molecules. (a) The monosaccharide mannose. (b) The aldehyde (2E, 4E), decadienal. (c) The aromatic hydrocarbon lamoxirene. (d) The hydrocarbon fucoserratene.

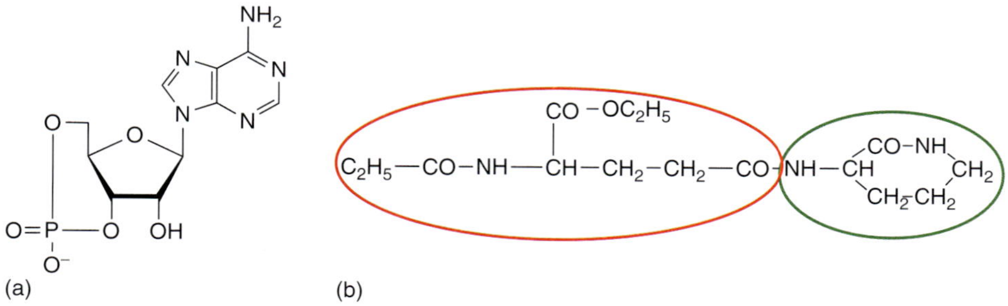

Figure 4.4 Molecular structures of signaling molecules for amoebozoa aggregation and differentiation. (a) cAMP and (b) glorin, a hydrocarbon and peptide-containing signaling molecule. The red oval encloses glutamic acid and the green oval encloses ornithine.

A well-studied fungal communication system is the signaling process that controls mating and meiosis. Two different haploid cell types (for this purpose, they can be thought of as "sexes") are involved in this process. Each produces a different signaling peptide and has different cell-surface receptors. One cell type (MAT**α** – for "mating-type alpha") secretes a peptide called α; in contrast, the second cell type (MAT**a**) secretes a different peptide, called **a.** The two peptides are produced by a single genetic locus [74], called *MAT*. This locus has both MATα- and MAT**a**-related genes [75, 76] (Figure 4.5).

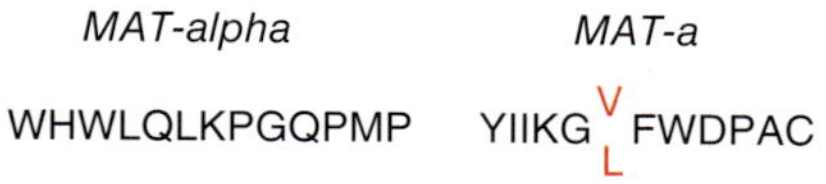

Figure 4.5 (a) Amino acid sequence of yeast-mating-type peptide MAT-alpha. (b) Sequence of yeast-mating-type peptide MAT-a. MAT-a can have either a valine or a leucine at position 6.

The **a** peptide binds to a GPCR on **α** cells, and the α peptide binds to another GPCR on **a** cells. The binding event leads to cell fusion and the production of a transient diploid cell that then divides by meiosis to produce two new haploid cells of different genotypes from the two originals.

Long-distance communication is adaptive for animals for which olfactory mechanisms are an important source of environmental information. Pheromones of aquatic animals are primarily non-volatile but water-soluble peptides (Figure 4.3). The pheromones of terrestrial animals consist of a diverse group of volatile hydrocarbons (e.g., Figure 4.6) and related chemicals. The behavioral effects of these compounds are mediated in part by receptors in the recipient animal, though receptor mechanisms of pheromone action (and subsequent intracellular effects) are poorly understood. To be considered as true pheromones, substances must act in a species-specific manner and they must evoke stereotyped behaviors [77].

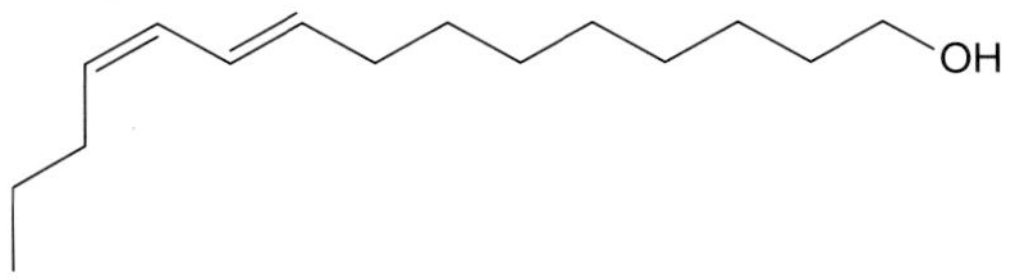

Figure 4.6 Molecular structure of bombykol, a volatile long chain alcohol used by female silkworm moths to attract males.

4.2.3.3 Invertebrate Pheromones

Serious research on pheromones began in the early 1900s when German biologists attempted to discover the basis for attraction between male and female moths. Evidence that a specific attractive substance existed was provided by Butenandt in the 1950s. He isolated the substance and named it bombykol. This was the first pheromonal substance isolated. He later determined its structure as a long-chain alcohol with two double bonds [78] (Figure 4.6).

4.2.3.4 Vertebrate Pheromones

The effects and importance of pheromones for animal communication have been known in popular culture for generations; to give one example, French fishermen have long used captured male lampreys as lures to catch females. The urine of the male contains one or more chemicals that attract females from great distances in sea water [4]. It is more difficult to apply the invertebrate criteria to mammalian chemical communication, even though the fact that most mammals are small, nocturnal, and have a highly developed sense of smell might make them ideal candidates to use airborne chemicals.

Doty [77] has recently questioned the role of typical pheromone mechanisms in mammals because the processes seem not to fit the normally expected stereotyped behavioral responses seen in insects. This is because the chemicals of communication in mammals are not species specific and because responses are often learned, rather than stereotyped. This is not to deny the role of scent in mammalian communication. Substances are present in urine, feces, and gland secretions such as sweat; their release is often accompanied by behavioral changes in other members of the same species. Lemurs and the new world monkeys are able to recognize the odors of individual troop members. Less research has been done on the other anthropoids, but secreted substances are suspected to be the basis for human mother–infant recognition and for synchronizing menstrual cycles in women who share living quarters [79]. The synchronizing substances are not known, and indeed researchers have become cautious in ascribing a specific pheromone concept to mammalian scent communication, preferring to use the idea that one or more compounds may generate "signature odors" [80] that are used in non-stereotyped social behavior.

The remainder of this chapter focuses on vertebrate hormone signal molecules, and treats them according to the receptor type on which they act.

4.3 Vertebrate Hormones and Transmitters

4.3.1 Peptide and Non-Peptide Agonists

The most fundamental classification scheme for vertebrate signaling molecules is one based on their chemical composition; we thus begin the discussion by dividing the hormones into two groups – the peptides and the non-peptides. The two are synthesized differently, transported within cells differently, act on different receptor types, and are disposed of in different ways.

4.3.1.1 Peptides

Peptides are proteins, each with a unique amino acid sequence that is based on a DNA sequence; thus, each peptide has a phylogenetic history that parallels the history of its gene. In principle, that history, and the relationships among peptide hormones themselves, should be recognizable from their amino acid sequences. However, it is apparent from the diverse amino acid sequences of peptide hormones (e.g., Table 4.1), that they belong for the most part to unrelated families. Only in a few cases (illustrated below and in Chapter 14) do the peptide hormones fall into recognizable families. Thus, because such an inclusive phylogeny of peptide hormones cannot be drawn, this chapter is organized instead according to the sources (different organisms, different organs) of these molecules.

4.3.1.2 Non-peptides

The non-peptides are diverse: catecholamines, quaternary nitrogen compounds, steroids, fatty acid derivatives, amino acids, nucleotides, and hydrocarbons. These are all synthesized by unrelated enzymatic pathways and thus the hormones themselves show no intergroup similarities. In the discussion below, the synthetic pathways and the various signaling molecules of each are described.

Table 4.1 Amino acid sequences of hypothalamic releasing hormones.

Releasing hormone	Sequence	References
CRH: corticotropin-releasing hormone (human)	pSEEPPISLDLTFHLLREVLEMARAEQLAQQAHSNRKLMEII.NH_2 [a),b)]	[81]
GnRH: gonadotropin-releasing hormone (mammalian)	pEHWSYGLRPG.NH_2	[82]
("chicken II")	pEHWSHGWYPG.NH_2	
GHRH: growth-hormone-releasing hormone (human)	pYADAIFTNSYRKVLGQLSARKLLQDIMSRQQGESNQERGARARL.NH_2	[83]
SS: growth-hormone-release-inhibiting hormone (somatostatin – human)	SANSNPAMAPRERK**AGCKNFFWKTFTSC** (C–C link [c)])	[84]
	(SS28 is the entire molecule; SS14 is the portion in bold type)	
TRH: thyrotropin-releasing hormone	pEHP.NH_2	[85]
PIF: prolactin-inhibiting factor	(Dopamine)	[86]

a) Pyro group at amino terminus.
b) Amide at carboxy terminus.
c) Disulfide link.

4.3.2
Peptide Hormones of the G-Protein-Coupled Receptors

The **GPCRs** themselves are discussed in detail in Chapter 7; the peptide hormones activating them belong to three separate groups:

- Hypothalamic peptides (**posterior pituitary** hormones and **releasing hormones**);
- **Anterior pituitary** peptides;
- Other peptides, including agonists for receptors in the brain, gut, blood, gonads, and vascular tissue.

4.3.2.1 Hypothalamic-Pituitary Axis

Hypothalamic Hormones These hormones are all short peptides (3–50 amino acids). Two types have been described, the posterior pituitary hormones and the releasing hormones.

The **posterior pituitary** is actually a continuation of the hypothalamus. Two different sets of hypothalamic neurons (**magnocellular neurosecretory cells**) produce two peptide hormones, oxytocin and vasopressin (the latter is also called ADH or antidiuretic hormone) in their cell bodies in the hypothalamus and release them near blood vessels in the posterior pituitary (Figure 4.7).

Oxytocin and vasopressin are nonapeptides (consisting of nine amino acids), have similar folding patterns, and have sequences that differ only at two sites (Figure 4.8).

The differences in sequence and structure are sufficient to make their binding properties different: oxytocin binds to receptors on uterine smooth muscle and causes contraction; vasopressin acts on receptors in vascular smooth muscle cells to cause **vasoconstriction** and it also binds to receptors on cells of the kidney collecting duct to increase water permeability, thereby promoting reabsorption and conservation of water [87]. This latter function is discussed in Chapter 6 with reference to the water channel (aquaporin) whose function is regulated by ADH.

Hypothalamic Releasing Hormones While the posterior pituitary is an extension of the hypothalamus, the anterior pituitary is a distinct tissue that does not have a neural connection to the hypothalamus. Instead, a connecting circulation, the **hypophyseal portal system**, is the communicating link. This short circulatory pathway carries releasing hormones from the axonal endings of the hypothalamic neurosecretory cells to the body of the anterior pituitary (Figure 4.9).

Four separate hypothalamic *stimulatory* releasing hormones control the synthesis and release of the anterior pituitary hormones discussed below.

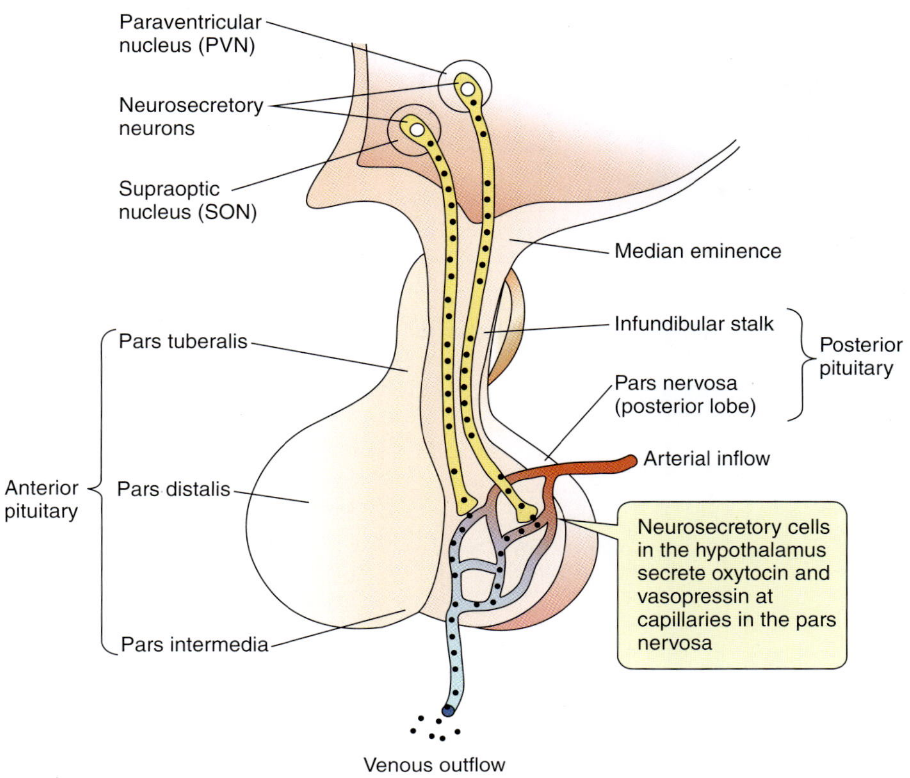

Figure 4.7 Posterior pituitary structure and locations of hypothalamic neurosecretory cells that release its hormones. Axons of neurosecretory cells extend from the hypothalamus to the posterior pituitary. Figure 15.7a of Ref. [87]. Used with permission.

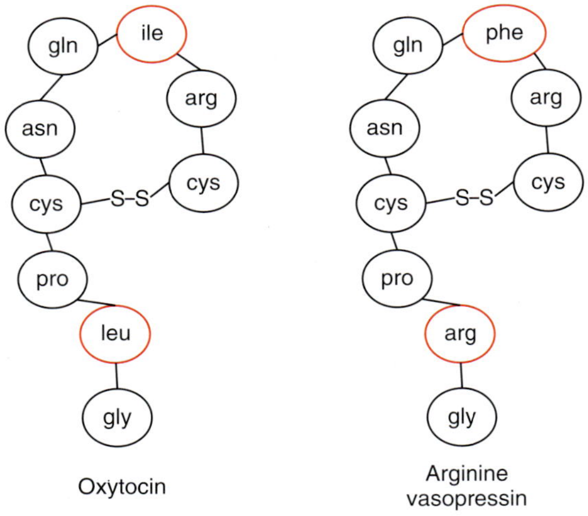

Figure 4.8 Amino acid sequences of the posterior pituitary hormones, (a) oxytocin and (b) vasopressin. The two-dimensional structures are very similar, but the two different amino acids (in circles) determine the different functions of each. Modified from Figure 3.14 of Ref. [62].

In addition, two *inhibitory* releasing hormones are involved, one a peptide and the other a catecholamine transmitter, dopamine. Each of the six substances acts on a specific cell surface receptor on a particular cell type of the anterior pituitary. The binding event modifies synthesis and release of the pituitary hormone.

The hypothalamic releasing hormones (RHs) now known are listed in Table 4.1, along with their amino acid sequences. These sequences are quite different, indicating the hormones do not form a single family of peptides.

Corticotropin-Releasing Hormone (CRH): This 41-amino acid releasing hormone stimulates release of adrenocorticotropic hormone (ACTH) from the pituitary. It is also found in other brain sites far removed from the hypothalamic and pituitary regions, where it causes release of β-endorphin and other products of the pro-opiomelanocortin (POMC) gene (see below). CRH thus is an important mediator of the response to pain and stress.

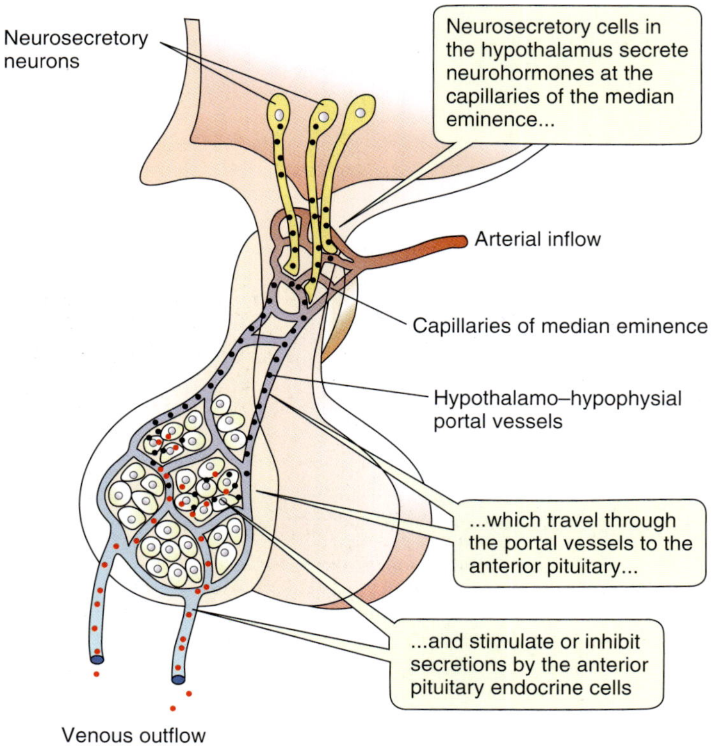

Figure 4.9 Structure of the anterior pituitary and its relation to hypothalamic releasing hormones. Axons of neurosecretory cells stop at portal vessels at the base of the hypothalamus. Figure 15.7b of Ref. [87]. Used with permission.

Gonadotropin-Releasing Hormone (GnRH): This decapeptide regulates the release of the two gonadotropins follicle stimulating hormone (FSH) and luteinizing hormone (LH) from the anterior pituitary. Most vertebrates have two forms of GnRH, one (differing in sequence in most species) that is responsible for LH and FSH release, and another (conserved across vertebrates and called *chicken II* because it was the second GnRH discovered in birds) that is involved in synaptic transmission within the brain [82].

Growth-Hormone-Releasing Hormone: GHRH is a 43-amino acid peptide that has a positive influence on growth hormone (GH) release; its synthesis and release are in turn stimulated by negative feedback from decreasing GH levels in the blood [87]. This feedback loop tends to maintain blood levels of GH.

Growth Hormone Inhibiting Hormone (Somatostatin): This substance is actually two related peptides, one of 28 amino acids and the other of 14 amino acids. Somatostatin 28 (SS28) is produced by cleavage of a 92-amino acid precursor, and SS14 is produced by further cleavage of SS28. About two-thirds of somatostatin activity in the brain comes from SS14. The somatostatins are released from neurons throughout the brain, and they also inhibit release of gastrointestinal (GI) hormones involved in secretion of acid, base, and digestive enzymes [88].

Thyrotropin-Releasing Hormone (TRH): In mammals, this tripeptide causes secretion of thyroid-stimulating hormone (TSH) from the anterior pituitary. It is also found in the gut of reptiles as well as in the brain and skin of amphibians. Its role in the amphibian brain is to release TSH from the anterior pituitary, as a part of metamorphosis. It is also synthesized in the skin and plays roles in both water/ion balance and defense from predation [85].

4.3.2.2 The Anterior Pituitary Trophic Hormones

The group of *hypothalamic* hormones just described is responsible for the control of release of the next group of peptide hormones – those originating in the *anterior pituitary*. The pituitary hormones are called tropins or **tropic** (or trophic) hormones because they,

Table 4.2 Anterior pituitary hormones

Hormone	Structure; receptor type	Function
Follicle-stimulating hormone	Dimer of glycoproteins, 92 and 118 amino acids; acts on GPCR	Causes development and maturation of gametes
Lutropin	Dimer of glycoproteins, 92 and 120 amino acids; acts on GPCR	Causes release of gonadal hormones; triggers ovulation
Thyrotropin	Dimer of glycoproteins, 92 and 118 amino acids; acts on GPCR	Causes release of thyroid hormone from thyroid gland
Adrenocorticotropin	Single polypeptide of 241 amino acids; acts on GPCR	Stimulates cortex of adrenal to release cortisol
Somatotropin	Single polypeptide 191 amino acids; acts on RK	Increases size of muscle cells; causes release of IGF-1
Prolactin	Single polypeptide 198 amino acids; acts on RK	Induces lactation

RK = receptor kinase and GPCR = G-protein-coupled receptor.
Colors indicate the two peptide families of the anterior pituitary.

in turn, cause the release of other hormones from target organs (liver, thyroid, adrenal cortex, gonads, and mammary glands). These peptides belong to three distinct families, illustrated in Table 4.2 (colored regions denote different receptor types: yellow is GPCR and purple is receptor tyrosine kinase (RTK)).

The Thyrotropin – Follicle-Stimulating – Lutropin Family: These hormones act on GPCRs (Chapter 7). They are all glycoproteins with heterodimer (α and β subunit) structure. The α subunit of 92 amino acids is identical in all three, and the β subunit has two large domains with homologous regions in all three [89].

Thyrotropin (TSH) acts in the thyroid to promote synthesis of thyroid hormones.

Follicle-Stimulating Hormone (FSH): This is the term for the peptide as it is found in females; in males, the same peptide is called interstitial-cell-stimulating hormone or ICSH. It promotes gamete production and sex hormone production in both sexes.

Lutropin (LH): This peptide is also identical in males and females, and has the same name in both. It promotes gamete and sex hormone production, and in females it is also responsible for ovulation.

Adrenocorticotropin (ACTH): This hormone acts on GPCRs (Chapter 7). ACTH is transcribed from a single gene, and the initial translated peptide is called *pro-opiomelanocortin*; it contains several domains that are later processed into separate hormones and neurotransmitters (Figure 4.10).

One of these domains contains the sequence encoding β-endorphin, an important peptide involved in the control of pain sensation. It belongs to a group of peptides (the opioids) found in the brain and elsewhere that regulates neurotransmission. These peptides are discussed in more detail below and in Chapter 7.

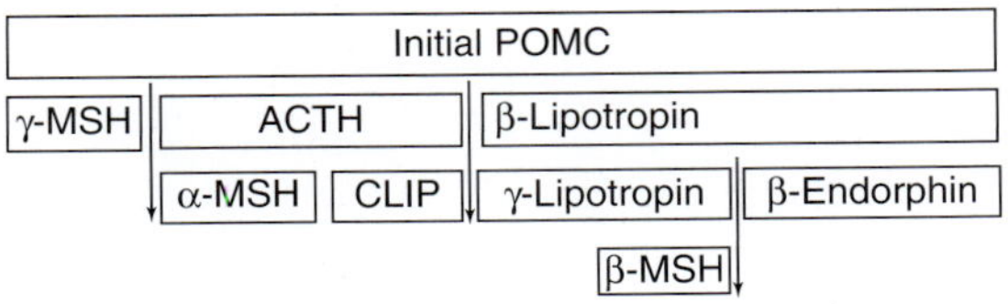

Figure 4.10 Post-translation processing of the pituitary peptide pro-opiomelanocortin. MSH: melanocyte-stimulating hormone and CLIP: corticotropin-like intermediate peptide. Modified from Figure 1 of Ref. [90]. Used with permission.

The Somatotropin–Prolactin Family: These hormones act on receptors of the growth factor superfamily (Chapter 8), and thus are discussed later in this chapter (Section 4.3.7).

4.3.3 Other Neural Peptides

4.3.3.1 Opioids

The **opioids** are important inhibitors of pain pathways in the brain. These substances are ligands for GPCRs that affect neurotransmitter release in pain pathways. They fall into three groups, the endorphins, the enkephalins, and the endomorphins. The peptides are shown in Table 4.3; four of them have the consensus YGGF at the N-terminus, and two are slightly different tetrapeptides.

The opioid peptides were first discovered in 1975 after earlier work had indicated that the opiate **analgesics** morphine and heroin acted on specific

Table 4.3 Opioid peptides.

Name	Sequence	References
Met-enkephalin	YGGFM	[91]
Leu-enkephalin	YGGFL	[91]
β-Endorphin	YGGFMTSEKSQTPLVTLFKNAIIKNAYKKGE	[92]
Dynorphin	YGGFLRRIRPKLK	[93]
Endomorphins	YPFF and YPWF	[94]

Table 4.4 Non-opioid brain peptides.

Name	Sequence	Action	Source
Substance P	RPKPNNFFGLM.NH_2	Transmitter at pain receptor neurons	[94]
Neurokinin A	HKTDSFVGLM.NH_2	Transmitter at pain receptor neurons; alternate splicing product of substance P gene	[94]
n-Acetyl-aspartyl-glutamate (NAAG)	nAc-DE	Modulator of transmitter release; control of gene transcription	[95]
Neuropeptide Y	YPSKPDNPGEDAPAEDLAR YYSALRHYINLITRQRY.NH_2	Learning and memory	[96]
Galanin	GWTLNSAGYLLGPHAVGNHR SFSDKNGLTS	Modulation of neuronal action potentials	[97]
Vasoactive intestinal polypeptide (VIP)	HSDAVFTDNYTRLRKQMA VKKYLNSILN.NH_2	Relaxes smooth muscle, inhibits T cell proliferation	[98]
Neurotensin	pELYGNKPRRPYIL	Dopamine signaling; increases vascular permeability	[99]
Orexin A	LGVDAQPLPDCCRQKTCS	Increases appetite and wakefulness	[100]
Ghrelin	CRLYELLHGAGNHAAGILTLG GSSFLSPEHQRVQQRKE SKKPPAKLQPR	Increases appetite	[101]

receptors in the brain (see Chapter 7 for details). The opioids and opiates both bind to the same receptor. The analgesic effects of the opiate drugs thus are explained by their fortuitous ability to bind to the body's own pain-reducing receptors. The sequence YGGF is found in the N-terminus of all these peptides, and is an important part of the receptor-binding domain.

4.3.3.2 Non-Opioid Transmitter Peptides

Several dozen peptides exist in neurons within the central nervous system (CNS); these peptides are usually released from the same neurons that release amino acid or catecholamine transmitters. However, patterns of peptide release are different from those of the other transmitters: the peptides are localized in vesicles, but the peptide vesicles are fewer in number and are located farther from the axonal ending. In addition, a higher frequency of neural impulses is required to cause peptide release. These patterns suggest that peptide release is less common and serves a modulatory role rather than a direct primary activator role in the brain [91]. Some of the important brain peptide transmitters are shown in Table 4.4.

4.3.4 Peptides from Non-Neural Sources

4.3.4.1 Digestive Tract Hormones

The amino acid sequences of two of these hormones (cholecystokinin (CCK) and gastrin) [102] are shown in Figure 4.11.

CCK is a peptide first discovered in the gut and later found in the brain. The name refers to its ability to cause activation of the gall bladder. In the digestive system, its action is to stimulate digestive enzyme secretion from the pancreas and bile release from the gall bladder (Figure 4.12).

CCK:

H_2N.KAPSGRVSMIKNLQSLDPSHRISDRDYMGWMDF.NH_2

Gastrin:

EGPWLEEEEEAYGWMDF.NH_2

Figure 4.11 Amino acid sequences of the gastrointestinal peptide hormones cholecystokinin (CCK) and gastrin. Underlining shows a short section of sequence identity.

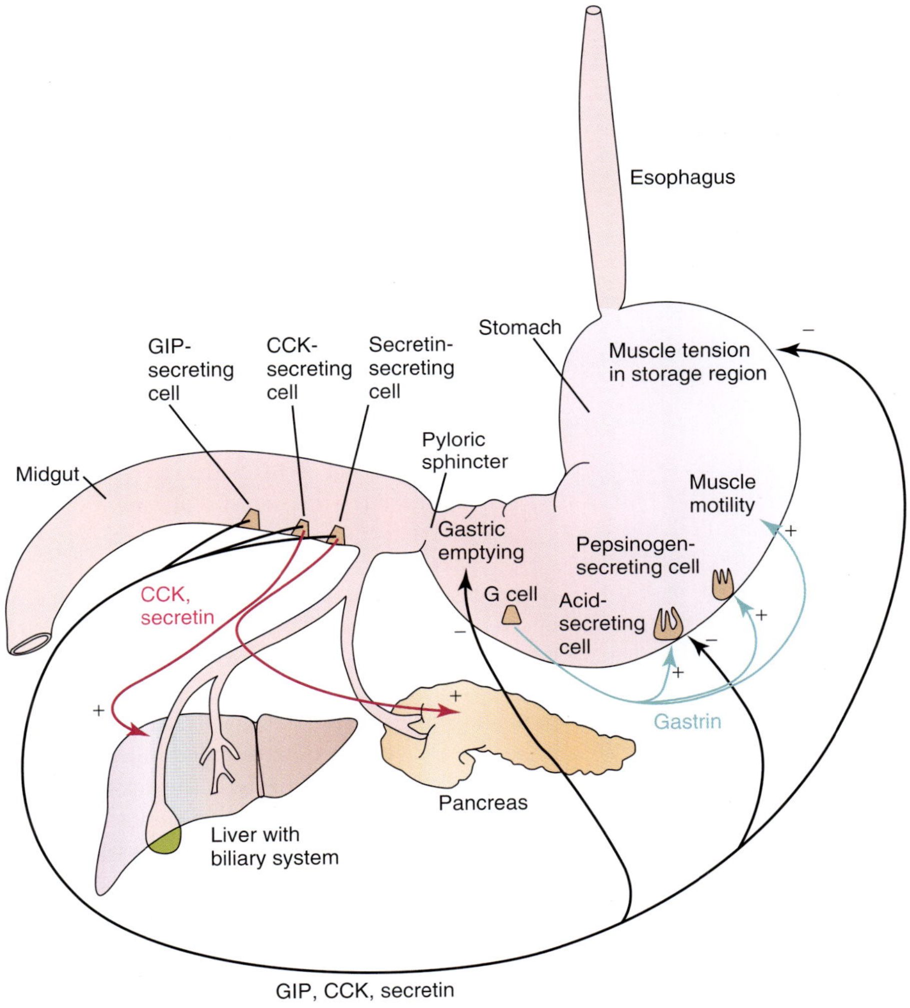

Figure 4.12 Feedback loops in production of CCK, secretin, gastrin, and gastric inhibitory peptide (GIP). Figure 5.24 in Ref. [87]. Used with permission.

Table 4.5 Peptide hormones related to glucagon.

Transmitter	Sequence	Source
Glugagon	HSQGTFTSDYSKYLDSRRAQDFVQWLMNT	[103]
Secretin	HSDGTFTSELSRLQDSARLQRLLQGLV	[100]
Orexin B	RPGPPGLQGRLQRLLQANGNHAAGILTMG	[100]
Orexin A	LGVDAQPLPDCCRQKTCS CRLYELLHGAGNHAAGILTLG	[100]

*Underlining indicates areas of sequence identity

These enzymes travel from the pancreas through specialized ducts into the lumen of the small intestine, where they carry out the digestion of food. In the brain, CCK is a widespread neuropeptide, occurring in a variety of neurons along with non-peptide transmitters such as gamma-amino butyric acid (GABA), dopamine, and serotonin. CCK thus acts in tandem with the non-peptides to modulate anxiety.

Gastrin is a peptide secreted into the blood from stomach cells when ingested proteins reach the stomach. It travels to other cells lining the stomach, where it stimulates the release of acid into the lumen (Figure 4.12). Gastrin and CCK share identical pentapeptide N-terminal regions.

The third GI hormone, secretin, is released into the bloodstream when acid enters the intestine as a result of stomach emptying (Table 4.5). Secretin operates on receptors in the pancreatic **exocrine** cells, and stimulates secretion of bicarbonate ion into ducts that empty into the small intestine. The bicarbonate neutralizes stomach acid and raises the pH of the intestines to a value that allows the functioning of intestinal digestive enzymes. Secretin has sequence homology with the hypothalamic transmitters orexin A and B (the hypocretins), involved in signaling states of appetite and awareness [100]. It also shares sequences with the pancreatic hormone glucagon, an endocrine hormone from the pancreas that opposes the actions of insulin by causing release of glucose into the blood.

Ghrelin is produced in the anterior part of the stomach; levels increase between meals and may thus be a stimulus to feeding via its activation of neuropeptide Y (NPY) signaling. Ghrelin shows action complementary to leptin, an adipose tissue peptide that suppresses appetite [101].

Glucagon is another member of this peptide family; it raises blood glucose by acting on GPCRs in the liver and muscle and activates enzymes for glycogen breakdown and glucose synthesis. It therefore opposes the effects of insulin (see below).

4.3.4.2 Hormones from Vascular Tissue

Endothelins are **vasoconstrictor** peptides that play important roles in vasomotor control. They are produced by the endothelial cells of blood vessels in several tissues and act on receptors that cause vasoconstriction. The three endothelins (ET-1, ET-2, and ET-3) have 21 amino acids and are nearly identical in amino acid sequence, though they show significant differences in binding and activation properties of the known endothelin receptors [104, 105]. These peptides are similar in sequence to the sarafotoxin group of snake venom toxins, which also bind to and activate the endothelin receptors in coronary vessels, causing vasoconstriction [105] (Figure 4.13 – venoms often target receptors, and this is an important aspect of their lethality; see Chapter 6 for discussions of the mechanisms of venom action).

4.3.4.3 Hormones from the Blood

Angiotensin II is a vasoconstrictor hormone produced in the blood when the enzyme renin is released from the kidney in response to reduction in blood pressure. Renin cleaves a peptide from a precursor, producing angiotensin I. This peptide is inactive, but is further cleaved by another blood-borne enzyme, angiotensin-converting enzyme. The final product, angiotensin II (eight amino acids in humans), has two important effects: vasoconstriction (elevating blood pressure directly by acting on angiotensin receptors) and release of vasopressin from the posterior pituitary (causing retention of fluid and thereby indirectly increasing blood pressure).

Bradykinin is a nine-amino acid vasodilator peptide formed in the blood by proteolytic cleavage of a protein involved in the blood coagulation pathway. It is

Et-1: CSCSSLMDKECVYFCHLDIIW

S-68: CSCKDMTDKECLYFCHQDVIW

Figure 4.13 Amino acid sequences of endothelin-1 and sarafotoxin-68, showing areas of sequence identity and disulfide bond locations.

produced during inflammation and acts on GPCRs to dilate blood vessels and airways, and to increase vascular permeability [106].

4.3.4.4 Peptide Hormones from Reproductive Tissues

Chorionic gonadotropin (CG) is a peptide released from the early embryo in mammals and from the placenta after implantation. Its action on ovarian cells converts the follicle into a secretory tissue, producing steroid hormones required for maintenance of pregnancy.

Relaxin is a hormone with sequence homology to insulin. It is produced by the ovary during pregnancy and softens the tissues holding the pubic bones together. This action allows the pubis to widen during birth. It is interesting that relaxin is an agonist on GPCRs (Chapter 7), while its close relative insulin is an agonist on **RTKs** (Chapter 8).

4.3.4.5 Hormones from Other Tissues

Bombesin is a peptide of 14 amino acids originally discovered in the skin of a European toad; its function in the toad is unknown, but it causes vasoconstriction when applied to exogenous tissue; its **endogenous** function may be either as a local vasoactive signaling compound or as a peptide used in defense against predation. Mammalian homologs of bombesin are neuromedin-B (NM-B) and gastrin-releasing peptide (GRP). NM-B is produced in the human CNS and activates genes for production of new brain cells.

Parathyroid hormone consists of 84 amino acids and is responsible for calcium- and phosphate-related regulation of the parathyroid gland. It acts on GPCRs to cause release of Ca^{2+} from bone, to absorb Ca^{2+} from kidney tubules, and also to increase synthesis of Vitamin D.

4.3.5 Non-Peptides Acting on G-Protein-Coupled Receptors

4.3.5.1 Transmitters Derived from Amino Acids

The actions of the adrenal gland hormones epinephrine (EP) and norepinephrine (NE) have been studied for over a century. These two substances, plus the brain transmitter dopamine (DA) (Figure 4.14), are called the **catecholamines** because of their structure, based on the aromatic molecule catechol.

These substances act both in peripheral tissues (the heart, blood vessels, muscle, and liver) and in the brain. Peripheral actions: EP and NE have long been recognized as part of the body's response to stress, exercise, and environmental stimuli such as temperature change. NE is the primary neurotransmitter of the sympathetic motor pathways that affect the smooth muscle of the gut, airways, heart, and blood vessels, and EP is the primary hormone released from the adrenal medulla.

The transmitters were first isolated from the adrenal gland (adrenalin = epinephrine; noradrenalin = norepinephrine). They act on related receptors (α- and β-adrenoceptors belonging to the G-protein-coupled superfamily); binding promotes the buildup of calcium within cells; the calcium acts as an

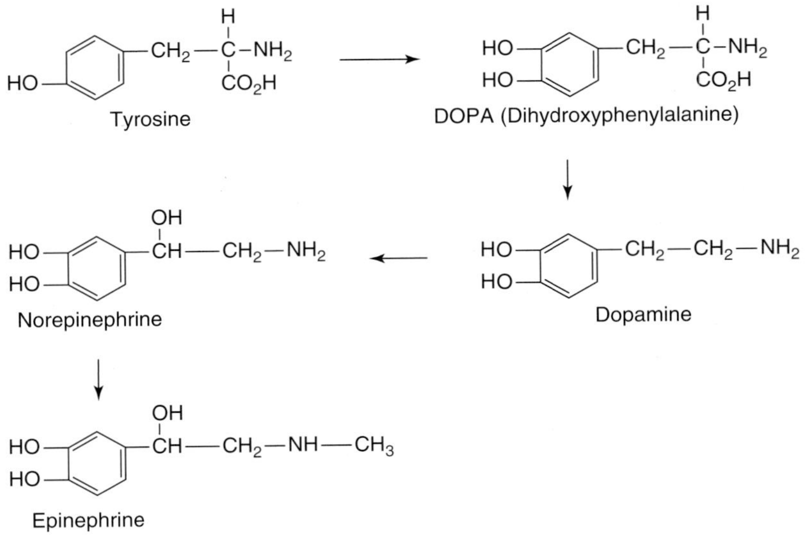

Figure 4.14 Synthetic pathway for catecholamine neurotransmitters and hormones. Dopamine, norepinephrine, and epinephrine serve as neurotransmitters.

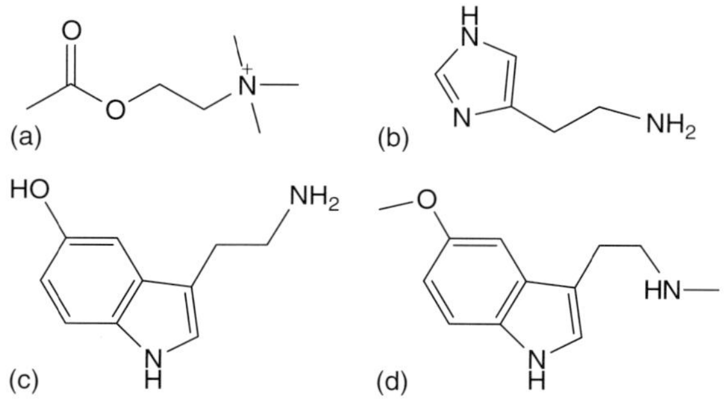

Figure 4.15 Molecular structures of (a) acetylcholine, (b) histamine, (c) serotonin, and (d) melatonin.

intracellular messenger to activate ion channels or enzymes. More about their receptor actions is found in Chapter 7.

Brain actions: Norepinephrine is an important neurotransmitter in brain pathways, especially in the hypothalamus and the reticular formation of the brainstem [107]. Other important transmitters are ACh, histamine, serotonin, and melatonin (Figure 4.15).

All four are derived from amino acid precursors (ACh from serine and histamine from histidine; serotonin and melatonin from tryptophan), and all are widespread in the animal kingdom. Acetylcholine and serotonin are unusual in that they act on two unrelated receptor types – the GPCRs of smooth muscle and the ion channels of nerves and skeletal muscle. These roles are explored separately in Chapters 6 and 7. Histamine is released by many types of cells, including brain neurons, but its best-known source is the mast cells of the immune system. It causes dilation of blood vessels, but at the same time it causes constriction of airways and gut and stimulates acid secretion in the gut. Melatonin is produced in the mammalian brain in response to visual inputs from the retina that signal decreasing daylight. It is thus is a sleep-promoting signal.

4.3.5.2 Transmitters Derived from Nucleotides

Several RNA nucleosides and nucleotides have been implicated as cell signaling molecules. Most are based on the structure of adenine and its association with ribose and phosphate to form nucleotides (Figure 4.16).

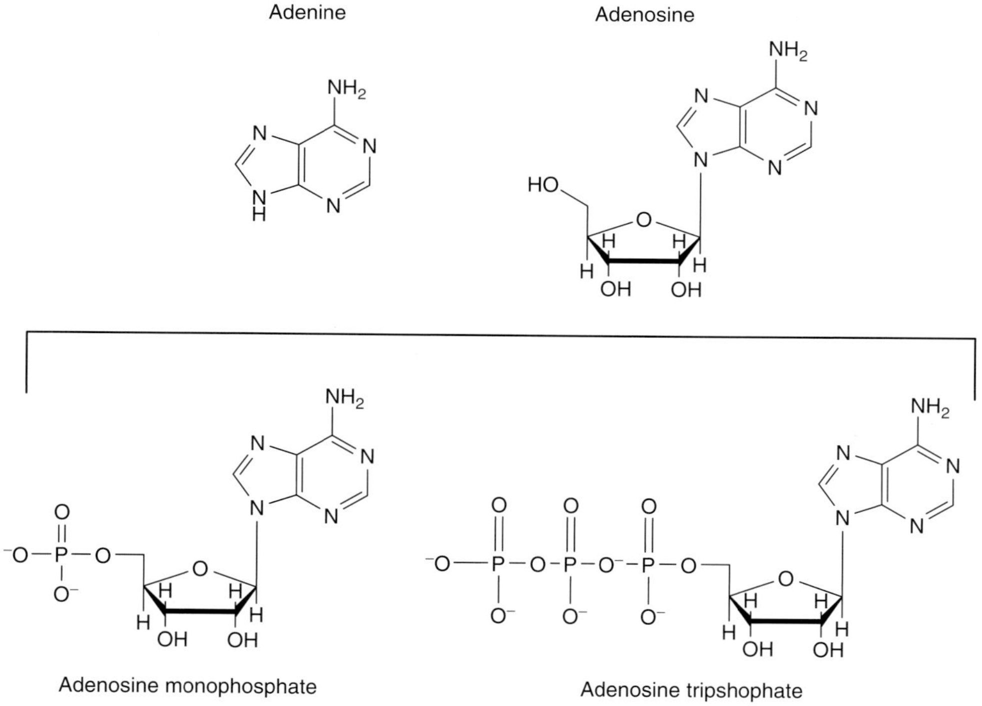

Figure 4.16 Molecular structures of nucleotide base alone (a), nucleoside (base plus ribose sugar – b), and nucleotide monophosphate (base plus sugar plus phosphate – c). These illustrations refer to the base adenine; similar molecules form for the other bases, guanine, cytosine, and uracil, and are based on the RNA version of each nucleotide.

Adenosine itself acts on GPCRs of the P1 (adenosine receptor) type (Chapter 7); ATP acts on two unrelated types of receptors – the P2Y type (also G-protein-coupled – Chapter 7) and the P2X type (a ligand-gated ion channel – Chapter 6).

4.3.5.3 Transmitters Derived from Membrane Lipids – Prostaglandins and Cannabinoids

Prostaglandins (PGs) are lipid-derived compounds that act as paracrine signaling agents in a variety of organisms, both prokaryotic and eukaryotic. The more general term for these compounds, eicosanoids, combines the PGs, PGI_2, and the thromboxanes. All are synthesized by the action of the enzyme phospholipase A_2 on cell membrane phospholipids to produce arachidonic acid. This product is then acted on by enzymes to generate the signaling eicosanoids (Figure 4.17).

The group name derives from the prostate, the **organ** thought to be the source of the first PGs discovered in 1935 in seminal fluid [108]. Though the seminal vesicles were the actual source for those secretions, the original name stuck. Subsequent research has shown that PGs are produced by all cell types. Though PGs are lipid soluble, they do not readily diffuse across cell membranes, and are released from cells by a unique mechanism involving a transporter of the ATP-binding cassette (ABC) family (Chapter 6).

The major signaling compounds in this group act on GPCRs of the **rhodopsin** class (Chapter 7).

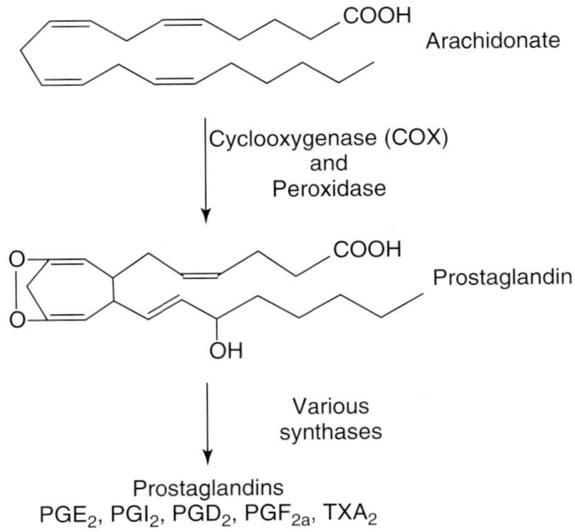

Figure 4.17 Synthesis of prostaglandins and related signaling compounds. COX: cyclooxygenase; terms with PG refer to prostaglandins; PGI_2 is also called prostacyclin; and TX refers to thromboxane.

The compounds are divided into three types and bear a two-letter PG designation plus a letter that describes which receptor each activates: PGD_2, PGE_2, $PGF_{2\alpha}$, PGI_2, and TXA_2 (thromboxane A_2). These act on D-type, E-type, F-type, I-type, and T-type receptors, respectively. Depending on the specific receptor activated, they cause smooth muscle constriction or dilation, alter neuronal function, regulate inflammation, and alter GI secretion [109].

A family of cyclooxygenase (COX) enzymes catalyze the reaction to form PGH_2, from which the remaining PGS are synthesized. COXs are inhibited by the **non-steroidal anti-inflammatory** drugs aspirin, indomethacin, and salicylate, and the inhibition of PG production by these drugs explains their analgesic effects.

More detailed structures of arachidonic acid and a representative PG are shown in Figure 4.18.

Cannabinoids in Plants and Animals This diverse group of compounds, known for their psychoactive effects in humans and other mammals, shows chemical similarities to the eicosanoids above, and act on two types of GPCRs in brain and peripheral tissues. Several dozen related compounds are produced by both animals and plants; not all have psychoactive effects. The animal compounds are called endocannabinoids and the plant compounds are called phytocannabinoids. The best-known phytocannabinoid is Δ^9-tetrahydro-cannabinol (THC), a product of the marijuana plant *Cannabis sativa.*

The actions of THC were initially thought to be via specific chemical reactions within cells, but Devane *et al.* [110] showed that it acted on a receptor, and that, as an analogy with opiate compounds, there must be natural ligands for these receptors. The best-studied natural ligands in mammals are anandamide and 2-arachidonoylglycerol, produced in several tissue types. Figure 4.19 shows examples of some phyto- and endocannabinoids.

The endocannabinoids are also synthesized from arachidonic acid derivatives, and thus share part of their biology with the prostanoid signaling molecules [111].

4.3.6 Transmitters of the Ion Channels

Over 90% of the receptors in the brain are fast ion channels that promote rapid neural activity or muscular action [91]. Ion channel transmitters are amino acids and their derivatives, never proteins. The main

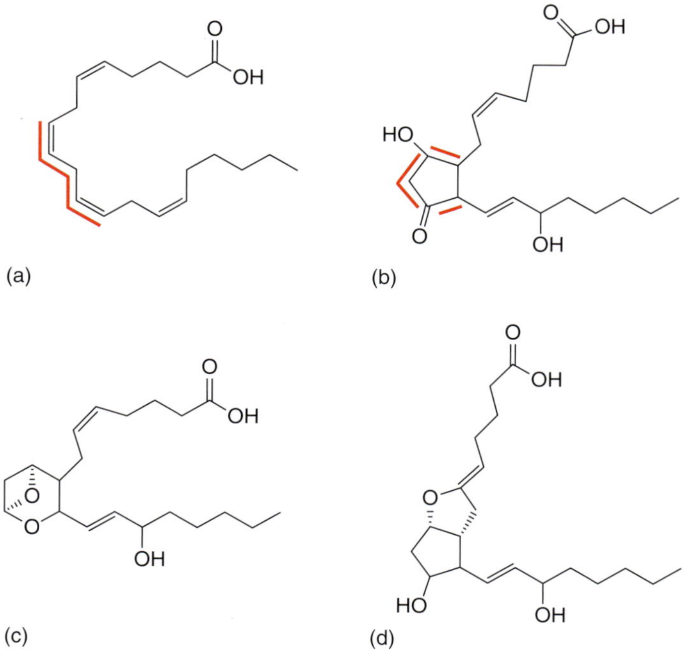

Figure 4.18 (a) Structure of arachidonic acid, from which the prostaglandins are synthesized by the action of cyclooxygenase. (b) Structure of an example of prostaglandins, PGD_2. The red lines represent equivalent carbon–carbon bonds, and in all PGs an oxygenated cyclic structure is formed between carbons 8 and 12. (c) Thromboxane A_2 (TXA_2), with cyclic oxygen on ring. (d) Prostacyclin (PGI_2), with cyclic oxygen between ring and chain.

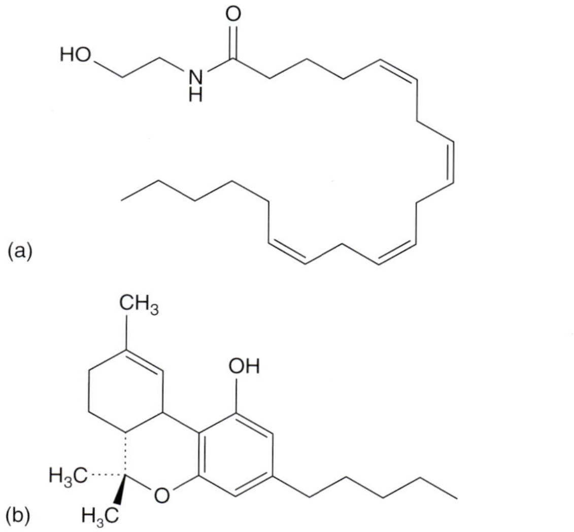

Figure 4.19 Molecular structures of cannabinoids. (a) Anandamide, an endocannabinoid. (b) Δ^9-Tetrahydrocannabinol, the active ingredient of the marijuana plant.

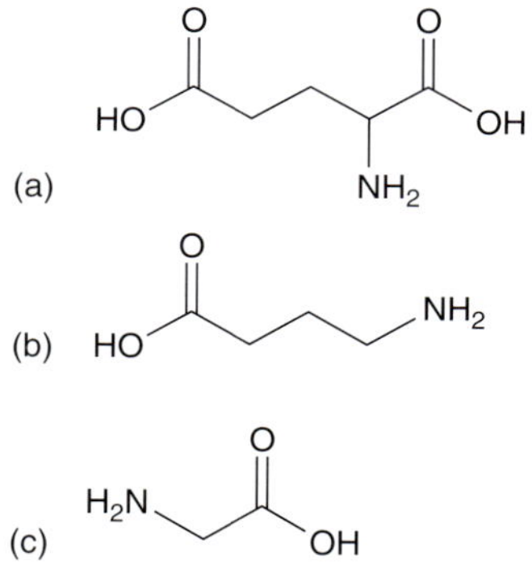

Figure 4.20 Molecular structures of amino acid transmitters. (a) Glutamic acid. (b) Gamma-amino butyric acid. (c) Glycine.

transmitters (Figure 4.20) are ACh, glutamate (both excitatory), GABA, and glycine (both inhibitory).

Excitatory Transmitters: Glutamate is the predominant transmitter in the brain, though ACh is used in certain pathways and aspartate has been proposed [91] as a possible transmitter.

Inhibitory Transmitters: GABA is the primary inhibitory transmitter in the brain, and acts primarily in the cerebral cortex and cerebellum. It is a metabolic breakdown product of the excitatory transmitter glutamate. Glycine is a major inhibitory neurotransmitter in the spinal cord and in the brain stem. The two transmitters act by increasing the conductance of chloride ions into neurons [91]. This causes a **hyperpolarization** of the cells as their membrane potential becomes more negative.

4.3.7 Hormones of the Receptor Kinases – Growth Factor Receptors

The hormones for growth factor receptors are always proteins, never amino acids or derivatives of fats or nucleotides. They often contain several residues of the amino acid cysteine, which form disulfide bridges and link portions of the peptide chains.

4.3.7.1 Insulin

Insulin is a protein of 51 amino acids consisting of two chains attached by three disulfide bridges. It is not synthesized in these two pieces, but rather in a single longer polypeptide called *preproinsulin*. The active form of the hormone results from the post-translational processing of preproinsulin, involving enzymatic removal of the N-terminal signal sequence and cleavage of a long peptide (the C peptide) to the final form consisting of two peptides joined by two disulfide links (Figure 4.21).

Responses to insulin are part of the "absorptive" process that occurs just after a meal, when glucose concentration in the blood is high. Increased glucose causes the β cells of the pancreas to secrete insulin into the blood [107]; in the liver, muscle, and fat cells, insulin binds to its receptor to promote uptake of glucose. The mechanisms of insulin action are discussed more fully in Chapter 8.

4.3.7.2 Insulin-Like Growth Factors

The liver secretes the somatomedins (the insulin-like growth factors IGF-1 and IGF-2), in response to GH in the blood. The somatomedins are similar in structure to insulin, including the presence of disulfide bridges

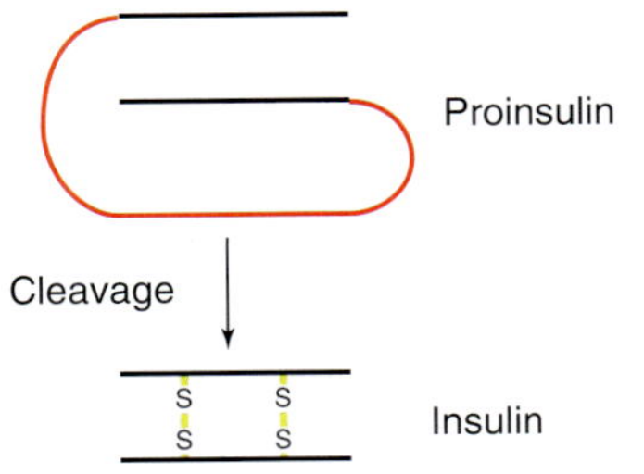

Figure 4.21 Overall structure and disulfide bonds of insulin. Proinsulin is synthesized as the red and black sequence. Cleavage of the amino acids at the black and red junctions results in the release of the nonfunctional C peptide (red); disulfide bonds connect the two short strands (black) that form the functional insulin.

connecting parts of the peptide chain. However, the somatomedins are single peptide chains because the C peptide is not cleaved [62].

4.3.7.3 Natriuretic Peptides

The heart produces and releases natriuretic peptides (NPs) in response to increased blood pressure. The NPs travel to the kidney, where they inhibit Na^+ reabsorption from the fluid in the kidney tubules. This inhibition promotes the loss of Na^+ and water from the body, resulting in reduced blood pressure [107]. In addition, the peptides act on vascular receptors to cause vasodilation.

4.3.7.4 Peptide Signal Molecules Important in Embryogenesis

Activin, Inhibin, TGFβ, and BMP: These peptides belong to the transforming growth factor (TGF) peptide class (Chapters 8 and 11). Activin and inhibin are produced by the gonads; they are both dimers, and their functions are mainly antagonistic to each other. Activin is a **homodimer** of two peptides with 116 amino acids, and inhibin is a heterodimer, one of which is an activin monomer. Activin activates gonadotropic functions in the gonads, while inhibin prevents release of gonadotropins from the anterior pituitary [112]. TGFβ and BMP (bone morphogenetic protein) are important signaling peptides during development.

Neurotrophins: During embryonic development, a major part of the differentiation of new cell, organ, and system types is directed by the actions of local modulators acting on cell surface receptors. Most of these modulators are peptides, and most also act on growth factor receptors (receptor kinases). Some important examples are fibroblast growth factor, epidermal growth factor, and the neurotrophins – nerve growth factor (NGF), brain-derived neurotrophic factor (BDNF), and neurotrophin (NT). These peptides regulate neuron survival during brain development, as well as the growth of neurite (dendrite or axon) projections on nerve cells. The general signaling functions of this class of peptides are discussed in Chapter 8 and their roles in embryonic development are discussed in Chapter 11.

4.3.7.5 Pituitary Gland Hormones – Somatotropin and Prolactin

Somatotropin (growth hormone – GH) stimulates growth in many tissues; it causes the release of somatomedins (IGFs – see below) from the liver. These hormones act further to promote growth in skeletal, muscle, and epithelial tissues. GH shows three areas of sequence homology with prolactin, and is nearly identical with the hormone placental lactogen from the placenta [62].

Prolactin and placental lactogen stimulate differentiation of the milk-producing tissues of the breast, and promote the synthesis of milk proteins. In this sense they are also GHs, though they have little effect on tissues outside the mammary gland.

4.3.8 Hormones of the Nuclear Receptors

The hormones and transmitters discussed above are all polar, and therefore water soluble. They thus remain outside cells and bind to cell-surface receptors. The following group of hormones consists of non-polar compounds that easily cross cell membranes. As discussed in Chapter 8, the binding of these molecules (steroids, certain vitamins, and thyroid hormone) to receptors within the cell activates transcription of new proteins.

Nuclear receptor hormones belong to several groups of similar molecules (Figure 4.22):

- Steroid hormones (estrogens, androgens, gluco- and mineralocorticoids)
- Non-steroids
  Thyroid hormone
  Vitamins A and D
  Insect metamorphosis (ecdysis) hormone.

4.3.8.1 Steroids

Steroids are found only in eukaryotes, and are involved in cell membrane construction, fat metabolism, and cellular signaling. The prototype, cholesterol, is not used as a hormone but is a crucial component of animal cell membranes (Chapter 3) and is the compound from which the hormones in this group are synthesized [62] (Figure 4.23):

The Adrenal Cortex: The cortex is the outer portion of the adrenal gland, and consists of two separate steroid-producing areas: the inner cortex releases cortisol when stimulated by the anterior pituitary peptide ACTH. The outer cortex releases aldosterone when stimulated by angiotensin II. Aldosterone acts on cells of the kidney to promote removal of Na^+ from the tubules and place it back into the blood.

The Gonads: Though the testis and ovary have different structures, they are homologous organs activated by identical hormones. The pituitary peptides FSH and LH both cause release of steroids and promote the maturation of gametes.

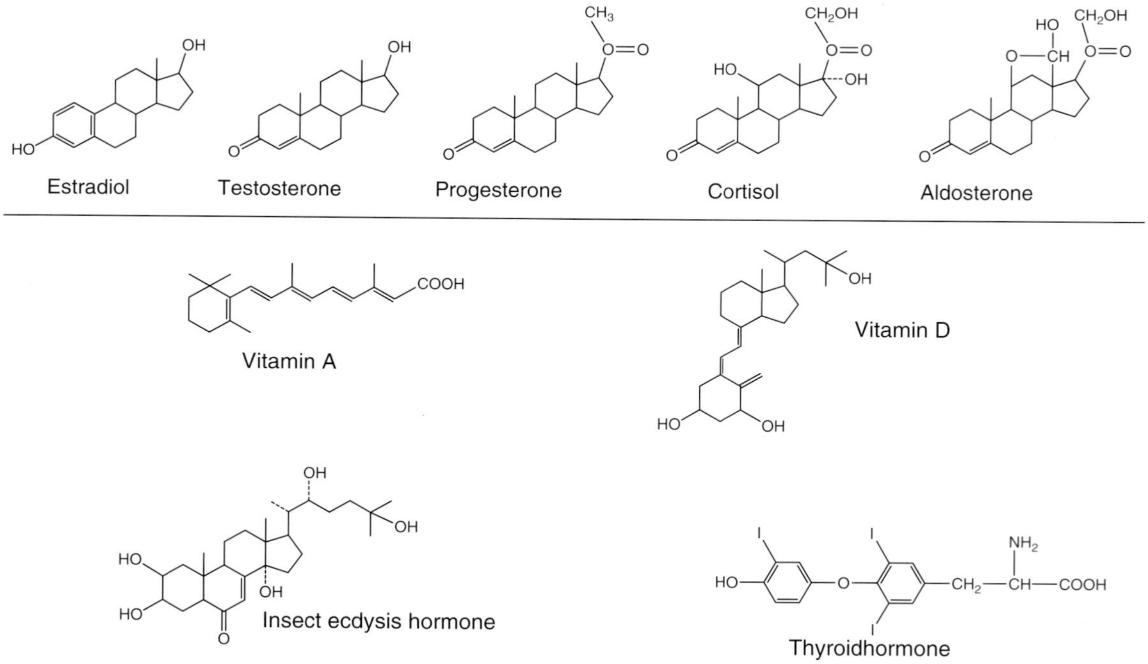

Figure 4.22 Molecular structures of hormones for nuclear receptors. Above: family 3 receptor ligands; below: family 1 receptor ligands. Receptor families are described in Chapter 9. Modified from Figure 1 of Ref. [113]. Used with permission.

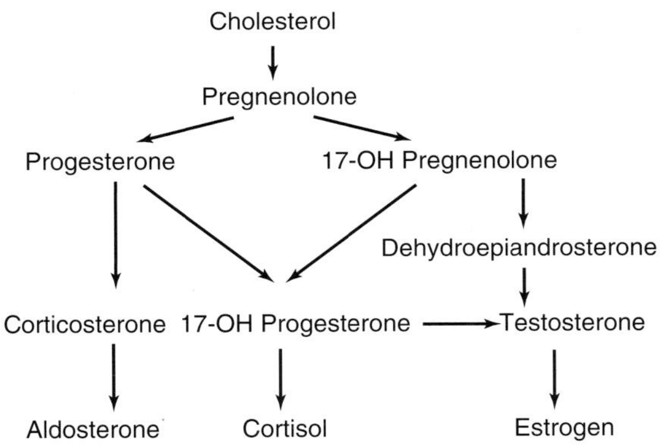

Figure 4.23 Synthesis of the major steroid hormones. Estrogen is the final steroid in the sex steroid pathway; its conversion from testosterone is catalyzed by the enzyme aromatase.

Though the steroid hormones are superficially similar to each other, important differences are present in the components of each ring that determine receptor-binding properties. Each of the slight variants shown in Figure 4.22 alters receptor-binding properties [62].

Steroid hormones are produced only in eukaryotes and are most diverse in animals, plants, and fungi. The steroids or steroid-like molecules seem to have originated as structural molecules rather than as signals, because bacteria, Archaea, and protists use them only to stabilize membranes. They are known to serve as signaling molecules only in plants and animals.

In vertebrates, steroid hormones are produced primarily by the adrenal gland and the gonads, though the placenta and brain are also sources of steroids and sites of their action. Thyroid hormone is produced by the thyroid gland in the neck of vertebrates, the homolog of the organ used in filter feeding in basal chordates; vitamin A is obtained in the diet, and vitamin D is both obtained in the diet and converted by sunlight acting on skin cells that contain the precursor steroid 7-dehydro-cholesterol.

The structure and function of the adult mammalian brain, including behaviors, is in large part determined by developmental processes that derive from the actions of steroid receptors as **transcription factors**. The developing brain is quite different from the adult brain, and those differences are established by the specific actions of steroids during development and in the first months after birth [114]. Though the gonadal steroids are associated in adults with one or the other gender – estradiol and **progesterone** with females, and **testosterone** with males – early development of the brains of both sexes relies on the action of *estradiol alone* on embryonic neurons.

The mechanism of normal male behavioral development appears to be the conversion of testosterone into estrogen in the neonatal hypothalamus, after which estrogen promotes the differentiation of the male hypothalamic and pituitary system (Figure 4.24).

Testosterone in external genitalia and ducts causes the production of male-specific structures that produce proper gonads, ducts, and secondary sexual features characteristic of males.

Estrogen thus appears to be the inducing steroid for male development; this is confirmed by the observation that estrogen added to female neonatal rats also masculinizes them. One might wonder why genetic females develop normally if estrogen is a masculinizing hormone. The reason is that no estrogen is available in the circulation of neonatal females, partly because little estrogen is produced in females at that early stage, and also because the little that is produced is bound and inactivated by α-fetoprotein, an estrogen-binding protein in the embryonic circulation [115].

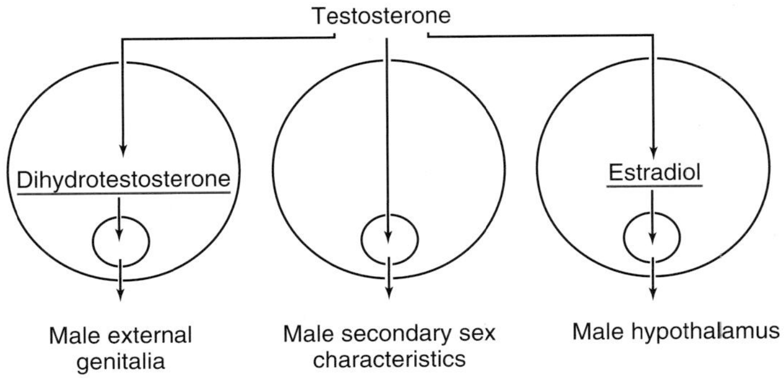

Figure 4.24 In different tissues of the male mammal, testosterone is either the source of the differentiating signal for male differentiation or the substrate for enzymes converting it to dihydrotestosterone or estrogen (red underlining). All three steroids have roles in male development.

Phoenix *et al.* [116] showed in 1959 that the behavioral effects of exogenous steroids were different during early development than during adulthood. This led to an important concept in psychology – the **organizational/activational** hypothesis of steroid action [114]. This idea proposes that in a biological male (XY), activation of Y chromosome genes causes embryonic production of testosterone. The testosterone acts as the "organizational" hormone to set up fundamental male structures and behaviors. Subsequent conversion of this *testosterone to estrogen* begins the process of development of the male gonads, brain, and reproductive tract. Because females (XX) do not have a Y chromosome, they do not make testosterone, and thus do not produce estrogen. The main consequence is that normal female development requires a steroid-free brain environment, and if either estrogen or testosterone is given to females during development, the result is sterility and lack of sexual responsiveness [114].

The *presence* of estrogen produces male gonads, brain, and reproductive tract and the *absence* of estrogen produces female gonads, brain, and reproductive tract. After birth, the "activational" aspect of this hypothesis comes into play, and gender-specific steroids affect male and female maturation and adult behavior [114].

4.3.8.2 Non-Steroid Nuclear Hormones

Thyroid Hormone: The thyroid is crucial for early development and growth, as well as for maintenance of adult metabolism. The gland is located in the neck, and produces several related hormones, all derived from the coupling of two molecules of iodinated tyrosine. The form most associated with thyroid activity is tri-iodo thyronine (Figure 4.22).

Vitamins A and D: Vitamin A is found in several forms in the body, including **retinal**, a component of the visual pigment **rhodopsin**. It is also involved in embryonic development, where it activates nuclear receptors to transcribe proteins involved in cellular differentiation (Chapter 11).

The term *vitamin* refers most precisely to substances that are required by the body but are not synthesized because of a lack of the proper enzymes. When Vitamin D was discovered [117], it was not known that it could be synthesized by humans. It is formed by the action of ultraviolet light on a form of cholesterol in the skin; the photoactivated compound then acts in the intestine to encourage uptake of Ca^{2+} from food. "Vitamin D" thus is actually a naturally occurring hormone, and its original name has been retained. Its continuing vitamin status results primarily from the occasional need to add it to the diets of people who live in regions lacking adequate sunlight.

4.4 Analgesics and Venoms as Receptor Ligands

We have earlier indicated the important roles of pheromones and other signaling molecules released from one organism and able to bind receptors on or in another organism. These substances are equivalent to intraorganismal hormones in the sense that they produce adaptive changes in physiology or behavior. But organisms have also evolved signaling molecules that pass to another organism, inhibit some physiological function in the recipient, and thereby promote survival in the organism releasing the signal.

In this category fall plant products such as curare (antagonizes the animal nicotinic ACh receptor – Chapter 6), strychnine (antagonizes glycine receptors and nicotinic receptors – Chapter 6), opiate drugs from poppies (activate opioid peptide receptors – Chapter 7), atropine (antagonizes **muscarinic receptors** – Chapter 7), and digoxin from *Digitalis* plants (inhibits Na/K pump in cardiac membranes – Chapter 6).

Animals such as predators also produce signals that affect other animals: the best known are toxins from venoms of snakes, spiders, and mollusks that, like their plant counterparts, alter function of receptors and channels in their prey. Later chapters introduce additional cross-species ligands.

5
Receptor Theory

> The receptor concept is to pharmacology as homeostasis is to physiology, or metabolism to biochemistry.
>
> *H.P. Rang [118]*

> Corpora non agunt nisi fixata
>
> *Paul Ehrlich [119]*
>
> THINGS DON'T WORK WITHOUT BINDING

5.1 The Materialization of Receptors

Chapter 2 showed how receptor thinking began: first, with the understanding that alterations in behavior are caused by molecular *chemicals* and not by spiritual "humors"; second, with the knowledge that the *concentration* and *structure* of these chemicals were the important features behind their action; third, with the postulate that such substances act on specific cellular receptors to exhibit their effects. As later chapters show, the receptor concept also means that certain drugs have powerful effects on some tissues and no effects on others, and can be effective in very small doses.

During the century after receptors were postulated, they remained largely mathematical abstractions rather than real objects because they had not been seen or directly manipulated, and their properties had to be inferred from pharmacological and physiological experiments. The idea of receptors thus emerged as a "black box" conceptual program in which investigators used chemical theory to determine how receptors might behave if they *were* real molecular entities. This conceptual program focused primarily on the physical chemistry of binding processes and to a lesser degree on biological processes.

As receptor structure became clearer from the 1980s on, pure mathematical approaches to modeling were combined with biochemical studies to determine the biochemistry of receptor actions. Today, a number of physical and chemical approaches are being used directly to visualize receptors: radioligand binding, X-ray crystallography, site-directed mutagenesis, and probe tagging (fluorescent and photoaffinity are prominent techniques. This research is providing the ability to define at least for some receptor types the molecular events that link receptor binding to cell activation.

5.2 Receptor Mechanisms

Four different superfamilies of receptors have been described (Chapter 1), of which three reside in cell membranes. Even so, the analytical approaches to receptor theory have been based almost entirely on the binding and responses of two receptor types – the ion channels of nerves and skeletal muscles (discussed in Chapter 6) and the G-protein-coupled receptors (GPCRs) of cardiac, smooth muscle, and glandular tissue (discussed in Chapter 7). These two types of receptors are the most widely distributed and give the most rapid responses, allowing the mathematics of agonist-receptor (AR) coupling to be most easily modeled.

For the purposes of this theoretical discussion, receptors can be viewed as large proteins that bind smaller activating molecules, or **agonists**, substances that act on receptors to initiate cellular processes such as neural activity, muscular contraction, metabolism, or protein synthesis. These agonists have been known from some of the earliest receptor studies [29]; that early work also compared the actions of agonists to other substances (**antagonists**) that opposed agonist actions. Thus emerged the idea that a cellular response was caused by a natural agonist (e.g., a neurotransmitter) that bound to a specific site on a "receptive substance," and that antagonists could also bind that same receptive substance to prevent the action of the agonist [29].

This binary concept of agonists as activators and antagonists as inhibitors guided research for several

Receptor Biology, First Edition. Michael F. Roberts and Anne E. Kruchten.

decades. It is still a useful way to think about receptor function, but recent theoretical and experimental work has produced a more complex view of cell activation. The remainder of this chapter explores the complexities of receptor activation.

5.2.1 Binding of Agonist to Receptor

In general, an agonist binds to a receptor to elicit a response. The initial hypothesis was that the binding of the agonist "caused" the receptor to acquire an active state that led to further cellular activation. As described in detail below, the current understanding is that receptor binding by agonists stabilizes an active state that the receptor arrives at spontaneously via normal molecular fluctuations [120]. This stabilization by the agonist allows the receptor to retain its active state long enough to initiate cellular processes.

For agonist binding to stabilize receptor active states, the bonds that form between the two must have two properties: they must be

- *selective*, so a given ligand makes a good fit to its receptor and binds only to that receptor, not to receptors of different ligands.
- *reversible*, so that ligands do not remain firmly bound to the receptor. This permits the cessation of receptor activation when the ligand is no longer being released.

Most of the antagonists used to inhibit agonist action also bind with **selectivity** and reversibility. We can thus use the term **ligand** for any molecule that binds a receptor, whether it activates a process or inhibits it. The chemical features of receptor binding are not unique to receptors, but involve the same types of bonds that characterize other interactions, such as those found in enzyme–substrate interactions.

5.2.1.1 Bonds

Ionic bonds form between atoms with opposite charges [121]. An example of ionic bond formation in receptor–ligand interactions is the association of the positive charge of the quaternary nitrogen in acetylcholine (ACh) with a negative charge on the ACh receptor. The strength of the ionic bond varies with the square of the distance between the two ions, and is a relatively strong attractive force for a charged ligand approaching an oppositely charged site on its receptor (Figure 5.1A).

Hydrogen bonds: hydrogen is a strongly electropositive atom, and when it is bound covalently to an atom with an electronegative character, such

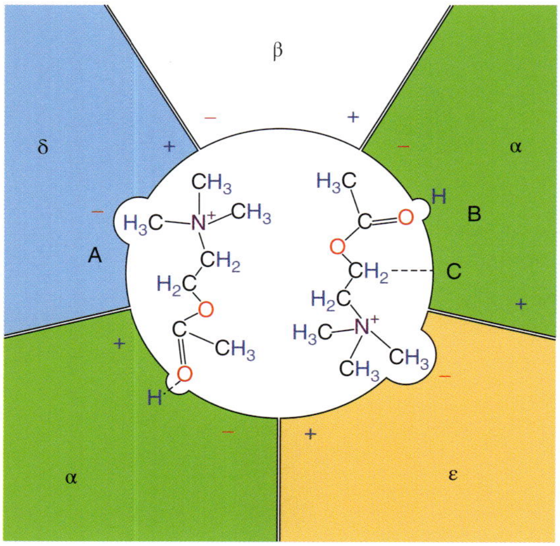

Figure 5.1 Top view of the nicotinic ACh receptor, showing the five peptide subunits (2α, β, δ, ε) in a circular arrangement around the central ion channel. Two ACh molecules bind, one to the α–δ pair (green and blue at left) and the other to the α–ε pair (green and orange at right). Binding is of three types: A (left): Ionic bonds between the quaternary N of the ACh and a negative charge on the δ and ε subunits. B (right): Hydrogen bonds between the carboxyl O of the ACh and an H on the α subunits. C (right): van der Waals bonds may also form between the intermediate portions of the ACh and nonionic amino acids of the receptor subunits. Only the β subunit seems not to be involved in specific bonding. Modified from: http://upload.wikimedia.org/wikipedia/commons/archive/5/5f/20091118095241%21Acetylcholine_nicotinic_receptor_binding.svg. Accessed 14 October 2014.

as oxygen or nitrogen, the hydrogen can share its electron with an adjacent electronegative atom, such as another oxygen or nitrogen [121]. ACh can form a hydrogen bond in association with its receptor: the carbonyl group oxygen is bound in this manner to a hydrogen ion on the surface of the receptor (Figure 5.1B). A single hydrogen bond is not sufficiently strong to hold a ligand to its receptor, but it forms a strong attractive force when acting along with the ionic bond.

van der Waals forces exist as attractions between adjacent molecules, and occur because even fairly neutral atoms have either a slight positive or negative charge. These charges promote alignments between adjacent molecules and act as weak attractive forces [121]. They may be too weak to assist the approach of a ligand to its receptor, but they act to stabilize the association once it has formed (Figure 5.1C).

It is now possible to see how a ligand approaches and binds to its receptor: in the case of ACh, the ionic bond at physiological temperatures is stronger than the disruptive influence of random thermal agitation; the ionic attraction thus guides the ligand to the vicinity of the receptor. Next, the hydrogen bond forms between the carbonyl group of the ACh and an –OH or –NH on the receptor, firming the association and holding the ligand close to the receptor. Finally, van der Waals forces develop between the neutral portions of ACh and neutral regions of the receptor.

This concept is still in development, as the actual atoms involved in most ligand–receptor interactions are known for only a few ligands (e.g., Figure 5.1). But the emerging picture is consistent with the idea that the two molecules must conform closely to each other: slight alterations in positions and charges of side groups on ligands alter binding. In the past few years, genetic and molecular methods have illuminated the amino acid sequences of many receptors, allowing researchers to develop models of their three-dimensional structure, the nature of the active site, and the properties of receptor interactions with cellular constituents. The term **affinity** is used to describe the result of all the above physicochemical binding forces; thus, high-affinity binding represents tightly fitting ligands and receptors. The role of affinity in ligand–receptor interactions is discussed next.

5.3 Binding Theory

5.3.1 Early Approaches to Understanding Receptor Action

Understanding the functions of the GPCRs has played a major role in the history of physiology and pharmacology: these were the receptors that initially defined the very existence of receptors. In 1905, Langley [29] proposed the existence of a "receptive substance" in muscle that mediated responses to certain substances. Hill [122] verified Langley's concept experimentally and began the quantitative analysis of how this hypothetical receptive substance might operate. Hill [122] and Clark [123] developed the earliest mathematical models of receptor binding. Their formulation (given below as Equations 5.1–5.5) shows the relation between agonist concentration and receptor occupation. We thus begin our discussion of receptor action with the *occupancy model.*

5.3.1.1 The Occupancy Model

The initial hypothesis of receptor action was based on the idea that the binding of agonist is adequate to activate the receptor. This hypothesis made several assumptions:

- Agonist and receptor interact via the law of mass action, standard for chemical reactions.
- All receptors are in the inactive state in the absence of agonist.
- It is not necessary to propose a specific cellular/biological mechanism for the activation of a tissue by an agonist – the mathematical description of binding alone is sufficient to explain receptor function.

As we show, the first assumption is reasonable; however, the others break down when the process of receptor activation is looked at carefully. The hypothesis nonetheless is a starting place for our theoretical development. This process occurs as agonist (A) is added to a solution containing receptors (R). The two combine to form the AR complex:

$$\mathrm{A} + \mathrm{R} \underset{V_2}{\overset{V_1}{\rightleftharpoons}} \mathrm{AR} \quad (5.1)$$

$$V_1 = k_1[\mathrm{A}][\mathrm{R}]$$

$$V_2 = k_2[\mathrm{AR}]$$

[A]	=	concentration of agonist
[R]	=	concentration of receptor
V_1	=	forward velocity
V_2	=	reverse velocity
k_1	=	forward rate constant ("on" constant)
k_2	=	reverse rate constant ("off" constant)

The hypothesis states that the rate of initial binding (V_1) depends on the products of the concentrations of the agonist and receptor, but that the rate of dissociation of the two (V_2) depends only on the concentration of AR. At equilibrium, the two velocities are equal ($V_1 = V_2$), and an equal number of ligand–receptor complexes are forming and dissociating per unit time. Thus, if we define

$$k_1[\mathrm{A}][\mathrm{R}] = k_2[\mathrm{AR}] \quad (5.2)$$

then

$$K_\mathrm{d} = k_2/k_1 \quad (5.3)$$

$$K_\mathrm{d} = \frac{[\mathrm{A}][\mathrm{R}]}{[\mathrm{AR}]} \quad (5.4)$$

K_d is the equilibrium or "dissociation" constant for AR binding. Its units are concentration (mol/l). Tight (high-affinity) binding occurs if the bound receptor [AR] is at a much higher concentration than the individual reactants [A][R]. Equation 5.4 contains variables we can measure (A, AR, and K_d) plus one we cannot measure (R). We thus wish to replace R by an equivalent expression: because the number of free

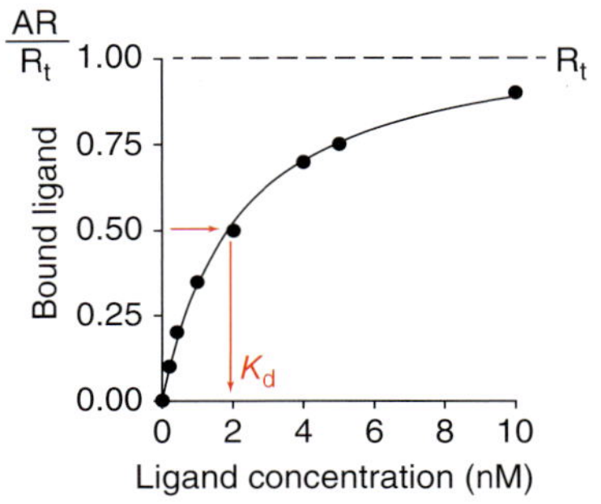

Figure 5.2 Graph of an experiment in which the amount of ligand bound to receptor (AR) is illustrated as a function of the ligand concentration. The K_d is about 2 nM. Agonist (ligand) concentration has a linear axis.

receptors R is equal to the total number of receptors (R_t) minus those bound (AR), we can set [R] = R_t − [AR], and rearrange, giving the final equation, which expresses the degree of binding (AR) as a function of the amount of agonist added (A) (Equation 5.5):

$$\frac{[AR]}{R_t} = \frac{[A]}{([A] + K_d)} \tag{5.5}$$

R_t	=	total no. of receptors
K_d	=	dissociation constant

This equation states that the amount of bound ligand (AR) as a proportion of the total number of

Box 5.1 Experimental approaches: isolated tissues

For the first century of receptor research, the main tool available to study receptor–ligand processes was the isolated tissue preparation (Figure 5.3), in which a muscle or other internal organ containing muscle tissue was placed into a physiological solution for measurement of its response to agonist.

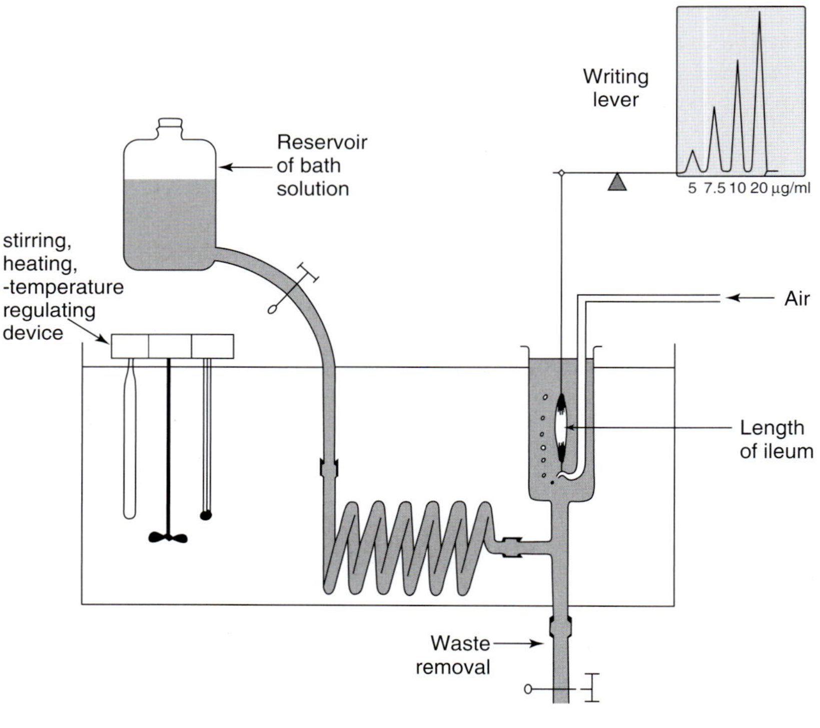

Figure 5.3 Apparatus for recording smooth muscle contractions. The tissue is within an organ bath at right, inside a larger bath that maintains temperature. Isometric tension is shown recorded at top right; increases in tension are caused by increasing concentrations of agonist, which are added to the organ bath in stepwise fashion. Tissue can be washed free of agonist by allowing solution (upper left) to replace organ bath solution. Stirring apparatus at left mixes large chamber. Figure 9.3 in Ref. [18]. Used with permission.

A known amount of drug was added and the response measured as a change in contractile strength. This approach produced data such as those shown in Figure 5.4. The data form a *dose–response* curve, a method of analysis pioneered by Clark [123] and by Ariëns [124].

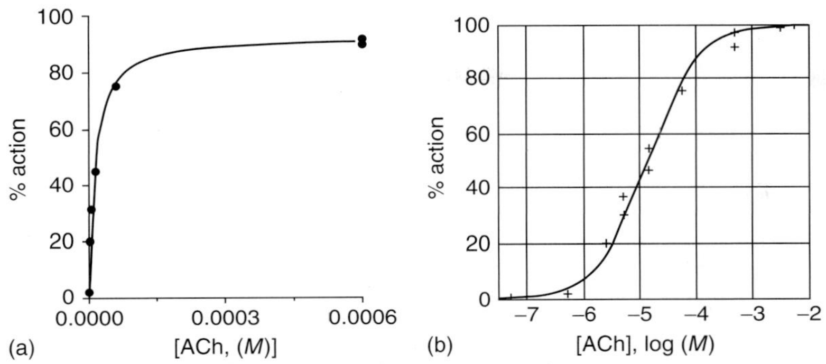

Figure 5.4 Response of tissue to hormone or neurotransmitter (agonist). Data from Clark [123] for action of ACh on frog rectus abdominis muscle. Same data plotted on (a) linear *X*-axis and (b) log *X* axis. Panel (b) modified from Fig. 6 in Ref. [123]. Used with permission.

binding sites (R_t) is a hyperbolic function of the concentration of ligand (A). The dissociation constant K_d is the concentration of ligand that gives 50% maximal binding. This holds because for

$$A = K_d$$

$$\frac{[AR]}{R_t} = 0.5$$

Figure 5.2 illustrates the relationship between agonist concentration and the fraction of receptors bound. This graph represents only binding of agonist to receptor, not the activation of the tissue that follows binding. Tissue activation will be dealt with below.

The form of the graph of response versus agonist resembles that for ligand binding alone (Figure 5.4), indicating a very tight coupling between binding and response; however, the cellular mechanism of that coupling is not given by either the graph or by Equation 5.5. Nonetheless, researchers have long used the dose–response curve as the basis for receptor modeling, and it will form an important part of the analysis we develop below.

Clark could only measure the contractile response of the tissue (e.g., Figure 5.4), and was not able to measure binding directly; he thus had to make simplifying assumptions about the steps between agonist binding and tissue activation. His most important assumption was that a linear relation existed between the two (no response at zero receptor occupation, leading to complete tissue response only at 100% receptor occupation – Figure 5.5, line A).

Clark thus thought that all agonists were alike in that *any* binding event led to a tissue response that was proportional to binding. In addition, he thought that the only important feature that related agonist binding to tissue activation is the dissociation constant K_d of agonist and receptor (Equations 5.4 and 5.5). Clark's hypothesis thus implied that all agonists with a given receptor occupancy produced the same tissue response.

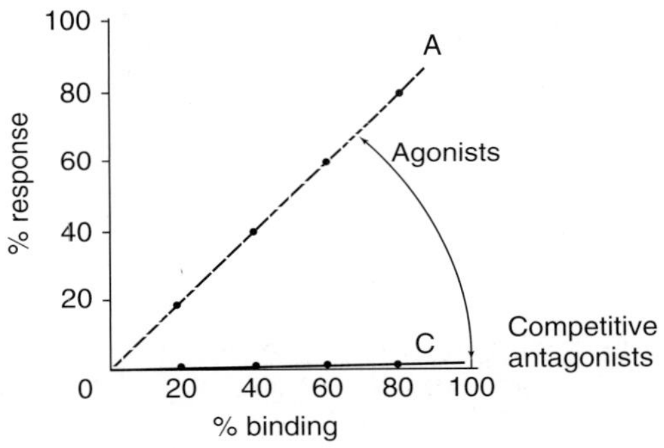

Figure 5.5 Clark's model of the relation between receptor occupancy (*X* axis) and response (*Y* axis). A: Linear relation between binding and response such that full response is reached only at full occupancy. C: The relation for a competitive antagonist, which may bind 100% of receptors without initiating a response.

Later researchers, in particular Furchgott [125] and Stephenson [126], questioned Clark's occupancy model on two grounds: first, not all agonists with equal occupancies were equally able to activate a given tissue, suggesting that besides affinity, another feature of receptor binding is important. Second, significant response could be achieved with only a small proportion of the receptors occupied [127], implying a nonlinear relationship between receptor occupancy and tissue response (Figure 5.6, line

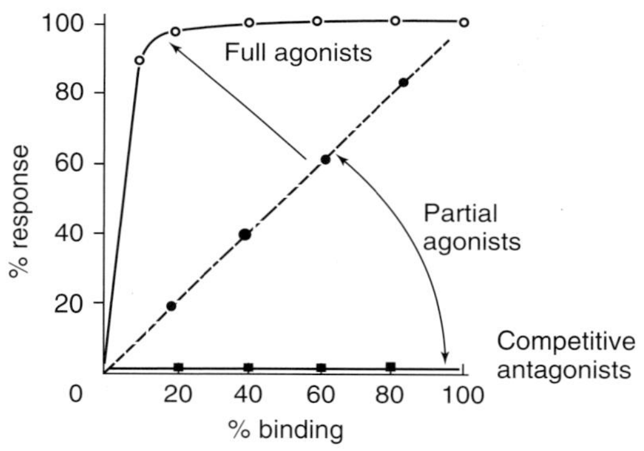

Figure 5.6 Early model of the relation between receptor occupancy (*X* axis) and response (*Y* axis). Partial agonists: Linear relation between binding and response (proposed by Clark) such that full response is reached only at full occupancy. Full agonists: significant response is reached when only a small proportion of receptors are bound. Competitive antagonists, which may bind 100% of receptors without initiating a response. Partial agonists may be defined as agonists with response patterns between 0 and the linear relation; full agonists by strongly hyperbolic response patterns.

labeled Full agonists). The nature of that nonlinear relationship produced the concept of differential agonist **efficacy** [120].

5.3.1.2 Processes That Follow Receptor Activation

Efficacy describes the ability of the receptor to activate the biochemical processes within the cell once binding has occurred. In the early days of theoretical modeling, efficacy was thought of in only three abstract ways: high-efficacy **full agonists** that activate cells at low concentrations; lower efficacy **partial agonists** that produce full activation only at high concentrations, and zero-efficacy **competitive antagonists** that bind to the same site as agonists but do not activate the tissue at all (Figure 5.6).

Antagonists can be natural (e.g., the plant substance atropine) or pharmaceutical (e.g., the drug propranolol). In both cases, they are important medically and can be administered at controlled levels to prevent unwanted effects of certain natural agonists. For example, atropine (muscarinic receptor antagonist) is used to counter the actions of ACh on salivation and gastrointestinal spasms, and propranolol (beta-adrenoceptor antagonist) is used to counter the **hypertensive** effects of epinephrine. Antagonists are also widely used in research for probing the binding properties of receptors.

In the sections that follow, we describe the current, more complex theoretical basis for ligand binding and how it is transduced into a cellular action; along the way we illustrate how the definitions of agonists and antagonists are being altered to account for the more nuanced interactions now known to occur between ligands and receptors.

5.3.1.3 Efficacy and Spare Receptors

Stephenson [126] hypothesized that in some cases, maximum effect can be produced when an agonist occupies only a small proportion of the receptors, and that different drugs have different capacities to initiate a response and consequently occupy different proportions of the receptors when initiating a response. His hypothesis thus proposes that an agonist has two properties, affinity and efficacy. Different substances have different binding affinities and different efficacies. Stephenson proposed that physiological substances had a spectrum of activities.

The concept that some compounds activate some tissues with a minimal amount of receptor binding also suggested that for specific AR combinations, a large proportion of receptors are "spare" in the sense that they could be eliminated without affecting the ability of the tissue to generate a maximal response. For example, if a ligand fully activates a tissue with 10% of receptors bound (Figure 5.6, line labeled Full agonists), then 90% of the receptors are not required to achieve full response. The concept of spare receptors became a part of the concept of high-efficacy full agonists.

Stephenson's idea of efficacy implied that researchers should also focus on the biochemistry of the cellular response to agonist. Efficacy could be seen as a single concept specifying the shape of the curve relating occupancy to activation: the higher the slope, the more an agonist could be considered a "full agonist" (Figure 5.6, line B). This analysis has a satisfying unity to it, as it appears to describe the functions of all possible ligands in one figure. But even this theoretical structure does not fully explain all examples of receptor function. We come now to new models that better account for the variety of receptor–ligand interactions and biochemical processes of cellular action.

5.3.2 Modern Approaches to Receptor Theory

5.3.2.1 The Two-State Model

In the 1950s, Koshland [2] developed a new way to model enzyme–substrate interactions. Rather than the static lock-and-key mechanism of Fischer, Koshland proposed that the active site accepts substrate in a reasonably tight bond (moderate affinity); the active site then is converted to a transition state

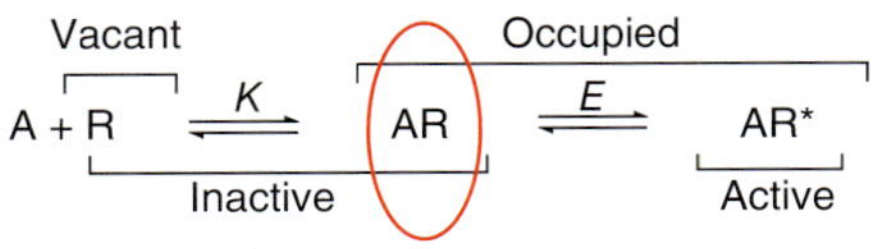

Figure 5.7 Del Castillo and Katz model for ion channel agonist binding and activation. *K* is the affinity of agonist binding. *E* is the efficacy of channel activation. This hypothesis shows the complex relationship between occupation and action, as inactive receptors may be vacant or occupied, and occupied receptors might be active or inactive. The portion within the ellipse is the occupied receptor converting from the inactive state to the active state.

(with a high-affinity bond for substrate), then cleaves the covalent bond holding the products. The main concept here is that binding causes an "induced fit" activation in the enzyme. A related model was applied to ion channel receptors by Del Castillo and Katz [128], who considered that receptors are inactive in the absence of agonist, but that binding causes a conformational change in the receptor; the pair remains inactive for a short while, then becomes active. This is the active state. Their model is given in Figure 5.7.

Del Castillo and Katz based their model on ion channel receptors (e.g., the muscle ACh receptor that causes Na^+ influx and depolarization) but it was also thought to represent other receptor mechanisms. A later version of the two-state model suggests that receptors exist in one of two conformations – R or R* – with the understanding that R* can exist even in the absence of agonist; the binding of agonist stabilizes the receptor in that state and promotes activity. At the time it was not known whether any cellular activity might occur without receptor binding. This idea is shown in Figure 5.8 (based on Equation 3.14 in Ref. [120]).

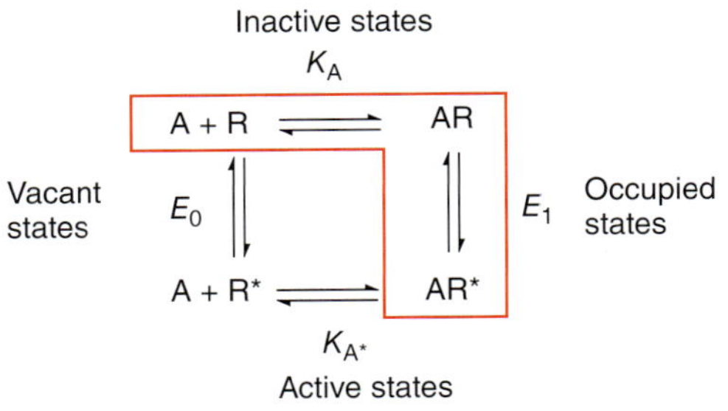

Figure 5.8 Two aspects of the two-state model of receptor activation. The boxed series represents classical hypotheses of affinity (K_A) and efficacy (E_1) according to the occupancy model and the Del Castillo and Katz [128] model. The remainder illustrates alterations accounting for active unbound receptors (E_0) and conformational change of receptor upon agonist binding (K_{A^*}).

This model suggests the existence of four separate receptor states, two relating to occupation and two relating to activity; these states are governed by four kinetic constants. The earlier classical models proposed that all unbound receptors are R, and are inactive. Agonist binding produces the occupied state AR, which only then converts to the active state AR*. In contrast, the two-state model proposes that R may change conformation to an active R* in the absence of A. The agonist A may bind to either state, but it has a higher affinity for R*. As the AR* conformation is the active one, the binding of A to R* stabilizes the active state, producing a cellular response.

Even though it considers spontaneous conformational changes, the two-state model does not reflect the types of interactions that occur between the receptor and **allosteric** proteins, for example, in the GPCRs. Thus, it has always been less inappropriate for the GPCR superfamily of receptors. Next we consider the implications of receptor theory for the complex interactions that occur between the receptors and these intracellular effectors.

5.3.2.2 The Ternary Complex Model

Research into metabolic processes in the 1950s and 1960s revealed the roles of intracellular molecules such as cyclic adenosine monophosphate (AMP) that were discovered to be products of GPCR action. When in the 1970s the connection was made between cyclic AMP and receptor action, it became important to develop a theoretical picture of the molecular interactions involved. The result was the 1980 ternary complex model [129], which describes the types of interactions between the receptor and its associated G protein.

In this model [129], the following interactions occur (Figure 5.9).

The agonist (A) binds to its receptor (R) to form the AR complex, which interacts with a membrane-bound G protein to form the agonist–receptor–G

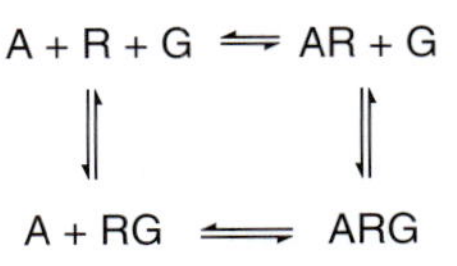

Figure 5.9 Ternary complex model. At top left, the agonist, receptor, and G protein are unbound. At top right, the agonist binds to receptor; at bottom right the G protein binds the others; at bottom left, the agonist dissociates. When initial state is reached again (top left); the receptor and G protein dissociate. The original figure of Ref. [129] used the letter X for what we now know is the G protein; it also illustrated coupling of the ARG to the effector, adenylyl cyclase.

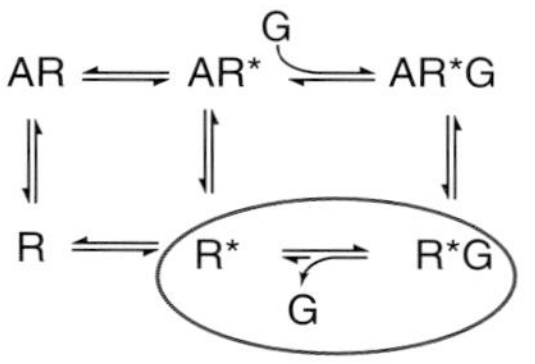

Figure 5.10 Extended ternary complex model. The four interactions at left form the ternary complex model (Figure 5.9); the extension involves incorporating the G-protein binding process along with its activating components (AR*G and R*G) at right. The R*G is not predicted to occur in the ternary complex model.

protein complex (ARG). In this formulation, the binding event activates the G protein to initiate one of many possible cellular events, including activation of an effector such as adenylyl cyclase (details in Chapter 7). This was the first receptor model expressly to invoke complex allosteric interactions that occurred in response to agonist binding. This hypothesis requires that activation of the G protein occurs only following agonist binding, and that binding creates the activated state of the receptor (R*). This concept was consistent with Koshland's [2] "induced fit" hypothesis for agonist binding.

Even though the ternary complex model explained the actions of many agonists in G-protein-coupled systems, by the late 1980s it was clear that it did not account for a newly discovered phenomenon: the intrinsically active receptor, noticed when mutated beta-adrenoceptors were constructed with the ability to take an active conformation (R*) in the absence of agonist. The R* interacted with G proteins in an unpredicted way, forming an active R*G complex (and thus able to initiate cellular responses even if no agonist is present).[1] Two groups – Costa and Herz [130] and Samama *et al.* [131] suggested that an extended ternary complex model is needed to account for this activity. The extension involves the direct activation of G protein by the unoccupied but active receptor R*, forming the R*G complex (blue oval in Figure 5.10 – based on Equation 3.19 in Ref. [132]).

Costa and Herz [130] also found that some ligands can bind to the R state of the receptor and alter the equilibrium between R and R* toward less R*, and therefore less R*G. This agonist action is a form of negative efficacy [132], is formally termed **inverse agonism** [130], and is shown in Figure 5.11.

1 This discovery, by the way, indicates that Ehrlich's dictum given in the epigraph to this chapter is out of date.

Inverse agonism is seen only when **constitutive** activity of receptors is present and cells have an intrinsic activity in the absence of agonist activation. During the years that theoretical studies focused on agonists as the sole activators of receptors, constitutive activity was thought to be an unusual occurrence. However, the models discussed in the previous few sections show that it is actually an occasional state of all unoccupied receptors, and that normal receptor activation involves the stabilization by agonists of a spontaneously arrived-at active state [120]. Once the concept of inverse agonism developed, investigators began seeking constitutive activity; it has been reported for native histamine receptors [133], so it may well be a widespread phenomenon.

5.3.2.3 Protean Agonism

Another consequence of recent theoretical and experimental advances is the discovery of variable agonism. A given agonist may activate or inhibit a tissue depending on the conformation of the receptor to which it binds. If an agonist A binds to an active receptor (thus forming AR*), it acts as a full agonist; if it binds to an inactive receptor (forming AR), it produces a lower efficacy complex that might even inhibit tissue function. This effect is part of the concept of **protean** agonism. The name refers to Proteus, one of the sons of the Greek sea god Poseidon, who frequently changed his appearance. If, in one tissue, an agonist bound its receptor and produced the state AR′ (somewhat lower efficacy than if it produced AR*), it could still be a positive agonist. If, in another tissue the same agonist bound a constitutively active receptor R*, it would reduce the efficacy and lower the overall tissue activity. A given agonist could thus be either a positive agonist or an inverse agonist. Inverse agonism is thus the lower half (reduced activity) of protean agonism. The study described above that defined constitutive activity of *in vivo* histamine receptors [133] also provided evidence for *in vivo* protean agonism.

The "protean" state was originally seen as a variable function of the same agonist, acting on the same receptor though on different final responses. It has since been extended to refer to the case where two different agonists act on the same receptor to produce two different effector outcomes. Figure 5.12 illustrates such an effect for two agonists and the dopamine 2 receptor (D2R). The receptor couples to both G_o and G_i; *S*-(−)-3-(3-hydroxyphenyl)-*N*-propylpiperidine (*S*-3-PPP) activates G_o and inhibits G_i (Figure 5.12a); in contrast, *n*-propyl norapomorphine (NPA) activates both G proteins (Figure 5.12b).

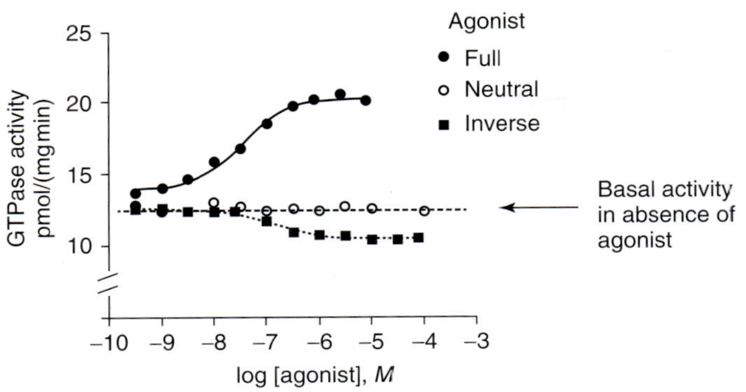

Figure 5.11 GTPase activity in isolated cell membranes in response to different ligands. Data show that receptor activity can be non-zero in the absence of agonist binding, and that different "agonists" have different modes of interacting with receptor. The neutral agonist in this model is equivalent to the competitive inhibitors described earlier; an inverse agonist creates a new type of inhibitory interaction. Based on data in Figure 1 of Ref. [130].

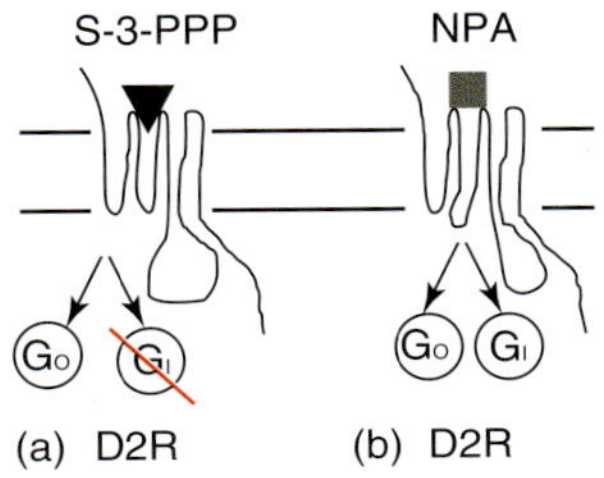

Figure 5.12 (a) *S*-(−)-3-(3-hydroxyphenyl)-*N*-propylpiperidine (*S*-3-PPP) binds to the dopamine D2 receptor to activate G_o, as well as to prevent G_i activation (red line indicates inverse agonism). (b) n-Propyl norapomorphine (NPA) binds to the same receptor, acting as an agonist of both G_o and G_i. Modified from Figure 1 in Ref. [134]. Used with permission.

5.3.2.4 Cubic Ternary Complex (CTC) Model

Interactions between G proteins and receptors may occur when each protein is in either its active or its inactive states; thus, R with G, R* with G, R with G*, and R* with G*. Each of these can also interact when the agonist is bound, and forms the basis of the cubic ternary complex model (Figure 5.13).

This model [135] includes all the possible protein–protein interactions as well as agonist–protein interactions and allows a complete description of G-protein receptor function. It is not known whether all the conceivable interactions actually occur in nature, but the model attempts to include all possibilities. Thus, the earlier ternary complex model is a subspecies of the cubic ternary complex (CTC), and differs primarily in that it does not include interactions between G proteins and inactive R. Future research will indicate whether, as with constitutive activity, some hitherto undiscovered active states can be identified.

5.3.3 Summary of Model States

The agonist actions of the preceding sections can be summarized by considering Figures 5.11 and 5.14 together.

Figure 5.11 illustrates full and inverse agonism, along with neutral (competitive) antagonism. To these "full" states can be added partial (positive) agonists and partial (inverse) agonists (Figure 5.14). The mechanism underlying these different agonist states is the differential affinity for states of the receptor. Thus, the receptor spontaneously alters conformation from R to R*; depending on which state it is in when it binds an agonist, it will be stabilized in that state. Activity or lack of it follows from the state that is stabilized. A full agonist has high affinity only for the R* state; a competitive antagonist has equal affinity for the R* and R states; an inverse full agonist has high affinity only for the R state. Partial agonists have greater affinity for one or the other of the R or R* states. A given substance may act as a competitive antagonist in some cases (e.g., where the receptor has little intrinsic activity) but may be an inverse agonist for a receptor with some intrinsic activity [137].

5.4 Visualizing Receptor Structure and Function

5.4.1 Determination of Receptor K_d

The conformation of a receptor changes with alterations in its function and with changes in the physical and chemical environment. Because conformation

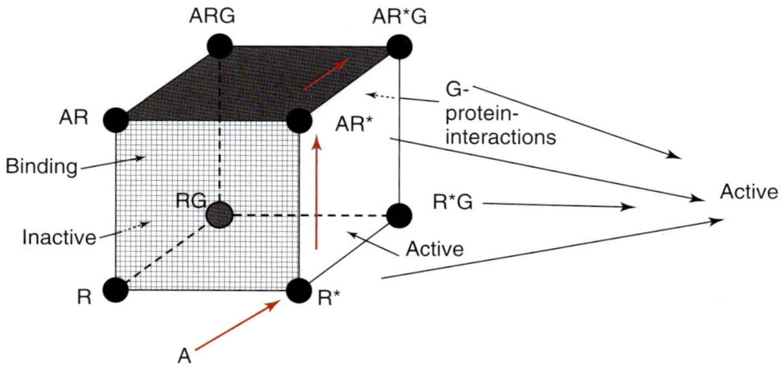

Figure 5.13 Cubic ternary complex model of G-protein-coupled receptor interactions. R: inactive receptor; R*: active receptor; A: agonist; and G: G protein. The front face and red arrows represent the extended ternary complex model from Figure 5.10. In the cubic ternary model, all states on the right side of the cube are potentially "active." Modified from Figure 2 in Ref. [135]. Used with permission.

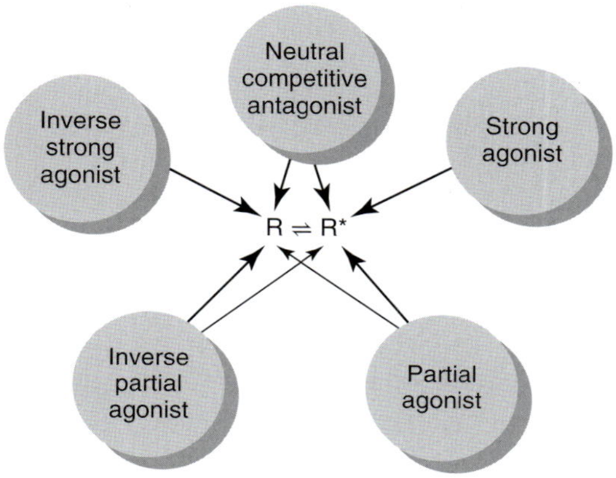

Figure 5.14 Explanation of the different states of agonism and antagonism. Receptor alters between R and R* states, and different types of agonists have different affinities for one or the other of those states. The thickness of the arrow is proportional to agonist affinity for the indicated state, leading to each circled agonist or antagonist description. Protean agonism is not shown, but would be the ability of a given agonist to be either an inverse or partial agonist on the same receptor. Figure 3 in Ref. [136]. Used with permission.

is a determinant of receptor K_d, one approach to measuring conformation is to measure K_d of ligands under different conditions. We discuss several methods, some using classical methods and some using modern instrumentation, that allow visualization of binding and activation of receptors.

5.4.1.1 Schild Analysis

When both agonists and antagonists compete for the active site; the agonist has efficacy to activate the receptor; competitive antagonists have zero efficacy, so binding is without effect. Antagonists thus occupy receptors and prevent the agonist from binding. The occupancy equation for agonists alone is

$$\text{Agonist receptor occupancy} = \frac{\mathrm{A}}{\mathrm{A} + K_\mathrm{A}} \tag{5.6}$$

A	=	concentration of antagonist
K_A	=	binding constant of agonist

In the presence of a competitive antagonist, agonist binding becomes:

$$\text{Agonist receptor occupancy} = \frac{\mathrm{A}}{\mathrm{A} + K_\mathrm{A}\left(1 + \frac{\mathrm{B}}{K_\mathrm{B}}\right)} \tag{5.7}$$

B	=	concentration of antagonist
K_B	=	binding constant of antagonist

The term $(1 + \mathrm{B}/K_\mathrm{B})$ in the denominator is the only difference between the occupancy equations for agonist alone and agonist plus antagonist [120]. Its effect is to shift the occupancy curve to the right in the presence of antagonists as compared to the curve with agonist alone. Thus, this suggests the existence of two occupancy curves, one with agonist alone and one with agonist in the presence of antagonist (Figure 5.15). In the presence of the antagonist, more agonist is required to achieve the same occupancy. We can use the variable d_1 to indicate the "dose" of agonist to achieve half-maximal binding with agonist

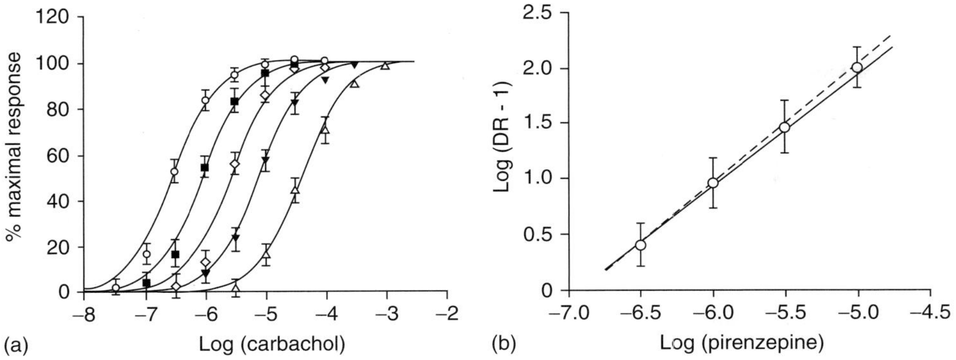

Figure 5.15 Schild analysis for the response of rat trachea to carbachol; competitive antagonist – pirenzepine. (a) Dose–response curves in the absence (open circles) and in the presence of pirenzepine from 300 nM to 10 μM. The antagonist causes concentration-dependent rightward shifts in the dose–response curve. (b) Use of Equation 5.8 to graph the Schild regression that allows calculation of pA_2. *Y* axis: log (dose ratio – 1) and *X*-axis: log [pirenzepine]. In this case, PA2 is approximately −6.9. Dotted line: linear best fit; solid line: line with slope = 1. Figure 6.6 in Ref. [120]. Used with permission.

alone, and the term d_2 to indicate the dose of agonist required in the presence of antagonist. We can also define a "dose ratio" $x = d_2/d_1$ to show the degree of antagonism. Because the term $(1 + \mathrm{B}/K_\mathrm{B})$ in the denominator is the only difference between the occupancy equations for agonist alone and agonist plus antagonist, that term is the mathematical expression connecting the dose ratio to B and K_B.

$$x = 1 + \frac{\mathrm{B}}{K_\mathrm{B}}$$

Rearranging and taking logs of both sides,

$$\log(x - 1) = \log \mathrm{B} - \log K_\mathrm{B} \qquad (5.8)$$

The graph of this equation gives us the value of K_B: if the ordinate is $\log(x - 1)$ and the abscissa is log B, the equation expresses a straight line with slope −1, and Y-intercept log K_B (indeed, because the slope is −1, both the *X*- and *Y*-intercepts are log K_B). The descriptive term often used for log K_B is "pA_2," because for a dose ratio x of 2, log $(x - 1) = 0$, which is the *X*-intercept of the line.

An example of Schild analysis is shown in Figure 5.15.

5.4.2
Visualizing Ligand Binding

This section describes methods used to study the binding of radiolabeled ligands to isolated receptors for the purpose of determining binding affinity, receptor numbers, and effects of physical factors on receptor function.

Several methods have been developed for observing ligand binding; all involve the use of a ligand containing a covalently bound radioactive atom. Radioactive forms of hydrogen (^{3}H), carbon (^{14}C), or iodine (^{125}I) are most commonly used for visualizing receptor binding (Figure 5.16).

The ^{125}I label has a very high degree of radioactivity (high specific activity) per molecule, so it allows researchers to locate bound receptors even at very low ligand concentrations. In contrast, the ^{3}H label has a lower specific activity, and thus cannot be used to label ligands that are present in low quantities [138]. Paton and Rang [139] were the first to apply the method, using a ^{14}C-labeled agonist to bind to guinea pig smooth muscle receptors. It soon became apparent that the binding affinity of most agonists is too low (K_d about 10^{-6} M) to give accurate estimates of binding, and the current procedure is to use labeled *antagonists* because of their higher affinity (K_d closer to 10^{-9} M) [140].

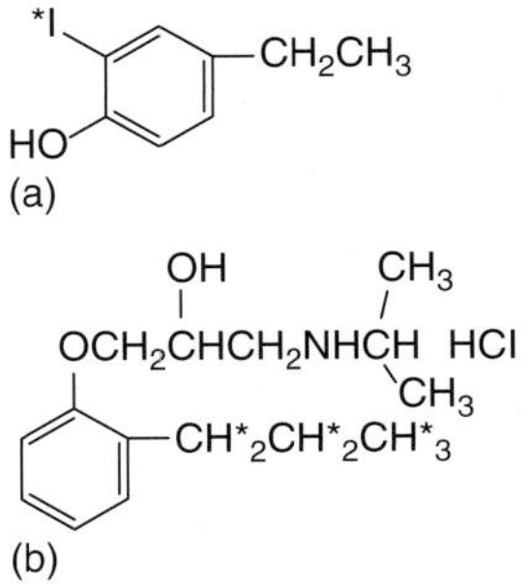

Figure 5.16 Radiolabeled compounds used for radioligand studies. (a) ^{125}I-labeled tyrosine as part of the insulin molecule. (b) ^{3}H-labeled dihydro-alprenolol, a beta-adrenoceptor antagonist.

5.4.2.1 Receptor Preparation

Receptor binding processes can be studied at several levels: by using whole cells, by obtaining isolated plasma membranes, or by using cell-free biochemical systems. To obtain a high concentration of isolated membrane-bound receptors, one fractionates a muscle or organ in buffer by grinding the tissue and centrifuging the solution at relatively slow speeds ($2000 \times g$) to pellet the denser cellular fractions such as nuclei and mitochondria. The membranes plus receptor reside in the supernatant at this stage, so that fraction is centrifuged at high speeds ($40\,000 \times g$) to obtain a new pellet consisting of the plasma membrane of the cell. The final pellet of membranes is resuspended in buffer for later study.

These methods are relatively crude in that they do not produce pure plasma membrane plus receptor. However, they are adequate with most tissues because the other membranes (e.g., endoplasmic reticulum, mitochondria) do not often carry receptors that might confuse binding studies. However, it is important for the experimenter to demonstrate that the preparation obtained is plasma membrane [140].

5.4.2.2 Equilibrium Binding Studies

To understand how ligand binding affects cell activation, researchers determine the values of the several binding parameters, such as the equilibrium dissociation constant K_d, and the total number of receptors R_t. This is done by using steady-state (equilibrium) binding experiments.

Receptors are **specific** sites for ligand attachment. However, any tissue or isolated membrane preparation contains sites of potential ligand attachment other than the receptors under study. These are **non-specific** sites that attract the ligand because of accidental conformity of shape or electrical charge. Thus, the **total** amount of binding of a radiolabeled ligand to a membrane preparation involves both specific binding to receptors and non-specific binding to non-receptor sites.

In experiments, total binding is achieved by incubation of a constant concentration of receptor with several different concentrations of ligand; nonspecific binding is achieved in parallel experiments by using the same concentrations of ligand and receptor, but by adding a high concentration of antagonist to all incubation tubes to prevent all specific binding.

At each concentration of ligand, the difference between total and nonspecific binding gives the specific binding of ligand with receptor (Figure 5.17).

From these curves, the parameters K_d and R_t can be determined: R_t is the maximum level reached for specific binding (graph at right; high ligand concentration), when all available receptor sites are occupied. K_d is the concentration of ligand needed to achieve half-maximal binding. Neither K_d nor R_t are actually obtained graphically; they are instead obtained from nonlinear regression of binding data because such methods are needed for experiments in which data are not distributed normally [141].

5.4.2.3 Competition Studies

The binding studies just described illustrate how to measure K_d of labeled *antagonists*. However, it is also important to determine how different **subtypes** of a given receptor might be distributed in the different tissues of an organism.

Determining Distribution of Receptor Types For this determination, a useful method involves **competition**, or **displacement**, experiments. In these experiments, one distributes identical membranes to several tubes. Each tube contains the same amount of membrane plus a constant concentration of radiolabeled ligand that can bind to many types of related

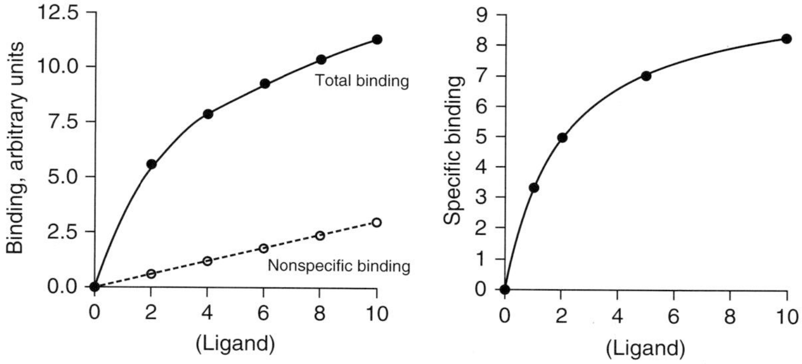

Figure 5.17 (a) Total and non-specific binding for an equilibrium binding experiment. (b) Specific binding (difference between total and non-specific binding). Both figures use arbitrary units of binding and concentration.

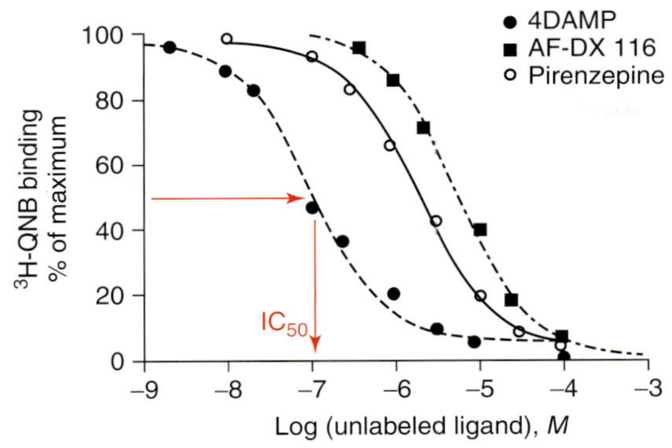

Figure 5.18 Competition study to measure K_i of antagonists in bovine tracheal smooth muscle. The vertical axis shows binding of a radiolabeled substance that binds to all types of muscarinic cholinergic receptors. The compounds listed in the legend are competitive antagonists that bind specifically to different types of muscarinic receptors. Data from Ref. [142].

receptor. Finally, the set of tubes contains increasing concentrations of an unlabeled ligand that is known to bind only to a specific type of receptor and not to others. If the receptor to which the unlabeled ligand binds is present in the tissue, the unlabeled ligand will bind those receptors and displace the labeled ligand. If that receptor type is not present, the unlabeled ligand will not be able to displace the label.

An experiment of this type is shown in Figure 5.18. The tissue is the bovine trachea; muscarinic receptors exist in several types (as of the time of the study, three were known: M1, M2, and M3). M1 receptors are bound with high affinity by pirenzepine; M2 receptors by AF-DX 116; M3 receptors by 4DAMP. To determine which of these receptors is most prevalent in the trachea, we [142] first labeled all the muscarinic receptors with ^{3}H-QNB (a nonspecific ligand). We then used the other three ligands one at a time to probe for the three types of muscarinic receptor. Because 4DAMP was able to displace ^{3}H-QNB most easily, we concluded that the most prevalent muscarinic receptor was of the M2 type, and that the M1 and M3 receptors were much less important.

The K_d of agonist binding can be computed from such an experiment. Competition curves (Figure 5.18) generate values of IC_{50}, the concentration of antagonist needed to displace 50% of the labeled ligand. This value is then combined with the concentration and K_d of the radiolabeled ligand alone (determined in other experiments) to calculate the K_i, (constant of inhibition) that is equivalent to the K_d. The method was pioneered by Cheng and Prusoff [143], who developed the following equation to calculate K_i:

$$K_i = \frac{IC_{50}}{1 + \frac{[A]}{K_d}} \tag{5.9}$$

K_i	=	the K_d of the antagonist being tested
IC_{50}	=	concentration of antagonist that reduces labeled ligand binding by half
[A]	=	concentration of *labeled ligand*
K_d	=	K_d of labeled antagonist (already known from previous experiments)

Munson and Rodbard [144] evaluated the Cheng–Prusoff calculation and determined that it provides an inaccurate estimate of K_i if the amount of bound radioligand is high compared to the total amount in solution. They recommend incorporation of the ratio of bound to free radiolabel in the equation, which becomes (Equation 5.10):

$$K_i = \frac{IC_{50}}{\left[1 + \frac{A\ (y_o+2)}{2K_d(y_0+1)} + y_0\right]} + K_d\left(\frac{y_0}{y_0+2}\right) \tag{5.10}$$

y_0	=	ratio of bound labeled ligand to free labeled ligand

If y_0 is small, Equation 5.10 reduces to Equation 5.9. If not, the Munson–Rodbard correction improves the accuracy of the K_i estimate considerably.

5.4.3
X-ray Crystallography of Native and Agonist-Bound Receptors

This is a method for determination of crystal structure of molecules from simple inorganic compounds to large organic molecules. X-rays are used as a probe to measure molecular size, bond angles, and lengths, based on the idea that a beam of X-rays will refract off planes within a structure, and will produce a pattern of dots on a photographic plate placed near the object. The process is similar to that seen when one allows sunlight to strike a crystal – the pattern of reflections on nearby walls can be used to determine the general structure of the crystal faces. Visible light can be used to see large-scale structures, but in order to visualize structure at the level of atoms, much shorter wavelengths of light must be used (the wavelength must be smaller than the structural detail one is looking for).

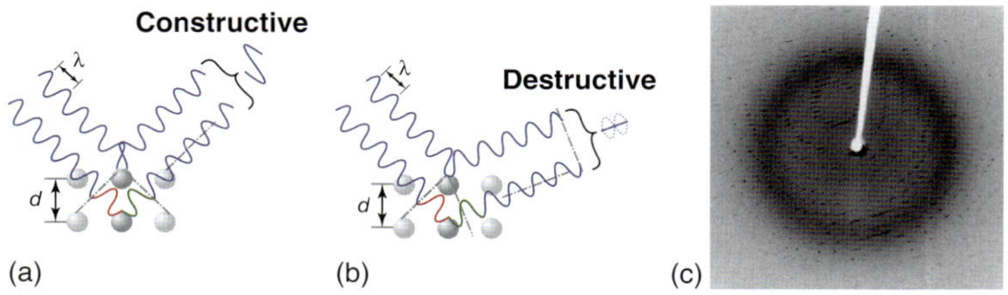

Figure 5.19 Interference patterns in an inorganic crystal, illustrating how constructive and destructive patterns arise from incident X-rays, allowing estimate of the atomic structure of the crystal. Structures of organic crystals are more complex, and are more difficult to determine. Panels (a, b) from http://en.wikipedia.org/wiki/Bragg's_law. Panel (c) is crystallography of 3Clpro, a SARS protease from http://en.wikipedia.org/wiki/X-ray_crystallography-. Accessed 11 September 2014.

X-rays (very high energy electromagnetic radiation with very short wavelengths) are thus the natural source of "light" for atomic structure analysis.

To use the method, one first obtains a crystal of the substance; for inorganic crystals this is fairly easy, but for organic molecules it is more difficult, and for membrane proteins such as receptors is even more so, as the receptors must be removed from the membrane without altering their functional structure, then placed in repeating crystalline arrays that also preserve functional shape. Once a crystal is obtained, it is placed on a holding device and X-rays directed at it while it is rotated. The rays scatter from electron-rich areas of the molecule, and are captured on recording devices arrayed around the crystal. A regular molecule has a regular array of electrons and provides a regular scattering pattern of X-rays. Thus, investigators determine the atomic and molecular structures from the pattern of X-ray scatter.

Images consist of dots on a photographic or digital surface and emerge from two processes – constructive and destructive interference. A dot is produced when X-rays scattering from adjacent atoms add together in phase (constructive) and a clear space is produced when X-rays are out of phase (destructive). Figure 5.19 shows how a regular inorganic crystal lattice produces interference patterns.

The constructive wave produces a bright spot (a "reflection") and results when two or more scattered beams are in phase (line 1 in Figure 5.19a); this occurs when the wavelength of the beam λ and its angle happen to bear a specific relation to the interatomic distance $\boldsymbol{d}$. Most beams have an angle such that a destructive overlap of waves occurs and the beams scatter out of phase (line 2 in Figure 5.19b). The few constructively produced spots must bear a specific relation to the crystal structure, and complex mathematical procedures allow investigators to interpret them to form a general idea of the molecular details of the molecule. Figure 5.19c shows one X-ray slice through an enzyme molecule (the white line represents the beam itself). The structure of the enzyme can be estimated by analysis of many such slices, each with a different pattern of reflections.

5.4.4 Probe Tagging (Fluorescent and Photoaffinity)

In addition to radiolabeled ligands, photoaffinity and fluorescent tags allow visualization of ligand binding.

Photoaffinity labeling involves use of a slightly modified high-affinity ligand that binds to the agonist site; this ligand has a reactive chemical group attached such that it spontaneously or after UV irradiation creates a covalent bond with the receptor at a nearby site [145]. An example of this approach uses an engineered version of epibatidine, a poison produced in the skin of a South American frog; this substance is a high-affinity agonist of the nicotinic ACh receptor (Chapter 6), and binds to the site that ACh normally occupies. The engineered version has an azide (N_3) group as part of the molecule, and this reacts with nearby groups on the ACh receptor; the permanently bound epibatidine-azide can be visualized and its binding location precisely located. Using this and related methods, researchers determined that agonist binding to this receptor involves the two alpha subunits as well as adjacent subunits [146].

5.5 Proteomics Approaches to Receptor Efficacy

Following agonist binding, receptor function (efficacy) depends on the assembly of a signaling complex involving recruitment of intracellular proteins such as enzymes or transcription factors [147]. An example is the recent work on *G protein-coupled receptor interacting proteins*, or GIPs. The many GIP interactions affect receptor synthesis, placement in the cell membrane, and coupling to cellular effectors. These interactions are being approached by the methods of proteomics (Chapter 3), and will be expanded in the following chapters, each of which deals with a different receptor class.

5.6 Physical Factors Affecting Receptor Binding

5.6.1 Temperature

Water molecules form hydrogen bonds with each other and with other molecules, such as proteins, in an aqueous solution. Hydrogen bonds from water molecules can weaken the degree of ionic and hydrogen bond attraction between receptors and their ligands by coming between and interacting with the charged portions of the two larger molecules.

Temperature affects ligand binding to receptors because it affects the magnitude of thermal agitation of a receptor and of its ligand; at high temperatures, agitation reduces the fit between the two molecules, decreasing the time a ligand remains bound to its receptor. These alterations in receptor binding, in turn, alter the ability of an agonist to activate a tissue; cooling from 37 to 23 °C increases the binding affinity of norepinephrine to the **alpha-adrenoceptor** [147, 148]. This increase in affinity is a major contributor to the ability of a cutaneous blood vessel to constrict at cool temperatures [149].

Hydrogen ion concentration of a solution affects the distribution of charges on the surface of a receptor, as well as the charge on a ligand. Each ligand–receptor interaction has an optimum pH for binding, and values of pH that differ from this optimum will reduce the ability of ligands to bind.

5.6.2 Relation of Agonist Affinity and Efficacy to Distance Traveled Following Release

Not all agonists have the same functions and properties. Some are hormones that travel long distances within the body; others are neurotransmitters that activate nearby neurons or muscles. Two interesting analyses give plausible explanations for the observation that binding affinities and efficacies of different agonists are roughly proportional to the distance they travel. Goldstein *et al.* [150] propose that because transmitters operate fast ion channels, they must bind and release quickly. High affinity (low K_d) means long receptor residence time (Equations 5.1–5.3), and is maladaptive for a fast-acting transmitter. However, it would be adaptive to couple a low affinity (short residence time) with a high efficacy such that maximal cell activity is achieved with few receptors being activated. This explains why neurotransmitters have low affinity but high efficacy, and is consistent with the proposal of Funder [151], who noted that affinity of blood-borne hormones for their receptors was a thousand times higher than affinity of neurotransmitters for their receptors. He suggested the difference reflected binding probabilities based on the distance traveled – a hormone was much less likely than a transmitter to find its receptor; thus, hormones should have high binding affinities.

Part III
Receptor Types and Function

Transduction: Introduction to the Four Receptor Types

Hormone binding is just the first step in the process of cell activation. Depending on the receptor type, many kinds of additional steps may be involved. The process by which the binding event is transmitted to the interior of the cell is called **signal transduction**, and though it was proposed theoretically by Langley in 1906, it has taken until the past two decades to unravel its varied molecular mechanisms.

The next chapter begins the second section of this book: a discussion of the receptors themselves, along with the mechanisms by which they activate cells:

- Ion channel receptors allow transmembrane ion exchanges – Chapter 6.
- G-protein-coupled receptors activate membrane enzymes – Chapter 7.
- Growth factor receptors activate an enzyme that is part of the receptor – Chapter 8.
- Nuclear receptors initiate gene transcription and protein synthesis – Chapter 9.

6
Transduction I: Ion Channels and Transporters

> The constancy of the internal environment is the condition for the free and independent life …
>
> *Claude Bernard [152]*

> "… you seem to me both in your appearance and in your power over others to be very like the flat torpedo fish, who torpifies those who come near him and touch him, as you have now torpified me … "
>
> *Plato: Meno [153]*

6.1
Introduction

Cell membranes separate the cytoplasm, where metabolism occurs, from the source of metabolic nutrients in the extracellular fluid. Thus, small molecules used for metabolism do not have easy access to cells. Indeed, because of the nonpolar nature of cell membranes, ions and other polar molecules are unable to pass into or out of cells without the aid of some sort of molecular device. Transmembrane (TM) movements are actually accomplished by two types of membrane molecules – transporters and channels.

Transporters belong to two functional types: one sort that translocates ions by passive **facilitated diffusion** (not requiring energy) and another sort that accomplishes **active transport** of ions by using energy to move ions *up* their electrochemical gradient. Both these transporters must bind the ion being moved, and thus are fairly specific for the ion species they transport.

In contrast, channels are passive flow conduits, and ions traverse them *down* their electrochemical gradients without binding to the protein. Channels are not as specific as are transporters to the ion species they allow through, and some channels allow K^+, Na^+, and Ca^{2+} ions to pass. A cation channel, however, will not allow an anion to pass. Channels also belong to two types: **voltage gated** and **ligand gated**. "Gating" refers to the mechanism of opening: voltage-gated channels open when a sufficient potential difference is applied to the membrane; ligand gating requires the binding of a neurotransmitter or other signaling molecule to the channel. The ligand-gated channels are thus receptors.

6.1.1
Family Relationships

Even though the transporters and channels do not show amino acid homologies [154], the two are sometimes treated as a single large superfamily of membrane proteins because of structural similarities in TM and cytoplasmic domains:

- Most consist of several polypeptide chains loosely associated in the membrane; those that are a single polypeptide consist of repeated identical or similar motifs.
- The polypeptides generally have an even number of membrane-spanning regions; this means that both the amino and carboxyl ends of the polypeptide are on the same side of the cell membrane, generally on the cytoplasmic side.
- The polypeptides are arranged in a ring. This ring is generally packed so that no space exists between the chains; but a small opening – a channel – may momentarily occur when the chains alter conformation.
- Conformation changes when one of several external events occurs: binding of an agonist, change in electrical potential, stretch, or change in pH.
- Some of these proteins are involved in the translocation of molecules across the cell membrane. The transported molecules are generally small (ions, neurotransmitters, or nutrients such as amino acids) and may be transported passively or by active transport.

Receptor Biology, First Edition. Michael F. Roberts and Anne E. Kruchten.

A reasonable scenario proposes that the transporter-like proteins evolved in the earliest bacterial cells as generalized **transducers** of environmental stimuli – osmotic shock, physical stretch, and pH changes [154]. Either these or different bacterial proteins evolved into pure voltage-sensitive channels for ions [155], as well as into ligand-activated transporters for food and into **ligand-gated ion channel** (LGIC) receptors [154].

Therefore, we discuss the transporters and channels in a single chapter, each section of which deals with the proteins that do appear to have adequate structural or sequence homologies to allow them to be considered together. This chapter organizes the transporters/channels into four types: (i) the small molecule channels, including the stretch and **osmotic detectors**; (ii) the **voltage-gated** cation channels; (iii) the transporters and voltage-gated anion channels; and (iv) the ligand-gated ion channels.

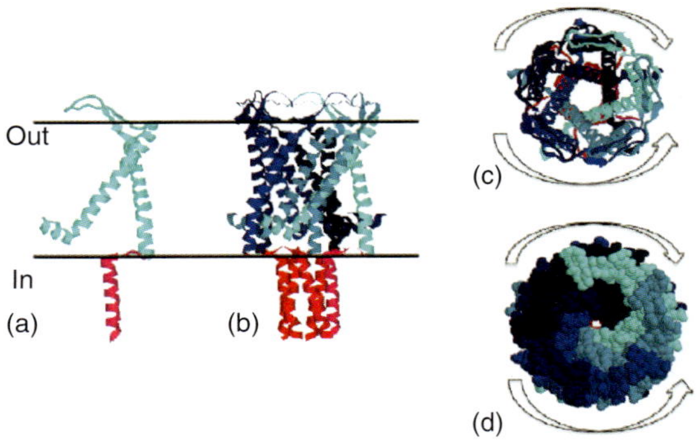

Figure 6.1 Mechanosensitive channel from *Mycobacterium*. (a) The monomer polypeptide consists of two transmembrane (TM) regions shown in blue and one cytoplasmic region in red. (b) The full channel protein is homopentameric, shown in shades of blue and red to contrast the individual monomers. When membrane is unstretched (b), channel is not open. (c) Top-down orientation of (b). (d) Membrane stretch acts in a rotational fashion to open the channel to 3 nm, allowing passage of ions or other osmotically important molecules. Images were created using Rasmol [8] from PDB ID 2OAR [156].

6.2 Small Molecule Channels

6.2.1 Osmotic and Stretch Detectors

The origin of membrane transport seems to have occurred nearly at the origin of cells themselves. Environmentally caused osmotic shock creates physical stress at cellular surfaces; in addition, fluids can stretch or compress membranes. To maintain cytoplasmic integrity, these changes must be countered by membrane proteins that can control cell volume. The earliest membrane proteins of this sort to evolve were the *mechanosensitive channels* of bacteria, today represented by several types of channels that detect and respond to osmotic shock and stretch. They are small peptides arranged in a ring of five identical units to form the channel (Figure 6.1).

Each monomer has two TM domains. When stretched, the proteins deform and increase conductance for ions and other small molecules. These proteins thus allow release or uptake of osmolytes when extracellular concentrations change [157]. The same properties that promote response to physical or osmotic shock (separate polypeptides, an intramolecular region for passage of molecules) are also well configured to promote ion passage, and the members of the next type, the ion channels, show some of the same structural features.

6.2.2 Voltage-Gated Cation Channels

It is uncertain why or how voltage sensing originated, but it is clearly adaptive for conducting tissues (nerves) and contractile tissues (muscles) among animals. Earlier hypotheses proposed that voltage-sensing channels originated with the animals, but recent work suggests that plants have channels homologous to animal K^+, Ca^{2+}, and Cl^- channels [158], indicating that these channels serve purposes other than simple movement. In contrast, the Na^+ channels may indeed be movement related: they appear not to be present in plants but they are present in the choanoflagellates, a sister taxon to the animals from which animals arose about 700 million years ago. These single-celled or colonial protists use flagella for movement and pseudopods for obtaining food. Both these structures may be early versions of the conductive and contractile tissues involved in animal movement.

6.2.2.1 History of Studies on Voltage-Gated Channels

Since the eighteenth century, it has been clear that nerve and muscle function involves electrical phenomena. Galvani first demonstrated the existence of the action potential in the 1790s by touching a frog's nerve ending to its muscle surface [159] and Berzelius proposed in the early 1800s that the source of the electrical eel's current was an "organic chemical process." Du Bois-Reymond discovered the action potential in 1848 [160]. Early in the twentieth century, Bernstein suggested that nerves had selectively permeable membranes and that differences in ion concentrations (specifically, K^+) across the membrane led to voltage differences between the cytoplasm and exterior of a nerve cell. He also suggested that ion movements were responsible for the action potential [161]. Soon after that, physiologists began measuring electrical potential changes on the outsides of vertebrate nerves (fibers containing hundreds of individual neurons), but rapidly appreciated that they needed single cells to measure ion movements more accurately.

The first intracellular voltages were measured by Kamada in 1934 in the unicellular protist *Paramecium* [162]. Within 2 years, Young [163] discovered the large-diameter squid neurons, and proposed the use of the squid giant axon to allow intracellular recordings for neural studies. Hodgkin and Huxley used this preparation to make recordings of action potentials from single cells and confirmed that the interior of the neuron was at a negative potential; they measured action potentials and showed that Na^+ ions played a crucial role in the process of nerve activation [165] (Figure 6.2).

By the 1950s, synaptic transmission was thought to involve channels, and specific channels for Na^+ and K^+ were predicted by Hodgkin and Huxley [166]. The methods of the time, however, did not allow measurements of single channel activity, as the electrical noise from equipment was much higher than the current characteristic of single channels. In the mid-1970s, Neher and Sakmann [167] developed the patch clamp technique, finally allowing the study of individual channels in cell membranes (Box 6.1).

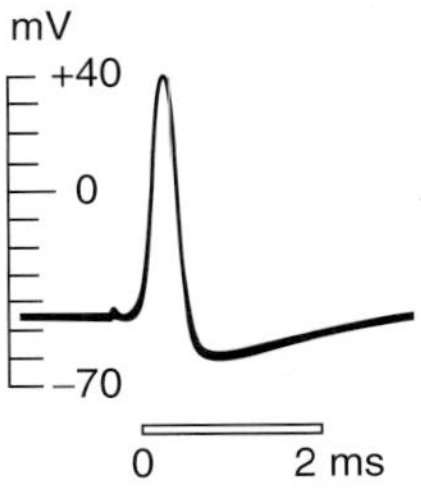

Figure 6.2 Example of recording of action potential from single neuron of squid giant axon. Rise in trace indicates opening of voltage-gated Na^+ channels. Modified from Ref. [165]. Used with permission.

Box 6.1 Visualizing Actions of Single Ion Channels

Ion channels are combinations of receptor sites and effector sites in a single membrane protein. Classical receptor theory involves equilibrium behavior of a population of receptors, not the behavior of individual receptors: over time and large numbers, the randomness of individual molecular transitions is averaged out. However, new insights into ion channel action can be obtained by measuring the responses of single ion channels in response to small amounts of agonist. Patch clamping allows the ion movements following single receptor binding to be visualized as deflections in a current trace.

A fire-polished tip of a glass microelectrode (1–5 μm) is applied to the surface of a cell whose extracellular matrix has been removed with proteases (Figure 6.3a). If gentle suction is also applied, the junction has a very high resistance (10^{11} Ω) so very small changes in current (the opening and closing of individual ion channels) can be followed (Figure 6.3b).

The cell can be studied in several ways: the membrane within the tip can be ruptured but the cell left intact, to produce a low-resistance pathway to the interior of the cell (whole-cell recording); as an alternative, the cell can be discarded and only a small vesicle of membrane left attached to the electrode. Each binding event causes the opening of a channel and the passage of a number of ions that creates a rectangular current pulse of variable size. Thus, ion channel recordings can give information about the open time associated with each binding event. With sustained exposure to agonist, ion channels show not a continuous open state but a combination of open periods separated by closed periods. As agonist concentration increases, channels are open a greater percentage of the time. Open times vary somewhat even for a given agonist at a given concentration, probably because of random molecular movements that affect conformation of the channel (Figure 6.3b). The summation of many individual recordings of patch clamp experiments gives an envelope of ion currents that resembles the whole cell recordings of earlier experimenters.

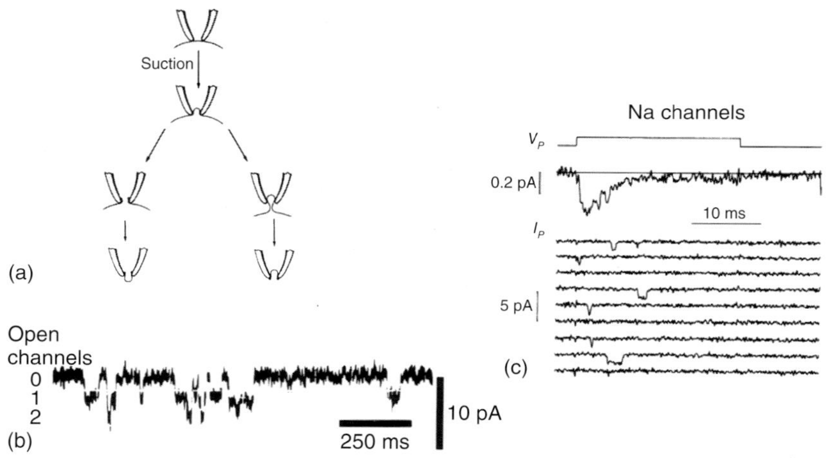

Figure 6.3 Patch clamp method and application. (a) Cell with microelectrode applied to surface. Suction produces high-resistance seal, and further pulling generates electrode contact with cell interior or patch. (b) Current recordings from a single patch that contains two channels. At each depolarization, one or two channels open. (c) An average of opening currents from many applications of 40 mV stimuli on the same membrane patch. Modified from Figures 2, 3, and 4 of Ref. [168]. Used with permission.

The patch clamp method is used both for ligand-gated and voltage-gated channels. Ligands are studied by placing them at appropriate concentrations into the fluid of the suction pipette; voltage-gated channels are studied by applying test voltages across the membrane with the patch in place.

6.2.2.2 Structure and Physiology of Ion Channels

In addition to Na^+ and K^+, Cl^-, and Ca^{2+} are motile ions that determine the voltage differences across the membranes of all cells. Specific channels have evolved for all four ions.

6.2.3 Potassium Channels

Because K^+ is the most abundant cation within cells, channels to control its concentration were likely the first to evolve and today are found in all organisms. Several types have been described, each having a fundamental subunit of two, four, six, or eight TM regions and each consisting of a fixed number of these subunits (Figure 6.4).

The 2TM K^+ channel evolved in prokaryotes; one organism, or even one cell type, may have several types of K^+ channels. The function of each type of K^+ channel depends on its size and the number of subunits, but all K^+ channels use the same five amino acids in the **selectivity filter** – TVGYG (Figures 6.5 and 6.6).

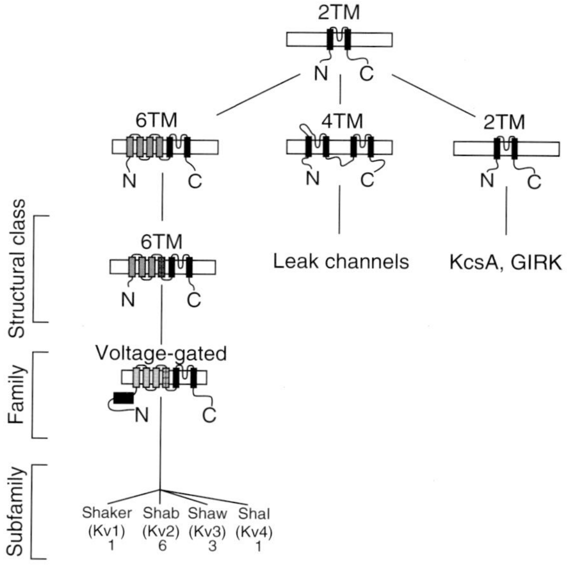

Figure 6.4 Diagram of potassium channel families and subfamilies. Three structural classes are shown defined by their subunit structure, having six (6TM), four (4TM), or two (2TM) transmembrane segments. The KcsA and GIRK channels consist of tetramers of 2TM channels. The 4TM channels control resting neuronal potentials. The colored TM regions at left represent a 4TM structure containing the **voltage-sensing domain**. The figures at left also show combinations of 4TM and 2TM domains to construct the large family of 6TM channels, only some of which are shown. For clarity, representatives of the 4TM and 2TM structural class are not shown. Modified from Figure 1 in Ref. [169]. Used with permission.

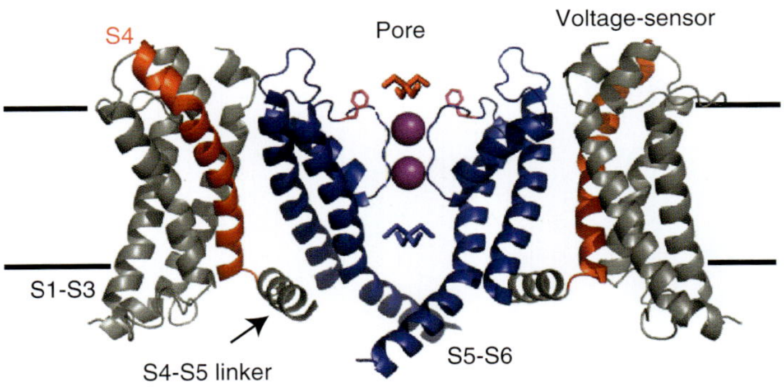

Figure 6.5 The voltage-gated K^+ channel. The front and rear 6TM domains have been removed. Two K^+ ions are visible in the channel, and the voltage sensors are shown in red. Modified from Ref. [170]. Used with permission.

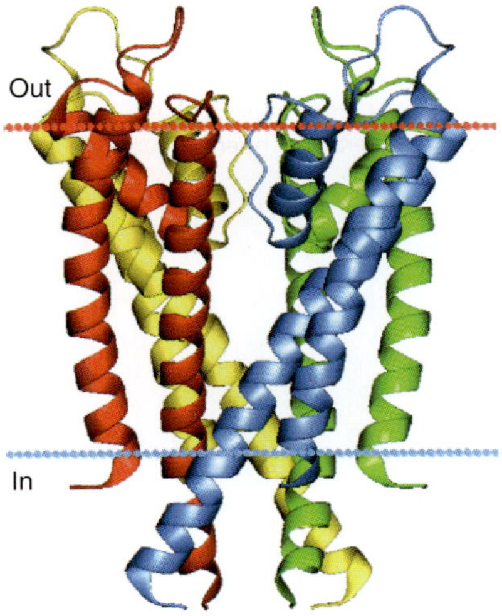

Figure 6.6 KcsA channel – organization. Each of the four peptides is shown in a different color. The selectivity filter is shown in the center as the thin bent domains of the peptides (all four peptides contribute to the filter, but only two of the peptides are visible in this diagram). Source: http://en.wikipedia.org/wiki/KcsA_potassium_channel. Accessed 9 January 2015.

The K^+ channels are classified by the number of TM domains (Figure 6.4). Even so, specific functions are associated with each TM format so we can also describe the channels in terms of their actions. The K^+ channels most commonly encountered in a general physiological context are of three kinds:

1) *Four or eight TM – two pore:* These are the K^+ passive leak channels responsible for maintaining the resting potential at about −70 mV in many types of cells. They form dimers, each consisting of four peptides in a ring. They appear to allow K^+ to cross the membrane virtually unimpeded; this free diffusion of K^+ ions (plus the higher concentration of K^+ inside cells than outside) controls the resting membrane potential, as predicted by the Nernst equation [164].
2) *Six TM – one pore:* The members of this family are voltage sensors, a feature allowed by four additional TM regions (Figure 6.4). They thus are the voltage-gated channels of eukaryotes (especially prominent in animal nervous systems) [154]. The channel closes when a portion of the cytoplasmic N-terminus of the TM4 domain moves into the pore to block ion movement (Figure 6.5a); it opens when an external voltage signal causes displacement of that same domain away from the pore. This exposes the central pore to the aqueous environment and allows K^+ to exit (Figure 6.5b) [170].

   These are the nerve channels that return membrane voltage to resting levels after an action potential. The earliest animal K^+ channel to be discovered was the squid axonal channel from this family.

   Several homologous genes form a variety of 6TM channels that are widely distributed in animals [161]. These form the group of related channels called the *shaker, shaw, shab,* and *shal* channels (Figure 6.4). They are found in nearly all animals and regulate neuronal voltage as well as non-neuronal cell functions.
3) *Two TM – one pore:* These channels (Figure 6.4) have provided a wealth of structural information on K^+ channels in general, as they are simpler than other K^+ channels and are more easily imaged with X-ray crystallography.

The KcsA channel, found in bacteria, is a tetramer of the 2TM structure. It was the first to be seen at the atomic level by X-ray crystallography [171]. The four subunits form a ring around a water-filled pore, in the middle of which is a selectivity filter that discriminates between K^+ and Na^+ ions (Figure 6.6); the combination of the long water-filled channel and the short selectivity filter is responsible for the ability of the channel to conduct 10^8 ions per second [171].

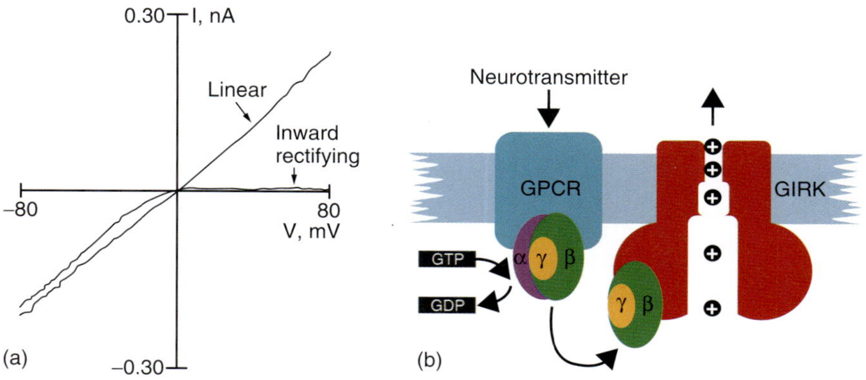

Figure 6.7 The GIRK channel, a tetrameric 2TM K^+ channel. (a) Graph of current (*Y* axis) and voltage (*X*-axis) relations of two types of K^+ channel; the linear channel remains open at negative and positive voltages; the inward rectifying channel closes at positive voltages. (b) Architecture of the channel (red) in the membrane; also shown is GPCR (e.g., muscarinic receptor) that activates the GIRK. The Gβγ protein binds to the channel following binding of ACh to receptor. Figure 1 in Ref. [173]. Used with permission.

The selectivity filter operates by pointing the oxygen atoms of the amino acid backbone (center of the channel in Figure 6.6) toward the pore. The distance between them is almost exactly the size of a K^+ ion; the selectivity of K^+ over Na^+ relates to the number of negatively charged oxygen atoms on the amino acids lining the pore [172].

The G-protein-linked **inward rectifying** channels (Chapter 7 and Figure 6.7) are also of tetrameric 2TM design.

This channel allows ions to pass only at negative membrane potentials and thus is largely responsible for keeping the resting membrane potential negative. When action potentials occur, the channel closes at positive membrane potentials and thus does not oppose the changes in membrane potential caused by inward movement of Na^+ ions. The rectifying block of the channel is accomplished by ions or other molecules such as polyamines that bind to the pore only at positive voltages and thus prevent K^+ movement (Figure 6.7b).

One type of inward rectifier operates with a G-protein-coupled receptor (GPCR), the ACh muscarinic receptor (Chapter 7). It has a $G_{\beta\gamma}$ subunit binding site on the C-terminus. When activated following ACh binding to the receptor, the βγ subunit causes channel opening, leading to hyperpolarization of atrial cells and decreased heart rate.

6.2.4 Sodium Channels

6.2.4.1 Bacterial Na^+ Channels

As for the K^+ channels above, the voltage-gated cation channels in vertebrates most likely derive from ancestral channels of bacteria. The channels are built on the same architecture as the K^+ channel: an even number of TM domains, a voltage-sensing domain, and a pore-forming loop (Figure 6.8).

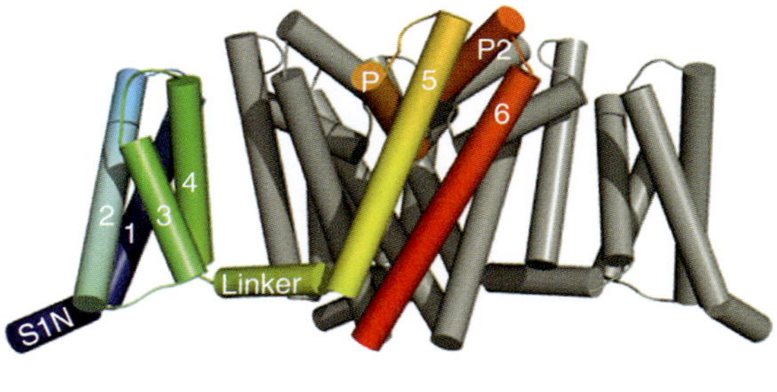

Figure 6.8 The bacterial voltage-gated Na^+ channel. The channel consists of four identical polypeptides, each with 6TM domains. Here, three of the four peptides are shown, and one is shown in color; TM domains 1–6 are numbered. The two helices of the pore loop (L) are indicated between TM5 and 6. Modified from Figure 1 of Ref. [174]. Used with permission.

6.2.4.2 Vertebrate Na^+ Channels

Neuronal Na^+ channels The role of Na^+ channels is primarily to produce rapid depolarization of the nerve or skeletal muscle cell membrane. This depolarization leads in nerve to propagation of an impulse along the axon, and in muscle it begins the process that leads to contraction. The channel consists of three polypeptides: α, β_1, and β_2. The α peptide is the largest; the two short β chains each have a single TM domain and long extracellular domains. The α chain has 24 TM domains (four groups of six in one long polypeptide – Figure 6.9).

Nine different α chains have been described in mammals, encoded by genes on four chromosomes.

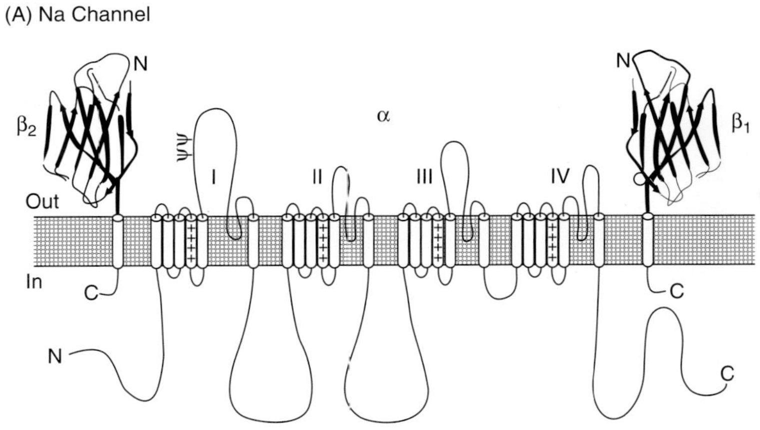

Figure 6.9 The vertebrate voltage-gated Na channel. Membrane architecture, showing its three proteins: α chain, β_1 chain, and β_2 chain. Figure 3.12a in Ref. [161].

Thus, they arose by duplication followed by mutation (see Chapter 14 for a discussion of receptor evolution). Their amino acid sequences are 75–95% identical, and their overall structure and function is similar [161]. The four units of the α chain form a channel flanked by the β chains. The role of the β chains is uncertain, as full activity can be achieved by reconstituting membranes with α chains alone.

The voltage sensor is the fourth TM segment (TM4 – shown with + labels in Figure 6.9). This domain contains several positively charged arginine or lysine residues that bind negatively charged amino acids on adjacent helices. In the polarized state (e.g., membrane potential −70 mV), the charges are situated such that the six TM domains of each unit are spread out and block the channel. When the membrane depolarizes, the four separate TM4 helices bind to different negatively charged amino acids and change position, allowing the TM6 helices to pull out of the channel and open it. These changes in conformation are sufficiently rapid that the channel remains open for less than a millisecond (Figure 6.3c).

Selectivity in this receptor is similar to that of the K^+ channel: the amino acids lining the pore (DDEE) expose electronegative oxygen atoms that attract Na^+ ions. However, as the pore is smaller (80% the size of the K^+ channel pore) and just the size of the Na^+ ion, these can pass easily without allowing K^+ ions through [172].

Epithelial Channels (ENaC) A related channel is the vertebrate epithelial Na^+ channel (ENaC), a 2TM channel occurring in heterotrimeric format. ENaC is found on epithelial cells of the airways, kidneys, and digestive tract. Renal sodium balance is important for control of blood volume and pressure. The channel is constitutively slightly open, but can be closed by diuretics such as amiloride [175]. This closure prevents reuptake of Na^+ from renal tubules, thus allowing water to remain in the tubules and to be excreted as urine, lowering blood volume and hence blood pressure.

6.2.5
Calcium Channels

Calcium channels are the primary transducers of electrical excitability [176]. They regulate Ca^{2+} fluxes into nerve cells for release of transmitter and into muscle cells for neuronal firing, muscle contraction, and other cellular functions such as secretion and enzyme activation. These two latter functions are discussed in Chapter 10, and the remainder of this discussion of Ca^{2+} channels focuses on their depolarizing actions. All the voltage-gated Ca^{2+} channels appear to have the same structure: five polypeptides (α_1, α_2, β, γ, δ – Figure 6.10).

The channels were defined physiologically, based on their rate of action and voltage dependence: L-type channels are activated at high voltage; they are slow to inactivate, and they produce large Ca^{2+} currents such as seen in the heart during ventricular depolarization (below). N-, P/Q-, and R-type channels are rapid acting and responsible for neurotransmitter release. T-type channels are transient regulators of cardiac pacemakers [176].

The differences between the functions of the several Ca^{2+} channels come primarily from differences in the α_1 peptide. The α_1 chain is homologous to the α

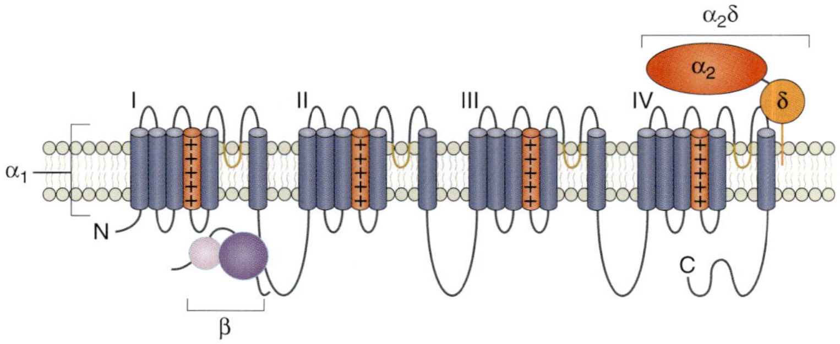

Figure 6.10 The voltage-gated Ca channel. It has five different peptides: α_1, α_2, β, γ, and δ (γ subunits are not shown). Cylinders represent TM domains; plus signs represent TM domains that form the pore and selectivity filter. Figure 2a of Ref. [177]. Used with permission.

chain of the voltage-gated Na^+ channel. Ten different α_1 genes are known in mammals, and occur on eight chromosomes. These also likely arose by duplication, but sequence homology among the duplicated genes is less than with the Na^+ channel genes. The genes for various L-type channels are closely related to the P/Q, N, and R channels, and the T channel is most closely related to the Na^+ channel (Chapter 14).

Ca^{2+} channels open on depolarization with a mechanism similar to that of the Na^+ channel, but 20–50 times more slowly. They thus facilitate more prolonged movements of Ca^{2+} that have a greater number of physiological roles than the relatively simple role of Na^+. Two examples follow:

L-type Ca^{2+} channels prolong depolarization in the heart: the initial event in the heartbeat is depolarization via Na^+ channels; this opens Ca^{2+} channels that remain open for 200 ms or more. During this period, the contractile process begins but the membrane is refractory to further stimulation. The long refractory period extends until the channel closes and the heart relaxes; this means that the heart cannot depolarize once it has begun to contract, and cannot generate sustained contractions (Figure 6.11). This is adaptive because sustained contractions, while useful in skeletal muscle, are lethal in the heart.

Another role for long Ca^{2+} currents in cells is to provide large amounts of Ca^{2+} for intracellular signaling: (i) at neuronal dendrites, the prolongation of depolarization increases neuronal excitability to incoming signals; (ii) at neuronal terminals where transmitter is released, Ca^{2+} provides the mechanism for transmitter release; and (iii) in endocrine cells, slow Ca^{2+} currents enhance enzyme activation via the Ca-binding protein calmodulin.

Calcium entry through its channel is similar to that of Na^+ through its channel: the amino acids of

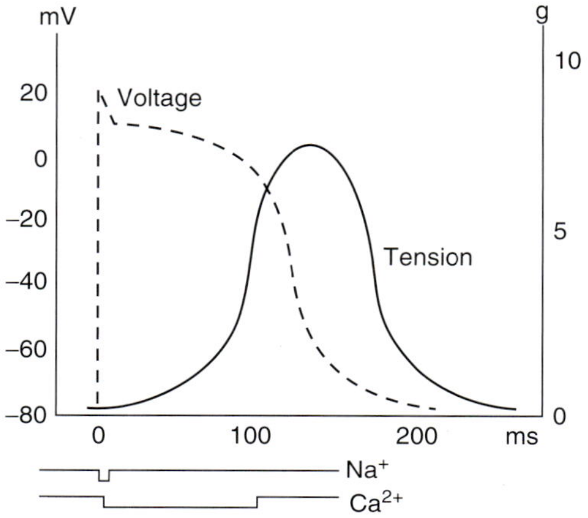

Figure 6.11 Changes in membrane potential and tension of a ventricular cell over a single heartbeat. Action potential (short Na^+ channel opening) initiates Ca^{2+} channel opening, which lasts 100 ms. The Na^+ channels cannot reopen while the Ca^{2+} channel is open, so the cell is refractory during that period.

the pore-forming TMs have exposed oxygen atoms that attract the Ca^{2+} at the pore constriction. The attraction is stronger than with the K^+ channel (amino acids TVGYG in the selectivity filter), as the Ca^{2+} and Na^+ channels both have E and D in the selectivity filter [172].

6.2.6 Non-Voltage-Gated Cation Channels – Transient Receptor Potential (TRP) Channels

These channels resemble the voltage-gated channels above but act independently of voltage changes. The transient receptor potential (TRP) channels form a large family (approximately 20% of all ion channel genes encode TRPs). They have six TM domains,

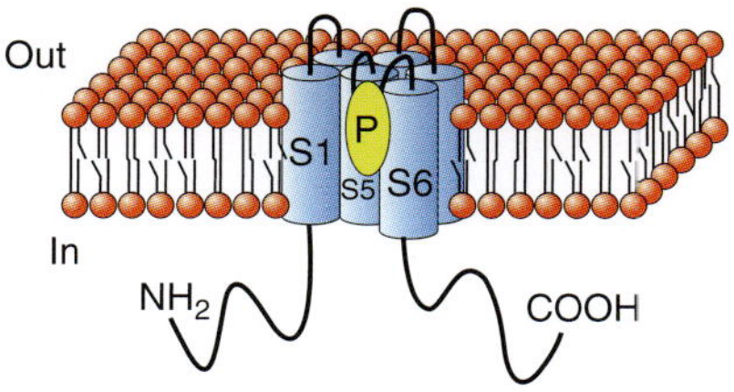

Figure 6.12 Generalized structure of members of the TRP Ca^{2+} channel family. Both N- and C-termini are in the cytoplasm. Six TM domains (S1–S6) form the channel. A pore-forming loop (P) lies between S5 and S6. Residues near the C-terminus may be involved in the binding of phosphatidyl inositol phosphate, a Ca^{2+} channel regulator. Modified from Figure 2 of Ref. [179]. Used with permission.

with both N- and C-termini intracellular and a pore-forming loop between the fifth and sixth TM domains [178] (Figure 6.12).

The TRP channel proteins fall into seven families and are found only in eukaryotes (primarily fungi, green algae, and animals). Functional channels consist of four peptides, and are generally homotetramers though some types form as heterotetramers [179].

TRPs are "polymodal" cell sensors that cause influx of Na^+ or Ca^{2+} ions following any of several different sensory inputs. An overview of the six subfamilies of TRP sensors is given in Table 6.1 [179].

These sensors are not generally ligand-binding receptors, but in a few cases ligands are known; for example, the cool-sensitive TRPM8 and warm-sensitive TRPV1 sensors also bind the plant compounds menthol and capsaicin, respectively. It is likely that this binding ability is responsible for the "cool" and "hot" sensations of these two food molecules [179].

Depending on whether the cell is excitable or non-excitable, TRP currents either promote further voltage changes (Figure 6.13) or modulate intracellular Ca^{2+} levels for control of enzyme function (Figure 6.13) [178]. For example, the ionic movements involved in the transduction of sweet, bitter, and umami tastes are mediated by TRPM channels, and sour by TRPP channels.

Table 6.1 Subfamilies of transient receptor potential channels.

Sub-family	Function
C	Neuronal growth, pheromone detection in mice
V	Warm sensation, Ca^{2+} absorption
M	Cold sensation; sweet, umami, and bitter taste transduction
A	Painful cold sensation
ML	Intracellular – lysozyme function
P	Sour taste transduction

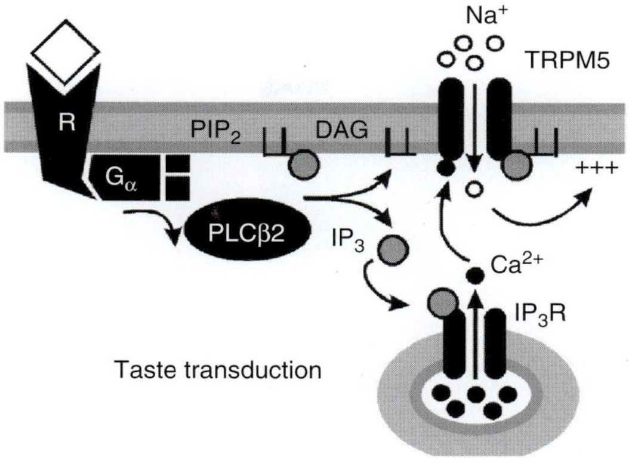

Figure 6.13 Functions of TRPM channel in taste transduction. Binding of sweet, bitter, or umami taste stimulus (diamond) to G-protein-coupled taste receptors (R) – discussed in Chapter 7 – leads to release of βγ subunits of the G protein, which activates PLCβ2, leading to IP_3 release. IP_3 activates its receptors (IP_3R), which release Ca^{2+} from intracellular stores. The Ca^{2+} opens TRPM5 channels, leading to an influx of Na^+ and depolarization of the cell. Figure 5 in Ref. [180]. Used with permission.

6.3 Transporters

6.3.1 Pumps and Facilitated Diffusion

The transporters all act by binding the transported substance on one side of the membrane and somehow transferring it to the other, where it is released. Some of the transporters (the *solute-linked carriers – SLCs*) are passive, meaning they do not transport a substance up its electrochemical gradient and do not require energy. Others (the *ATP-powered pumps*) are active transport molecules and split ATP to provide energy for the concentration of substances on one side of the membrane or the other. The various molecules show no amino acid sequence homology, and it is not known whether they are derived from a common ancestor.

6.3.1.1 The SLC Proteins

These proteins couple the "uphill" movement of a small molecule such as an amino acid or neurotransmitter to the "downhill" movement of Na^+. The sodium gradient is created by the action of the

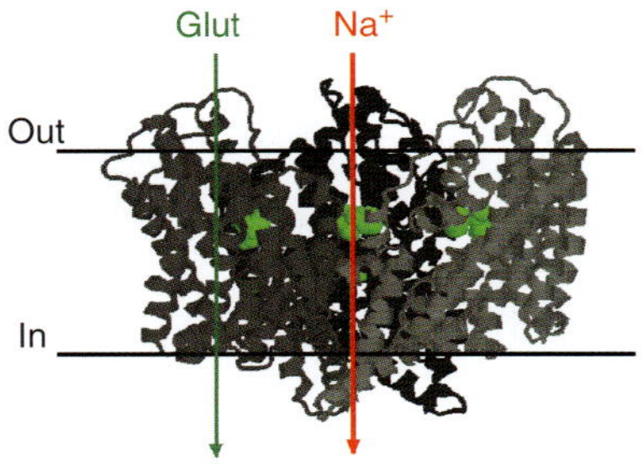

Figure 6.14 Model of the glutamate transporter. The molecule forms a large basin in the extracellular side of the membrane; this basin is filled with water molecules, which allows the glutamate (green spheres) to be halfway through the membrane before binding to the transporter. Glutamate transport is coupled to Na^+ diffusion down its gradient. The transporter is a homotrimeric protein, and the three subunits are shown in shades of gray. The movements of Na^+ and glut are not meant to be stoichiometrically accurate. Images were created using Rasmol [8] from PDB ID 2NWW [181].

Na^+/K^+ pump (the P class pump below). Thus, the SLC proteins require the prior establishment of a gradient of Na^+ to drive the **cotransport** of the amino acid. The Na^+ binds to a site on the exterior of the protein; this creates a high-affinity amino acid binding site in adjacent residues. An example is the amino acid transporter Glt_{ph} (Figure 6.14).

This is a prokaryotic homolog of the eukaryotic Na^+-coupled glutamic acid (glutamate) transporter. The glutamate is transported against its electrochemical gradient by the diffusion of Na^+ down its own gradient. The transporter is a homotrimer, and each subunit can transport one glutamate. In eukaryotes, this transporter is part of a group called the *excitatory amino acid transporter*s (*EAAT*s), which are trimers coupled to a P-type ATPase (below) that pumps the Na^+ needed to form the Na^+ gradient used by the transporter to move the amino acid. The pump moves 3 Na^+ in and 1 K^+ and 1 H^+ out to power the gradient to move 1 glutamate into the cell.

The SLC family also comprises the transporters for the monoamine neurotransmitters (MATs) norepinephrine (NE), dopamine (DA), and 5-hydroxy tryptamine (5-HT or serotonin). They are homologous to the transporters for octopamine and tyramine, which serve as MATs in the insects [182].

6.3.1.2 The Pumps

Three major types of ATP-powered pumps are known, the P class, the F/V class, and the ABC (ATP-binding cassette) transporters. Ions are transported by the P and F/V class pumps, and small molecules by the ABC transporters. The P-type pumps were discovered by Skou [183] in the 1950s and are the primary Na^+/K^+ pumps in all organisms, often accounting for a third of their energy consumption (Figure 6.15).

The V-type (Vacuolar) pumps are proton pumps in prokaryotes (also in eukaryotic **organelles**, the "vacuolar" compartment); they use ATP bond energy to pump protons across membranes for acidification of **endosomes** and **lysosomes**, as well as for volume control in mitochondria.

F-type ATPases sometimes do the reverse: they are the ATP synthases found in mitochondrial and chloroplast membranes and use a proton gradient to drive synthesis of ATP. In eukaryotes, this process is coupled to the oxidative phosphorylation portion of metabolism, and these molecules are the primary source of ATP. In prokaryotes, they may function as V-type ATPases to use ATP energy to create a proton gradient.

The ABC transporters are large protein complexes consisting of two to four polypeptides. The overall configuration resembles that seen in Figure 6.15c, but detailed structures are different in the importing transporters and the exporting transporters. The importers are tetramers (two identical TM peptides and two identical ATP-binding cytoplasmic peptides – Figure 6.16a) and the exporters are homodimers (TM and cytoplasmic domains are separate and are encoded by two genes – Figure 6.16b).

Translocation occurs as an active process via splitting of ATP in the cytoplasmic domain. In both bacteria and eukaryotes, the two intracellular peptides generally remain separate until ATP binds to the intracellular nucleotide-binding domains. This binding event brings the cytoplasmic domains together and thereby alters conformation of the TM protein, promoting its ability to bind and transport a molecule.

The ABC transporters were retained within each of the eukaryotic taxa that evolved from bacteria. Bacterial members of this family are either importers or exporters, but only the exporters evolved in eukaryotes. In addition, the eukaryotes modified some proteins from this class to become DNA repair proteins or enzymes involved in protein translation. One branch of the family in bacteria became the cell membrane Cl^-/H^+ transporters, and is present in eukaryotes as a Cl^-/H^+ **antiporter** in intracellular compartments. A member of this latter group became the **cystic fibrosis** transmembrane conductance regulator CFTR (below), and from this transporter the

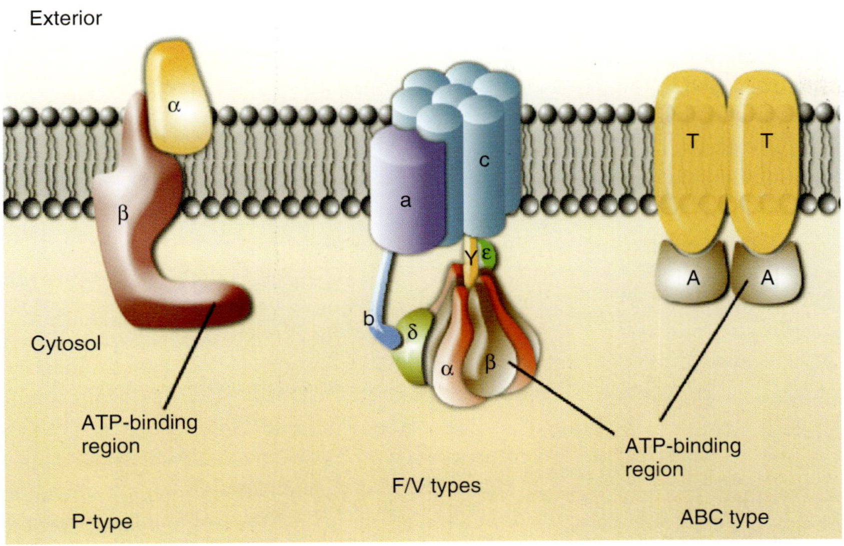

Figure 6.15 Models of the four classes of transmembrane active transport pumps. Left: P-class pump (H^+, Na^+/K^+). Center: F and V-class pumps. Right: ABC transporter (small molecule pump). Source: http://cc.scu.edu.cn/G2S/Template/View.aspx?courseType=1&courseId=17&topMenuId=113300&menuType=1&action=view&type=&name=&linkpageID=113471. Accessed 25 January 2015.

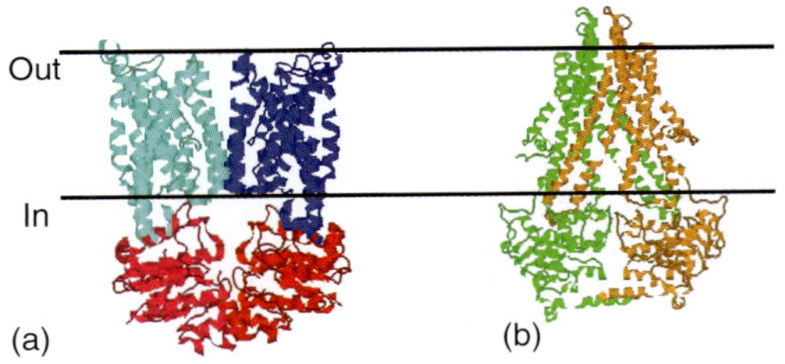

Figure 6.16 Structures of two bacterial ABC transporters. (a) BtuCD ABC Importer. The turquoise and blue subunits are separate TM polypeptides; the pink and red subunits are separate ATP-binding polypeptides, giving a tetramer as the functional transporter. (b) The Atm1-type ABC exporter. The TM domain is fused to the ATP-binding domain in a single polypeptide, and the functional transporter consists of two such polypeptides (one is green, the other is brown). Images were created using Rasmol [8] from PDB ID 1L7V [184] and PDB ID 4MRN [185].

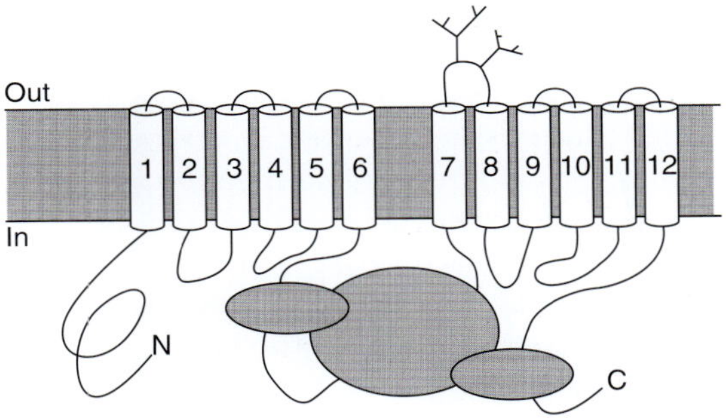

Figure 6.17 The CFTR (cystic fibrosis transmembrane conductance regulator) contains 12 TM domains, two nucleotide-binding regions (NBD1 and 2 - the two small ellipses), and a regulatory region. Figure 5.17 of Ref. [161]. Used with permission.

pure Cl^- channels (below) evolved by loss of the H^+ transport function of the antiporter [155]. The CFTR is an ATP-powered Cl^- channel (Figure 6.17), and is the only member of the ABC family to function as a channel.

The CFTR is a Cl^- channel that is normally closed, but opening it allows diffusion of Cl ions followed by water movement. Secretion of ions and water from the airways produces a mucus secretion that allows removal of irritants. Mucus secretion from epithelial cells is initiated by activation of GPCR stimulation in response to airway irritation. Agonists include epinephrine (EP), vasoactive intestinal polypeptide (VIP), and adenosine (ADO) (Chapter 7), all of which activate receptors coupling to G_s and cyclic adenosine monophosphate (cAMP). Binding of agonist to its receptor causes synthesis of cAMP, which activates a cAMP-dependent protein kinase A (PKA). The kinase then phosphorylates the R(egulatory) subunit of the CFTR, allowing ATP to bind to the N(ucleotide) B(inding) D(omains) 1 and 2; these steps promote unblocking of the channel and the diffusion of Cl^- out of the cell.

Once the channel is open, Cl^- moves down its electrochemical gradient. Movements of Cl^- are also

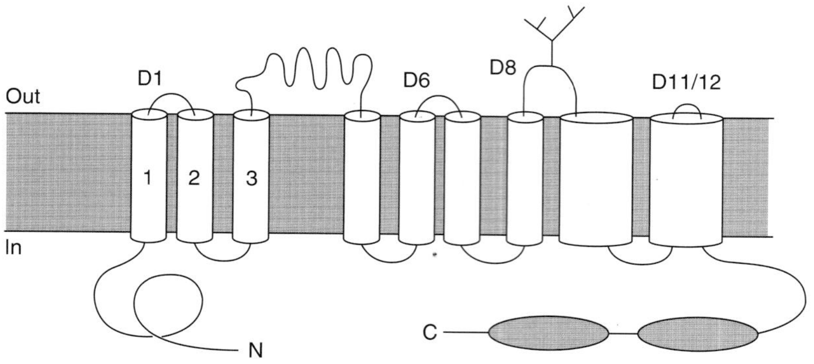

Figure 6.18 The ClC channel with 12 TM domains; the functional channel consists of two such units, and each is a separate Cl– channel. Figure 5.17 of Ref. [161]. Used with permission.

important for the proper movement of water, ions, and proteins into and out of cells, and mutations in the CFTR compromise fluid regulation in the airways and gastrointestinal (GI) tract. The primary disease of this protein is cystic fibrosis, and its mechanism is covered in Chapter 12.

6.3.2 The Chloride Channel

Chloride channels are involved in several cellular functions: volume regulation, transport of small molecules, and membrane voltage control. This large class of proteins (the ClC family) thus includes pure channels as well as Cl^-/H^+ transporters. It was first discovered in the electric ray *Torpedo* and soon identified in other eukaryotes and bacteria. Mammals have at least nine types of ClC channels, which function as homodimers, each peptide having 12 TM domains and forming its own Cl^- channel (Figure 6.18).

Recent research has indicated that the ClC family is related to, and perhaps derived from, the ion transporters such as the ABC family. Though the ABC transporters are ATP-powered ion pumps, the ClC channels handle Cl^- in the same way as do the ABC family members, and unlike the way that cation channels operate [155]. For example, the K^+ channels are tetramers of 2- to 6TM peptides, while the Cl^- channels are homodimers of larger 12TM peptides. The channel configuration is very different, with the K^+ channels forming a linear pore with negatively charged amino acids lining the pore, while the two ClC proteins form an hourglass-shaped structure with positive charges near the centers of the two molecules, thus guiding Cl^- ions through the channel [155].

6.4 Major Intrinsic Proteins

This is a large and poorly understood class of membrane proteins of largely unknown function, though two members – pores for water and for glycerol transport – have received recent attention.

6.4.1 Water Channels

These channels are called *aquaporins* because they allow the TM movement of water and they control the permeability of many types of animal and plant cell membranes.

Physiologists have known for a century that water can cross lipid membranes at a rate faster than one would expect from its polar nature, but only in the past two decades has the primary mechanism for this permeability been determined: a class of membrane proteins that form channels with many of the same structural and functional properties of the ion channels.

These channels are multimeric proteins as are many ion channels; each monomer has an even number of TM domains (generally 6–10 – Figure 6.19a, b), and the channel is formed by the association of four such monomers (Figure 6.19c).

Though aquaporins spontaneously form tetramers, each monomer can function as a water channel, and the structure has thus been described as four channels per tetramer (Figure 6.19c). Each of the several TM domains surrounds its own channel lined with polar amino acids through which water molecules move single file.

Mammals express 11 different aquaporins, one or more in each of several different tissues. A partial list

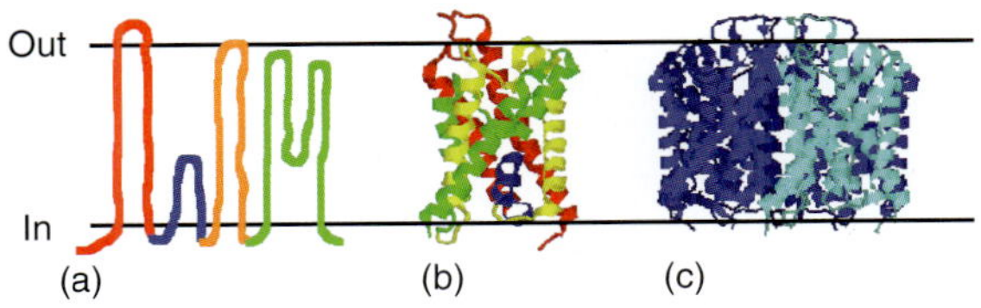

Figure 6.19 Model of an aquaporin. (a) One monomer; (b) the arrangement of the TM domains of the monomer; and (c) four peptides in the membrane. This figure represents the glycerol channel, but water moves through aquaporins in the same manner. Images were created using Rasmol [8] from PDB ID 3NE2 [186].

Table 6.2 Types of aquaporins in mammalian tissues.

Cell type	Aquaporin channel type	Function
Red blood cell	1	Osmosis
Brain	4	Osmolarity sensor
Salivary gland	5	Saliva production
Kidney		
Proximal tubule	1	Concentration of urine
Distal tubule	2	Action of ADH
Distal tubule	3	Water transport to blood

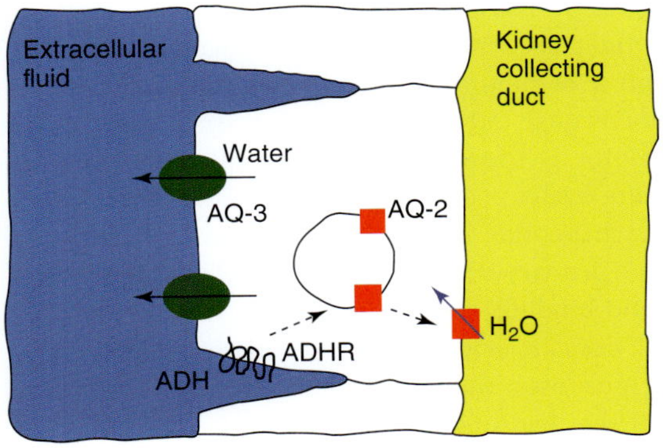

Figure 6.20 Mechanism of action of ADH on water balance in the kidney. Kidney epithelial cell shown in white, with basolateral surface at left and apical surface at right. Aquaporin-3 is always present on basolateral membrane but aquaporin-2 is stored in vesicles within epithelial cells. ADH binds to its receptor on basolateral side, causing aquaporin-2 to be placed on apical membrane. This permits flow of water from tubule to blood.

of the channels and their functions [187] is shown in Table 6.2.

An interesting example of the role of aquaporins in mammals is shown by the combined role of these channels and antidiuretic hormone (ADH) in controlling water balance. Two kinds of channels are found in the kidney distal tubule [187] (Figure 6.20).

Type 3 channels are permanently located on the basolateral membrane (the side away from the tubules containing urine); type 2 channels are present within these cells but not normally expressed on the cell surface. ADH is a peptide from the posterior pituitary (Chapter 4). It is secreted in response to water loss or increased blood Na levels in the kidney distal tubule cells. It binds to a GPCR on the basolateral membrane of tubule cells and causes the cells transiently to express the submerged channels on the apical surface (Figure 6.21). This expression allows water to pass from the tubule to the cells by osmosis; and because of the already-expressed channels on the opposite side of the cell, the water passes through the cells into the kidney matrix.

6.4.2 Glycerol Transporters

These transporters are homologous to aquaporins, and promote the movement of linear sugars such as glycerol [188]. Glycerol channels are nearly identical to water channels, but have a slightly larger diameter channel and lower polarity amino acids lining the channel.

6.5 Ligand-Gated Ion Channels

Ion channels operated by ligands are present only in bacteria and in **metazoan** animals [154]. Their absence in other eukaryotes makes it likely that animals obtained the channels via a unique event of lateral gene transfer at about the time of metazoan origins [189]. The bacterial channels bind small molecular weight **chemotaxic** molecules such as amino acids; the ion movements caused by binding alter cellular behavior for attraction or avoidance. The animal channels show the same ability to bind small molecules, and it is therefore likely that amino acid neurotransmission in animals is homologous with bacterial chemotaxic sensation. Three main classes of LGICs have been described: the "cys-loop" receptors, the **ionotropic** glutamate receptors, and the ATP-gated receptors.

6.5.1 Four-TM Domains – the Cys-Loop Receptors

These receptors possess a unique structure in the extracellular domain – a cysteine–cysteine **disulfide** link that creates a loop near the agonist-binding site

Table 6.3 Ligand-gated ion channels.

	Domains	Functions
Receptors with 4TM		
Excitatory	Cation-specific channels	
	nAChR	Neuronal, skeletal muscle depolarization
	$5HT_3R$	Neuronal depolarization
Inhibitory	Anion-specific channels	
	$GABA_AR$	Neuronal hyperpolarization
	GlyR	Neuronal hyperpolarization
Receptors with 3TM		
Excitatory	NMDAR	Neuronal depolarization
Excitatory	Non-NMDAR	Neuronal depolarization
Receptors with 2TM		
Excitatory	P2XR	Smooth muscle contraction Neuronal modulation

(Figure 6.24a). This feature is found only in the animal LGICs, and suggests that the channels came from a single common metazoan ancestor that originated the loop after obtaining the channel gene from bacteria. These channels also fall into three main structural types based on the number of TM regions each peptide subunit has (Table 6.3). Subclassification is based on whether the channels are selective for cations (Na^+, K^+, Ca^{2+}) or for anions (Cl^-).

6.5.1.1 The Four-TM Channels for Cations

Within this class, the cation-selective channels are the nicotinic ACh receptor and the serotonin ($5HT_3$) receptor.

Nicotinic Acetylcholine Receptor (nAChR – ACh-Gated Na^+ Channel) Nicotinic receptors were among the first receptors ever studied, being the target for curare in Bernard's studies in the 1850s (Chapter 2). The term "nicotinic" refers to early experiments showing that a low concentration of nicotine activates them in the same manner as the usual agonist, ACh. Similar versions of the receptor are also present on many types of nerves in the central nervous system (CNS) and in autonomic ganglia.

The receptor is a combined agonist-binding site and ion channel formed by five separate polypeptides, each with four membrane-spanning regions. These are present on all skeletal muscle cells, as well as on the **electroplax** structures of the various species of electric fish (e.g., electric eels and electric rays – Box 6.2) and in many cells of the nervous system. They are responsible for the ion flows into muscle cells that lead to depolarization and contraction, and, in the case of the electric fish, the delivery of hundreds of volts of electricity for stunning prey.

The five peptides of the receptor are of four types: two α, one β, one γ, and one δ (Figure 6.22b).

Though the amino acid sequences of the four peptides are different, they are related by descent from a common ancestor, and form similar three-dimensional structures in the membrane. The receptor is shown in a slightly different format in Figure 6.23.

Each peptide becomes a membrane protein with three primary domains: the extracellular N-terminal ligand-binding domain consisting of β sheets, the TM domain, and the intracellular domain [9]. For our purposes, the ligand-binding sites, the β sheets, and the hydrophobic TM helices from the second transmembrane domain (TM2) are the most important. These helices form the hydrophobic gate inside the channel, and directly above them in the extracellular portion are the β sheets. In the inactive state, the β sheets are organized such that the TM2 helices lie close together in the center of the channel (Figure 6.23). Two binding sites are present, one between the α–δ subunits and the other between the α–γ subunits. Following the release of acetylcholine from nerve endings, two molecules bind to these sites. Binding of agonist causes rotation of the β-sheets, and this leads to rotation of the TM2 helices of each subunit away from the center of the channel, increasing the distance between the helices so Na^+ ions can flow through it [9]. These movements

Box 6.2 Electric Fish Acetylcholine Receptors as Shock Generators

As shown by the quote from Plato at the beginning of this chapter, people have known for over 2000 years about the power of the electric fishes. Not until the last 300 years, though, were electrical phenomena in organisms subjected to systematic study.

All cells are negatively charged inside compared to the outside. When muscle cells are activated by nerves, their membrane potentials change by 100–150 mV and generate a few microamperes of current. Most organisms are adapted to using these small amounts of voltage and current in operating their muscles. In fact, these voltages are easily measured and can be detected on the surface of animals as the electromyogram or electrocardiogram. However, some species of fish have organized the muscles of their trunks such that hundreds of thousands of cells lie in series, allowing the ion flows to add to several hundred volts and to generate currents of 30 A. These particular muscle cells do not contain contractile elements; their structure is devoted to the generation of electricity. The electric eels (*Electrophorus*) and the electric rays (*Torpedo*) are the most powerful of the electric fish.

The electric ray has two large electric organs, one on each side of its head. The organs are modified muscle cells called **electrocytes**; the cells are arranged in disk-shaped stacks and are innervated only at the underside of each cell. This arrangement produces a battery-like structure where the voltages and currents of the individual cells can be added together. The cells have a very high density of nAChR on the innervated side of cells in each layer of the stack. When the cell is not discharging, the voltage gradients across the cell cancel so no net external current flows (Figure 6.21).

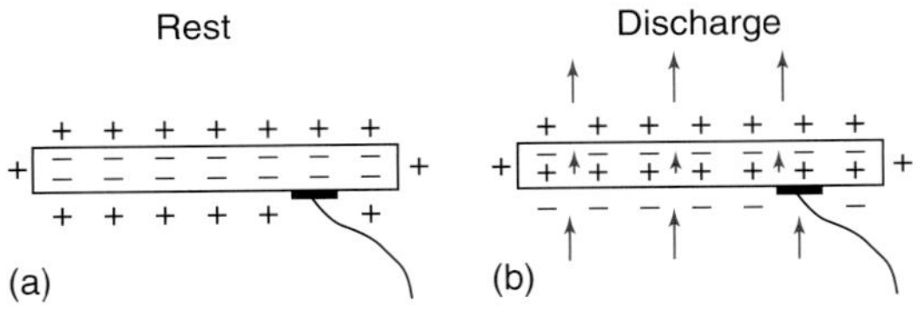

Figure 6.21 Single electrocyte from a stack of discs of an electric fish, for example, *Torpedo*. Ventral side only is innervated. The electric organ consists of stacks of thousands of such electrocytes, each with its motor neuron, in vertical arrangement. (a) At rest, outside of cell positive, inside negative; the +−−+ pattern from ventral to dorsal produces no net current. (b) During discharge, depolarization of the ventral side only reverses ventral polarity so the pattern is −+−+. This allows current from each cell to add to that of adjacent cells.

When ACh is released simultaneously at each electrocyte, it produces a depolarization of 100–150 mV on the ventral side of each cell. Because the dorsal side of each cell is not innervated, that side retains its polarity; the reversal of polarity (Figure 6.21b) at the ventral side along each of thousands of stacked cells allows voltages and currents to sum, producing a large amplification of the individual cellular contributions [187].

of Na^+ cause local depolarization that spreads to adjacent voltage-gated channels and eventually to channel-by-channel depolarization of the cell.

Differences between muscle-type and nerve-type nAChRs are subtle; the muscle type has the stoichiometry $(\alpha_1)_2\beta_1\delta\gamma$ (meaning it has two α_1 subunits, one β_1 subunit, and δ and γ subunits). The nerve-type receptors also have a pentameric structure, but in general have less diversity of subunit types, consisting either entirely of α subunits, for example $(\alpha_7)_5$ or of only α and β subunits, for example $(\alpha_4)_2(\beta_2)_3$.

The bacterial channels closest in structure and function to LGICs are ELIC (*Erwinia ligand-gated ion channel*) from *Erwinia chrysanthemi* and GLIC (*Gleobacter ligand-gated ion channel from Gleobacter violaceus*). Both are homologous to the ACh receptor [190], and are homopentamers with five binding sites in the same positions that bind ACh in animal receptors. The natural ligands for the two are not known, but are likely molecules of roughly the same size as ACh. The ELIC protein is homologous with the acetylcholine-binding protein (AChBP) secreted

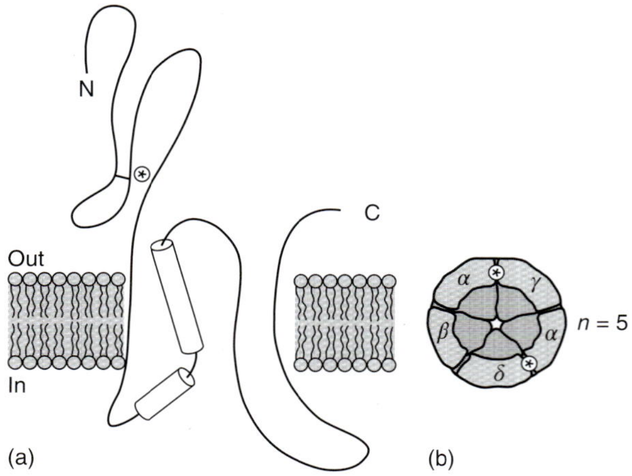

Figure 6.22 The muscle-electroplax nicotinic Ach receptor. (a) Diagrammatic representation of one peptide with four TM domains. The two cylinders represent the M2 transmembrane helix responsible for controlling ion permeability. The star indicates the cys-loop that gives its name to this family. (b) Top view of the five peptides in the membrane. Each of the five channel-forming units at the right has the general TM arrangement of the peptide in part A. Ligand binding occurs in two regions, between α and γ and between α and δ. Figure 6.1 of Ref. [161]. Used with permission.

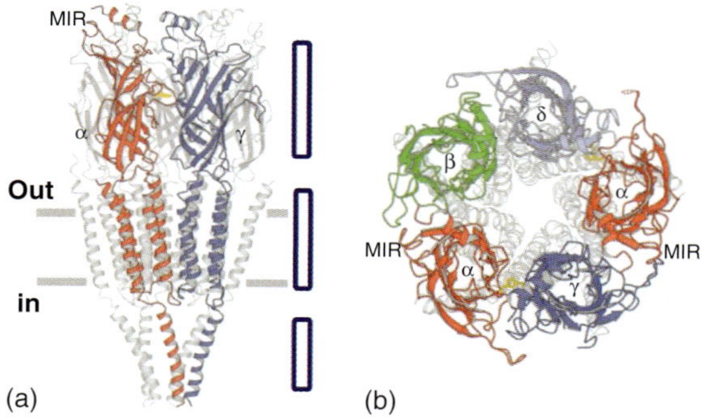

Figure 6.23 Structure of the nAChR of muscle. (a) View from side. Each subunit has four TM domains consisting of β sheets and α helices. Both the β sheet and M2 helical domains are involved in ion permeability. Open bars at right show the three receptor domains. (b) View from extracellular aspect showing the five subunits. Modified from Figure 3 of Ref. [9]. Used with permission.

by some eukaryotes into the synapse and used as a synaptic modulator of ACh [190].

Ionotropic Serotonin ($5HT_3$) Receptor These are similar in sequence to nAChR, with pentameric structure in the membrane. Five different subunits occur, and functional channels are made with at least one subunit A, with the remainder from the other subunits. They are present in brain pathways, where they influence cognitive processes; they are also present in the intrinsic nervous system of the gut, where they are implicated in vomiting and smooth muscle contraction. These receptors are also present on presynaptic neuronal membranes, and modulate transmitter (especially dopamine) release.

6.5.1.2 The Four-TM Channels for Anions

Gamma-Amino Butyric Acid (GABA) Receptor, Type A This receptor, also called the *$GABA_A$ receptor*, is the most widespread inhibitory receptor, and is found in all parts of the brain. It consists of five separate polypeptide chains forming a ring (Figure 6.24).

The five separate peptides are of three kinds – generally two α, two β, and one γ. The receptor can be built in different ways, as at least six α, four β, three γ, one δ, and a ρ peptide are known from molecular cloning experiments [195]. The sequences of the peptides are similar, and each has four membrane-spanning regions.

When gamma-amino butyric acid (GABA) molecules or GABA-like compounds activate it, the channel opens and allows the influx of Cl^-. The increase in negative charge forces the cell membrane to become more negative, and thus less likely to depolarize, or fire. Activation of many $GABA_A$ receptors can produce sedation. One hypothesis for the action of inhalant anesthetics such as halothane proposes that they act by binding the $GABA_A$ receptor on neuronal cells, causing Cl^- influx, hyperpolarization, and sedation [196].

The receptor seems to be a binding target, not only for GABA from brain neurons but also for alcohol,

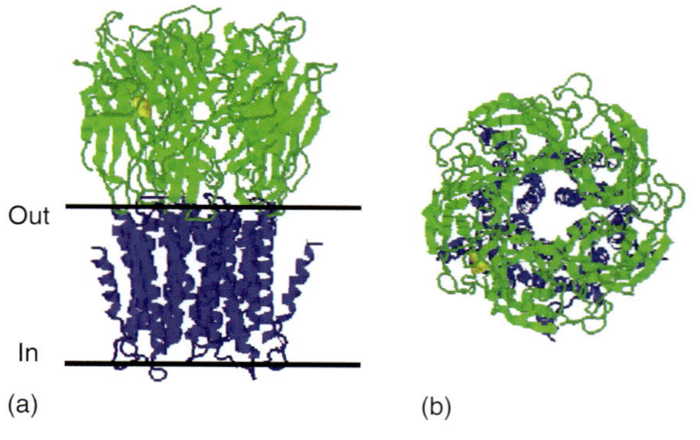

Figure 6.24 Orientation of the $GABA_A$ receptor in the membrane. (a) Side view and (b) top view. This receptor is a Cl^- channel opened by the binding of GABA; binding site for benzodiazepines, compounds that increase the sensitivity of the channel to GABA, are near the central channel. Images were created using Rasmol [8] from PDB ID 2YOE [194].

Box 6.3 Cation Receptors and Channels as Targets for Venoms

A wide variety of predators use venoms that act as nAChR antagonists; interfering with agonist binding either causes the death of prey or prevents their escape. The best-studied venoms are those of the snake family Elapidae (cobras, kraits, mambas, and taipans) and the marine snails of the genus *Conus*. The primary elapid toxin is a peptide of about 70 amino acids that is a competitive antagonist of the neuronal and muscle AChR. Mambas also have a peptide (dendrotoxin) that binds and inactivates the neuronal voltage-gated K^+ channel. The result of inactivation is a prolongation of the action potential and increased convulsive muscle activity [166]. The predators themselves are not affected by their own venom peptides, as their AChRs have altered sequences that reform the active site so it does not bind their toxins. In addition, mammals such as the hedgehog and the mongoose that prey on cobras are immune to the toxins, as they have similarly modified AChRs [191].

Cone snails produce hundreds of small peptides (each about 12 amino acids). The major paralytic peptide is α-conotoxin, antagonist of the muscle nAChR; additional toxins target voltage-gated Na^+ channels, K^+ channels, and Ca^{2+} channels [192].

Tropical frogs of the family Dendrobatidae release neurotoxins from their skins as defensive measures. The main toxic substance is batrachotoxin (BTX), which binds to and opens neuronal and muscular Na^+ channels, inactivating them in that position; the channels remain open, producing an inappropriately increased activity that causes paralysis [193].

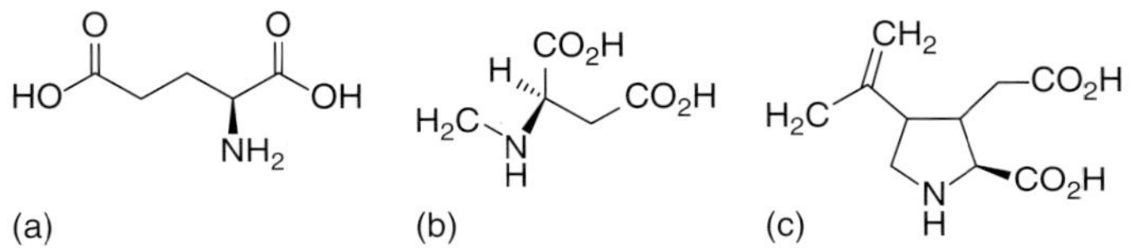

Figure 6.25 Molecular structures of (a) glutamate; (b) NMDA, and (c) kainate.

barbiturates, steroids, and the group of tranquilizer drugs (the **benzodiazepines**), developed in the 1960s to combat anxiety [195]. The benzodiazepines bind to an allosteric site (Figure 6.25; not the same site as the agonist) and increase the probability that the channel will open when GABA binds; the barbiturates and steroids prolong open time once it has opened. Benzodiazepines, barbiturates, and steroids thus accentuate the tranquilizing actions of $GABA_A$ receptors. In contrast, another set of psychoactive compounds, represented by bicucculine, antagonizes the $GABA_A$ receptor and increases neural excitability [161].

The Glycine Receptor is also a Cl^- Channel It contains five polypeptides chains, each with 4TM helices [197]. It is most prevalent in the spinal cord and brainstem and differs from the $GABA_A$ receptor in not being a target for the benzodiazepines. The amino acid sequences of the separate peptide chains are similar to those of the $GABA_A$ and *nAChR*, and different from the various glutamate receptors.

Possible Histamine Receptors Histamine has long been known to be an agonist of GPCRs (H_1 to H_4 subtypes – Chapter 7). One of those, the H_2 receptor, also couples indirectly to a Cl^- channel via the second messenger cAMP. Recent evidence also suggests that in mammals, histamine activates increased Cl^- conductance via a class of LGIC. Little is known about the receptors, and they have only recently been described; so far they appear to belong to the same family as the glycine and $GABA_A$ Cl^- channel receptors [198].

Amine-Gated Cl^- Channels The nematode *Caenorhabditis elegans* possesses several different ligand-gated neuronal Cl^- channels activated by the amines dopamine and 5HT [199]. This is an unusual situation, as the only other ion channel known to be activated by 5HT is a cation channel (above), and dopamine is otherwise known only to activate GPCR. It is possible that other amine-gated anion channels exist, and that these new receptors may provide important information on the behavioral repertoire of animals.

6.5.2 Three-TM Domains – Ionotropic Glutamate Receptors

6.5.2.1 Glutamate-Gated Channels

This class of excitatory LGICs contains receptors that are activated in the brain by the amino acid glutamate. The receptors are divided into three types – the *N*-methyl-D-aspartate (NMDA) receptors, the kainate receptors, and the AMPA receptors, depending on whether, in addition to glutamate, they respond to the artificial agonists NMDA or to kainate (KA) and alpha-amino-3-hydroxy-5-methyl-4-isoxazole-propionate (AMPA – Figure 6.25 and Table 6.3). The most useful classification scheme simply lumps the latter two receptor types, and thus distinguishes between NMDA receptors and non-NMDA receptors.

The NMDA and non-NMDA receptors all have four peptides, each of which has three TM domains [5] (though a partially reentrant additional loop also occurs – Figure 6.26a). The various peptides that can combine to form the receptor are similar in amino acid sequence and structure, but they differ markedly in function between NMDA and non-NMDA types.

6.5.2.2 N-Methyl-D-aspartate (NMDA) Receptor

This receptor is an excitatory Ca^{2+} channel found in many regions of the CNS, particularly in the **hippocampus**, cerebral cortex, and cerebellum. In vertebrates, it is not found in peripheral nerves outside the CNS. The conformation of the receptor in the membrane is as a ring of four peptides of two different types – generally NR1 and NR2 (Figure 6.26a).

This channel functions in a manner different from that of the ligand-gated Na channel. The resting receptor (Figure 6.26b) has a Mg^{2+} ion bound within the channel, blocking Ca^{2+} ion flows; this Mg^{2+} ion is not dislodged simply by ligand (glutamate) binding to the channel. Two other conditions must be met simultaneously: a glycine must first bind to an allosteric site on the NMDA receptor and the membrane must have been depolarized by another type of channel. Once the NMDA channel opens, it remains open for tens of milliseconds (in contrast to less than 5 ms for the ligand-gated Na channel). This complex activation pattern means that the NMDA receptor is not a simple depolarization-causing channel as the Na^+ channel is. Because it requires two simultaneous signals, it is a molecular *coincidence detector* that integrates signals from depolarization and transmitter release (initiated elsewhere) [161]. As discussed in Chapter 13, these actions are involved in **learning** and in the consolidation of **memories**.

6.5.2.3 Non-NMDA Receptors

These glutamate receptors mediate fast ion movements in CNS neurons. They are activated by ligand binding alone [161], and appear not to be involved in coincidence detection and learning. They thus are simple ligand-gated Ca^{2+} channels.

6.5.3 Two-TM Domains – ATP-Gated Receptors (P2X)

These receptors are 2TM peptides (Figure 6.27) that combine as trimers to make Na^+ and Ca^{2+} channels that bind extracellular ATP.

Though ATP is not normally thought of as an extracellular molecule, it is released from many types of cells [201], and it is coreleased with NE from sympathetic nerves in the periphery and from **adrenergic** nerves in the brain. Neuronal P2X receptors are sources of neurotransmission in the central, enteric,

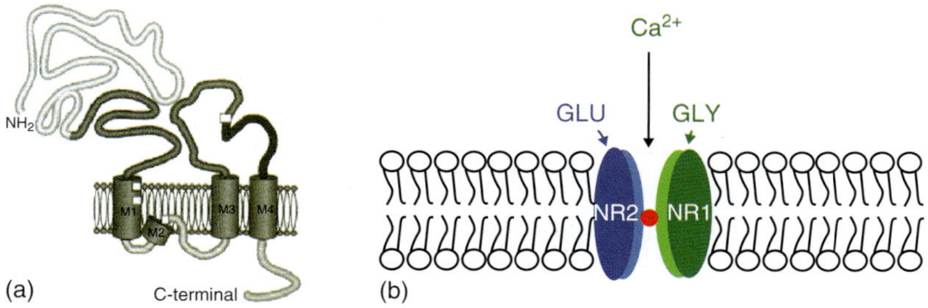

Figure 6.26 Generalized views of the NMDA receptor. (a) One peptide with three full TM domains and a partial TM domain. Each receptor consists of two of these peptides and two other similar peptides. Extracellular black areas contain the agonist binding site. (b) All four subunits (combined in one cartoon) in their membrane architecture; blue color represents the agonist (glutamate) binding unit (NR2); green represents the allosteric (glycine) binding unit. Red circle is the Mg2+ ion in the unstimulated channel. Panel (a): Figure 3.9b of Ref. [200]. Used with permission.

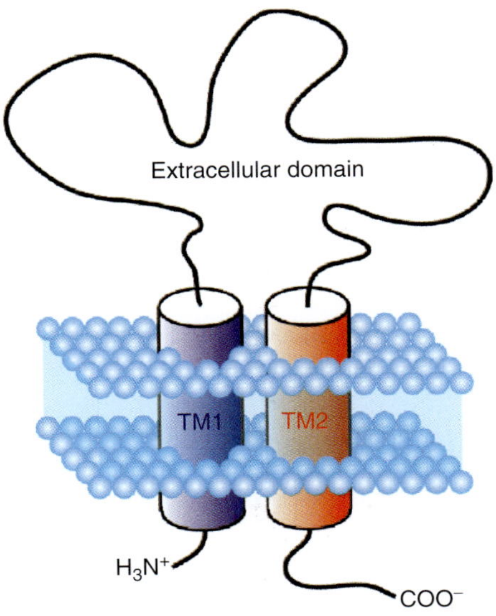

Figure 6.27 P2X peptide orientation in cell membrane. Note that the N- and C-termini are both intracellular. Three such molecules form a single channel, and simultaneous binding of three molecules of ATP is required to open the channel. Source: http://en.wikipedia.org/wiki/P2X_purinoreceptor. Accessed 2 January 2015.

and peripheral nervous systems. Their location on presynaptic terminals also allows them to modulate synaptic transmission [202] (Table 6.3).

The N-terminus and the C-terminus are on the cytoplasmic side of the cell, and have phosphorylation sites for intracellular kinases. The TM1 domain is involved in gating, and the TM2 domain makes up part of the ion pore. The ATP-binding site is on the extracellular portion between the two TM domains. The external domain is complex, with 10 cysteines that form 5 disulfide links, plus sites for glycosylation, all involved in ATP binding.

Seven different peptides of the P2X receptor are known, and most can combine in homotrimeric or heterotrimeric formats. P2X receptors appear to be present only in eukaryotes, and primarily in the unikont clade (amoebozoans, fungi, and animals). Though recently described in a single-celled chlorophyte alga [203], they are not yet known from other eukaryotes nor from prokaryotic taxa.

In vertebrates, these receptors are widely distributed, and subtypes $P2X_1$ through $P2X_6$ produce responses in neurons, glia, epithelia, endothelia, bone, muscle, and blood-producing tissues. The responses to P2X activation involve modulation of neural activity, detection of inflammation, and contraction of skeletal, cardiac, and smooth muscles [201]. $P2X_7$ is unusual in that in addition to ion channel action, it may also form a pore rather than a channel; this pore allows large molecules (up to 900 Da) to pass through; it thus alters cell integrity and is implicated in inflammation and cell death.

P2X receptors appear to be unrelated to other LGICs, but are among the oldest eukaryotic channels, and evolved at or near the origin of eukaryotes. Though they are seen most prominently in multicellular organisms with nervous and circulatory systems, it is significant that they are present in animals with no neural or muscle tissue [203], where they regulate ion fluxes and cell volume. This presence suggests that these cellular processes may have been important in signaling processes that allowed origin of the multicellular state.

7
Transduction II: G-Protein-Coupled Receptors

> ... strip the model down to only the basic inputs, minimize the complexity and build the simplest model. ... if I am going to screw up, at least I will do so with a lot less work.
>
> *Aswath Damodaran [204]*

> Do not use a simple model when a complex model will do.
>
> *J.C.J. Nihoul [205]*

7.1
Introduction

The G-protein-coupled receptors (GPCRs) comprise the single largest receptor superfamily in eukaryotes. The receptors control visceral functions such as cardiac and smooth muscle contraction, cellular metabolism, hormone secretion, and sensation. Their ligands are diverse: neurotransmitters, amino acids, peptide hormones, pheromones, light-sensitive molecules, and odorant/taste molecules. Their name comes from the intracellular proteins they activate after binding agonist – these **G(uanine nucleotide binding) proteins** go on to affect intracellular effectors such as enzymes and ion channels. The receptors are biologically and medically important, as over 800 genes (4% of the human genome) have been identified as actual or potential GPCR encoders, and close to 50% of all drugs manufactured by the pharmaceutical industry target these receptors [206].

The receptors are integral membrane proteins that cross the cell membrane seven times from the cytoplasmic side to the extracellular side; they have an extracellular N-terminus and an intracellular C-terminus. Ligand binding occurs either within a central well between three of the helices (Figure 7.1b) or externally to the N-terminus (Figure 7.1c).

DNA sequencing does not shed light on the earliest origins of these receptors, though because of their seven transmembrane (7TM) structure, they very likely were derived during the evolution of eukaryotic phyla from chemosensory receptors [210]. *Bacteriorhodopsin*, a similar membrane protein in the prokaryotic domain Archaea, uses solar energy to pump protons in the process of ATP synthesis. The rhodopsins are not true receptors because they do not bind ligands; their relationship to receptors is shown by the 7TM loops and by (in animal rhodopsins) their activation of a G protein [91]. The connection is further shown by the presence of a small light-sensitive molecule called **retinal**, attached covalently to one of the amino acid residues in precisely the same location that activating ligands attach in the true receptors (Figure 7.2).

The receptors also form reversible associations with other cellular proteins. The best-studied are the G proteins, but many others are involved in activating the cell following agonist binding. A short list gives a sense of the complexity of receptor interactions:

- G(uanine nucleotide-binding) protein
- Effector – enzyme, **ion channel**, or transcription factor
- GEF (guanine nucleotide exchange factor) – promotes exchange of guanosine diphosphate (GDP) for guanosine triphosphate (GTP) after ligand binding to initiate activity.
- GAP (GTPase activating protein) – terminates activity by hydrolyzing GTP to GDP.
- PKA (protein kinase A) – terminates activity by phosphorylating receptor.
- GRK (G-protein-coupled receptor kinase) and arrestin – GRK phosphorylates the intracellular portion of the receptor, providing a binding spot for arrestin, which prevents further G protein activation. The combination thus terminates activity.
- Other receptors – some GPCRs always work as dimers (see next section).

Until the 1990s, the GPCRs were thought to function exclusively as monomers; in 1993, Maggio *et al.* [211] first identified the ability of two receptors

Receptor Biology, First Edition. Michael F. Roberts and Anne E. Kruchten.

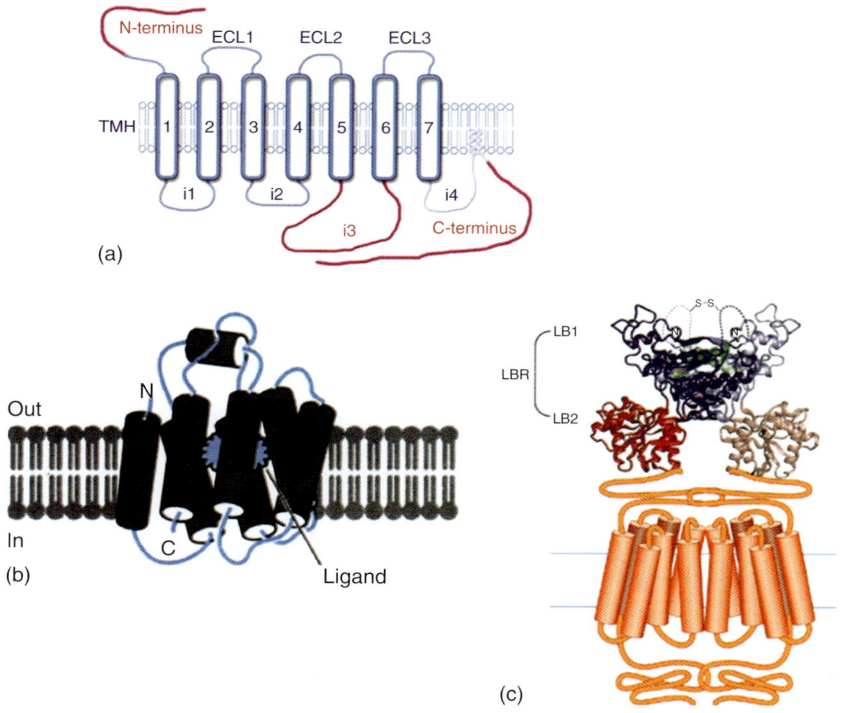

Figure 7.1 Overall structure of G-protein-coupled receptors, shown as cylindrical transmembrane domains alternating with linear cytoplasmic and extracellular domains. (a) Seven transmembrane (TM) segments (1–7); three extracellular loops (ECLs), four intracellular loops (i1–i4); i3 couples to G proteins. N-terminus is extracellular and C-terminus is intracellular. (b, c) Members of two families of G-protein-coupled receptors. (b) Family A receptor, showing central ligand-binding pocket (using TM domains H3, H5, and H6). (c) Family C receptor with long extracellular N-terminus for ligand binding. LBR: ligand-binding region. (a) From Figure 1 of Ref. [207]. (b) From Figure 4 of Chapter 6 in Ref. [208]. (c) From Figure 1 of Ref. [209]. Used with permission.

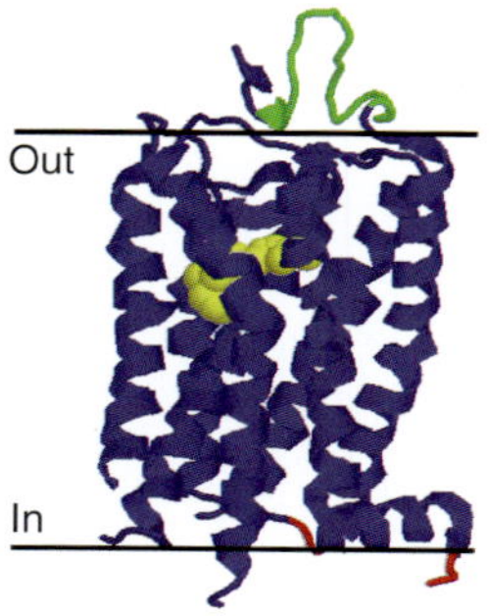

Figure 7.2 Model of G-protein-coupled receptors, based on bacteriorhodopsin. Hydrophobic helices form the membrane domains. At top is extracellular space; at bottom is cytoplasm. Rhodopsin binds covalently to a molecule of retinal (yellow) in the same location as ligand binds in true receptors of this superfamily. Images were created using Rasmol [8] from PDB ID 1F88 [10].

to function as dimers, and subsequent work has shown that dimerization is a common feature of the action of certain members of the superfamily, for example, the gamma-amino butyric acid type B receptor ($GABA_BR$). It is not clear, however, that dimerization is required for full function of other GPCR members. Dimerization seems to be involved in receptor maturation, cell surface expression, and receptor **internalization** [212].

7.1.1
Receptor Function

GPCRs are often termed **metabotropic** because their functions generally involve metabolic transformations [91]. Their transduction mechanisms are generally slower than those of ion channels (Chapter 6) and consist of three steps:

- Hormone or transmitter binding;
- Binding-induced activation of a membrane-bound **transducer** protein – the G protein;
- The activation by the G protein of one or another of these processes:
  Stimulation or inhibition of adenylyl cyclase and production of the second messenger cyclic adenosine monophosphate (cAMP).

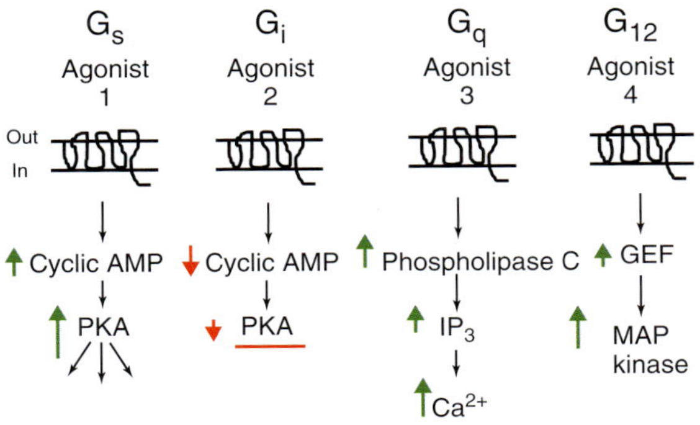

Figure 7.3 Principal signaling systems using G-protein-coupled receptors. PKA: cyclic AMP-dependent protein kinase and IP_3: inositol trisphosphate. Examples: agonist 1 stimulates glycolysis; agonist 2 inhibits glycolysis; agonist 3 activates gastrointestinal secretion; and agonist 4 activates transcription.

The activation of soluble enzymes that break membrane phospholipids down to produce other second messengers, inositol 1,4,5-triphosphate (IP_3) and **diacyl glycerate**

The activation of membrane ion channels (Figure 7.3).

7.1.2 Sensory Transduction

Cell-membrane receptors are known primarily as hormone or neurotransmitter signaling systems, but recent work has established them also as transducers of several of the body's senses: vision, taste, and smell are receptor processes different from hearing and touch in that they use photons or the binding of small molecules to proteins to alert an animal of details of its surroundings. We first briefly give an overview of chemo- and visual-sensory receptor mechanisms, and then place these particular systems within the context of GPCRs.

7.1.2.1 Chemoreception in Non-Mammals

Cyclic AMP Unicellular eukaryotes such as the slime mold *Dictyostelium discoideum* use cAMP as a signal for aggregation and formation of a temporary multicellular state. cAMP acts as a pheromone, being released from starving cells and binding to cell-membrane receptors of other cells. The binding event promotes aggregation and the formation of a complex structure that forms spores for release and colonization of new areas. The receptors for cAMP form their own family along with other chemoreceptor families [213].

Insect Gustation and Odor Sensing Insects use taste and smell as important cues for locating food or avoiding predators. The genomes of the fly *Drosophila melanogaster* and the mosquito *Anopheles gambiae* have been sequenced, allowing researchers to discover dozens of genes for GPCR-like proteins that are probably taste or odor detectors. These receptors seem to have no clear homology with other GPCRs [213].

7.1.2.2 Chemoreception in Mammals

Taste Mammals can discriminate five tastes – sweet, savory (also called *umami*), bitter, salt, and sour. Receptors for sweet, umami, and bitter are GPCRs, while receptors for sour and salt are ion channels, covered in Chapter 6. The bitter receptors belong to Family A, and the sweet and umami receptors belong to Family C, the families being distinguished by sequence differences and by the length of the N-terminus (Figures 7.21 and 7.23). Taste reception originates in sensory cells on the tongue and palate, and travels to the brain via sensory nerves (Figure 7.4).

Odors For most organisms, the sense of smell is the primary method of obtaining information about the environment; humans can recognize 10 000 different scents, and though this ability to discriminate is not as extensive as that of many mammals, it is enough to influence our physiology and behavior [215].

The initial detection of odors occurs in ciliated cells at the olfactory epithelium on the roof of the nasal cavity directly beneath the cerebrum of the brain. The olfactory bulb, a projection of the brain containing olfactory nerves, makes intimate contact with the thin bone at the roof of the nose that separates brain from nasal cavity.

The olfactory epithelium is divided into four large regions, each with a characteristic range of neuron/**odorant receptor** types. Each type of olfactory neuron seems to project to a small region of the olfactory bulb and hence to a particular region of the brain (Figure 7.5).

These distributions of receptor-type and projecting neuron suggest that odorant processing begins in the nose, is further modified in the olfactory bulb, and is completed in the brain. The receptors for odor belong to Family A (the rhodopsin-like receptors).

Vision The visual sense is also mediated by a GPCR, the protein rhodopsin; this particular receptor is important to describe at this point because it is the prototype of the largest family of GPCRs, the Family A receptors. Proteins acting as visual detectors have

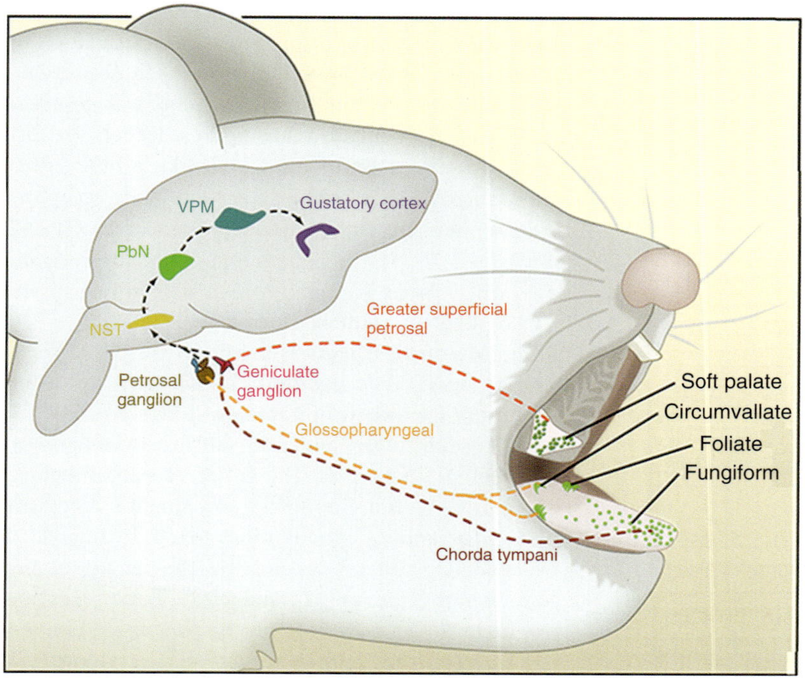

Figure 7.4 Taste buds are distributed on the tongue and soft palate. Three types of papillae are found in mice: circumvallate, foliate, and fungiform. Three nerves carry taste information from the taste receptor cells to the brain stem, then to the thalamus (VPM) and the primary gustatory cortex. Figure 1 of Ref. [214]. Used with permission.

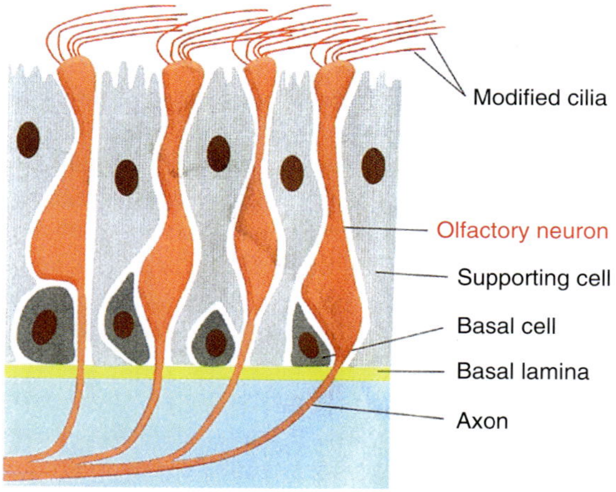

Figure 7.5 Neural connections between olfactory neuron cilia (top) and axon to olfactory bulb of the brain (bottom). The cilia are exposed to olfactory stimuli and contain receptors in their cell membranes. Figure 15–46 in Ref. [311]. Used with permission.

evolved many times in prokaryotes and eukaryotes. All are based on different ways of using the rhodopsin molecule to transduce light energy, and some of these detectors have evolved the complex anatomy called *eyes.* Five general light-transduction systems exist:

1) Pumps and channels in unicells
   a. *Archaea:* bacteriorhodopsin in cell membranes pumps H^+ for ATP synthesis.
   b. *Eukaryotic Algae:* channelrhodopsin in *Chlamydomonas* that uses light to open a H^+ channel that controls phototaxis.
2) Visual reception in metazoans
   a. *Molluscs:*
      Bivalves: several designs (Figure 7.6)
      Octopus and Squid: right-side-out camera eye
   b. *Insects:* **ommatidial** compound eye
   c. *Vertebrates:* inside-out camera eye.

The visual pigment rhodopsin, whether acting in unicells or vertebrates, is always a 7TM protein with a covalently bonded **chromophore**, retinal or a related vitamin A derivative. The chromophore binds in the same position within the transmembrane (TM) domains as do many GPCR agonists in binding to their receptors (compare Figure 7.1b with Figure 7.2). Though the rhodopsin chromophore is permanently bound, it alters conformation when a photon strikes it, and this change in conformation activates the rhodopsin. Retinal in its active state acts like an agonist, and for this reason it has been suggested that the rhodopsin-retinal complex evolved from an earlier chemoreceptor [210].

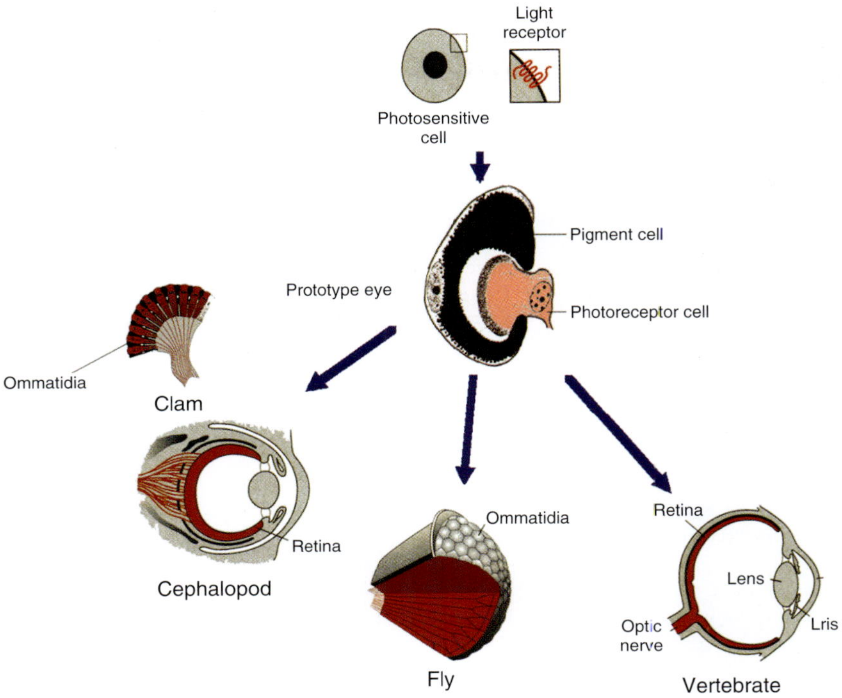

Figure 7.6 The evolutionary history of eyes. Top: Single-celled eukaryote with 7TM protein such as rhodopsin as photodetector (red) but with no further focusing or signaling function. Middle: In metazoans, a possible ancestral light-gathering and focusing cell leading to the multiple types of eyes found in animals. At lower left, two molluscan eye types; center: insect; right: vertebrate eyes. Modified from Figure 8 of Ref. [210]. Used with permission.

Gehring [210] has proposed that the eyes of animals are homologous in that they are all controlled by the same developmental gene, *Pax6*, which encodes a transcription regulator of the many structural genes encoding eye formation. The differences among animal eyes reflect levels of complexity of the rhodopsin-containing cells and neural connections with a nerve net or a central nervous system (CNS). This concept led to the evolutionary pathway shown in Figure 7.6.

7.2 Families of G-Protein-Coupled Receptors

The GPCRs of vertebrates are divided into five families according to sequence homologies within the families. Other animal phyla also express GPCRs that are in one or more of the same families as vertebrate receptors; however, some non-vertebrate eukaryotes have GPCRs that do not fit into the vertebrate receptor families. We discuss the evolution of the superfamily as a whole in Chapter 14; an overview of the families is given in Table 7.1 [206].

7.3 Transduction Mechanisms

7.3.1 Discovery of Receptor Control of Metabolism – Cyclic AMP and G Proteins

7.3.1.1 Components of the Process of Metabolic Activation

The hormones epinephrine (EP) and glucagon (GL) have long been known to increase blood glucose. The mechanism was not understood until researchers studying cellular enzyme function began to look for the signaling function – the receptor/agonist mechanism – that was responsible. Sutherland and Cori [217] determined that the enzyme **phosphorylase** was the rate-limiting enzyme for conversion of glycogen to glucose, and its concentration was affected by EP and GL. Rall *et al.* [218] showed that phosphorylation of the enzyme controls its function. Krebs and Fischer [219] showed that ATP was required for enzyme activation, indicating that ATP serves in a role distinct from its action as the energy source for metabolism. These data together indicated that

Table 7.1 Families of G-protein-coupled receptors.

Family	Name	Examples of receptor types
A	Rhodopsin-like	
	Adenosine receptors	
	Melatonin receptors	
	Histamine receptors	
	Serotonin receptors	$5HT_{1,2,4,5,6,7}$
	Amine receptors	Adrenoceptors, muscarinic ACh
		Dopamine, melatonin, nongenomic steroid receptor
	Rhodopsin	
	Some tastes	Bitter
	Some neuropeptide receptors	Endothelin, GnRH, ADH, OXY, NPY, CCK, TRH, GH, opioid, SST, ACTH, angiotensin, bradykinin
	Odorant receptors	Olfactory substances, FSH, TSH, LH
	Prostanoid receptors	Prostaglandins
	Cannabinoid receptors	Cannabis and endocannabinoids
B	Adhesion and secretin-like receptors	Secretin, calcitonin, GHRH, VIP, CRH, PTH, SEC, GL
C	Metabotropic glutamate receptors	Glutamate, GABA
	Some tastes	
	Calcium-sensing receptor	
F	Frizzled/smoothened receptor	Developmental receptors for Wnt and hedgehog signaling
D	Fungal pheromone receptor	a and α protein
E	cAMP receptor	Cyclic AMP
	Insect gustatory receptor	Taste
	Nematode chemoreceptor	Chemical sensing

GnRH, gonadotropin-releasing hormone receptor; ADH, antidiuretic hormone; OXY, oxytocin receptor; CCK, cholecystokinin receptor; TRH, thyrotropin-releasing hormone receptor; GH, growth hormone; SST, somatostatin receptor; ACTH, adrenocortical tropic hormone; FSH, follicle-stimulating hormone; TSH, thyroid-stimulating hormone; LH, luteinizing hormone; GHRH, growth-hormone-releasing hormone; and SEC, subepithelial cell.

enzyme phosphorylation is an important metabolic regulatory process, and that hormones somehow activate the phosphorylation process. At that time the nature of the hormone receptor and its mechanism of action were still not known.

7.3.1.2 Discovery of Cyclic AMP

Rall *et al.* [218] also discovered the agent responsible for phosphorylase action. It was cAMP, produced from ATP by a previously unknown enzyme that was somehow activated by hormone. The cAMP-producing enzyme is now known to be **adenylyl cyclase**.[1] These data suggest that two separate processes are involved in the control of metabolism: hormones activate adenylyl cyclase to produce cAMP, and cAMP activates the metabolic enzyme phosphorylase, which breaks glycogen down to form glucose.

1 Sutherland named it adenyl cyclase, but proposed [219] that chemically more correct terms might be adenylyl cyclase or adenylate cyclase. All three terms have been in use at one time or another; the current term is adenylyl cyclase.

By the mid 1960s, Sutherland's group showed that adenylyl cyclase was present in the plasma membrane rather than in cytoplasm or nucleus; this location near the interface of the cell with the environment suggested that the cellular response to agonist may involve production of an intracellular messenger. Sutherland formalized this new concept by describing cAMP as a **second messenger**; he thought of the extracellular hormone as the **first messenger** that worked by binding to a receptor on the cell membrane. This mechanism quickly came to be seen as a very important general activation system for cells, because many cell types could be activated by hormones that cause formation of cAMP.

7.3.1.3 Discovery of G Proteins

Rodbell had determined that guanine nucleotides (GDP and GTP) were involved in the regulation of adenylyl cyclase and proposed that these nucleotides bound to an as-yet unknown membrane protein [220]. Both this protein and GTP were required for hormone activation of cAMP, suggesting they acted

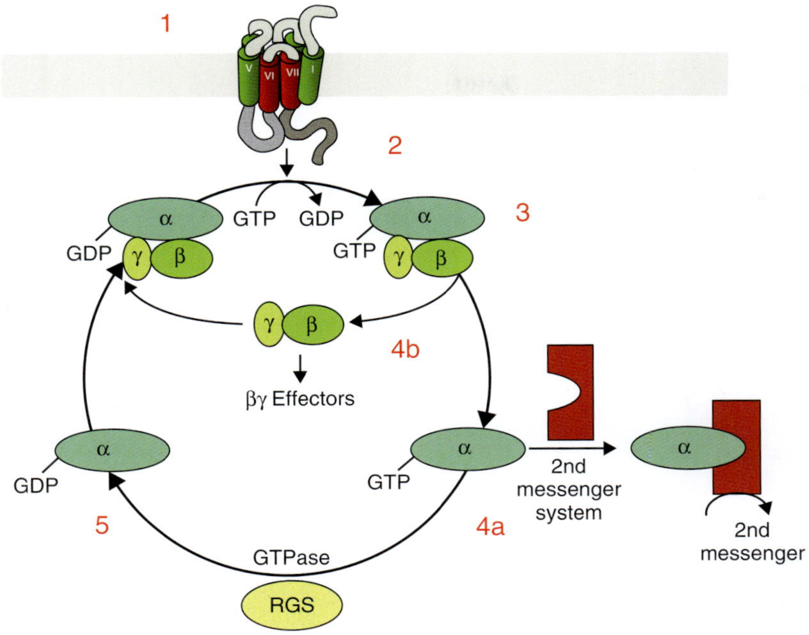

Figure 7.7 Transduction of hormonal signal via G protein. (1) Hormone binding alters receptor conformation. (2) G protein activation begins when (3) GDP is replaced by GTP. (4a) Alpha subunit activates effector proteins via second messenger and (4b) β–γ subunits activate effectors. (5) Hydrolysis of GTP to GDP deactivates G protein. Modified from Figure 3 of Ref. [224]. Used with permission.

at a post-receptor step. Cuatrecasas [221] suggested that hormone binding to a receptor initiated a lateral movement of the receptor–hormone complex within the membrane, and that this mobile receptor was able to bind to another membrane protein to initiate cell activation. Cassel and Selinger [222] showed that receptor binding of hormone causes dissociation of a molecule of GDP that had been bound to a membrane protein, to be replaced by GTP. Activation of adenylyl cyclase occurs at this GTP-binding step, and deactivation occurs when GTP is subsequently hydrolyzed to GDP. Thus, the cyclic alternation of GDP and GTP as components of a transducing membrane protein was step that caused activation and deactivation of the enzyme.

This transducing protein was finally isolated by Gilman's group [223] in 1980 and named *guanine-nucleotide-binding protein*, or G protein. An overview of the receptor–transducer–effector activation is shown in Figure 7.7.

7.3.2
Actions of G Proteins

The G proteins are heterotrimers consisting of α (39–46 kDa), β (37 kDa), and γ (8 kDa) subunits. Investigators have isolated about 20 different α peptides; at least five distinct β subunits and 12 different γ subunits have been described [224]. If all possible combinations of α, β, and γ subunits occurred, close to 1000 different G protein combinations would be possible. It is highly unlikely that these subunits are actually used in all combinations, and the actual number of functional associations is not known; however, new interactions are being discovered on a regular basis.

The molecular structure of G proteins began to emerge in the mid-to-late 1980s, following determination of their amino acid sequences [225]. The current view of their structures is given in Figure 7.8.

The proteins were first thought to be intermediaries between only the β-adrenoceptor and adenylyl cyclase, but evidence for their wider importance soon emerged [220, 227]. In the early days of G protein research, the β and γ units appeared not to be directly involved in signaling, but recent work has shown they are implicated in as many processes as the α subunits [224]. In particular, the β–γ units may affect the activation state of the receptor by binding to regulatory kinases [228], and may also activate ion channels [229]. In some cases, the α and the β–γ subunits may activate the same effector. When this happens, the actions are additive and independent [200]. In addition, β subunits may also operate in

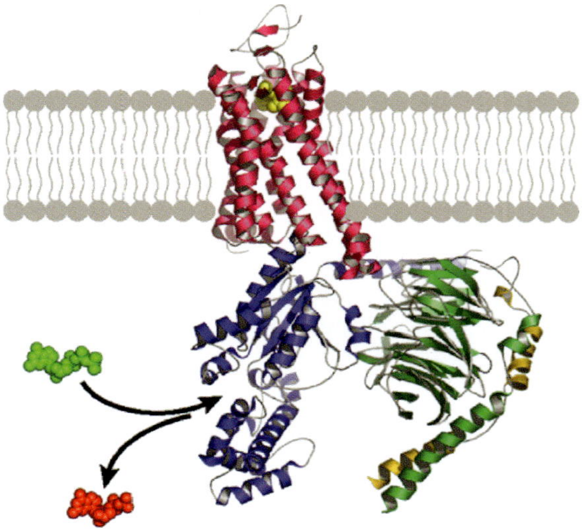

Figure 7.8 GPCR (pink) in membrane with bound ligand (yellow) and associated G protein. The α subunit (blue) exchanges GDP (red) for GTP (bright green); β subunit (dark green); and γ subunit (gold) shown at right. Part of Figure 1 of Ref. [226]. Used with permission.

isolation, on effectors different from those acted on by α subunits [230].

7.3.2.1 G-Alpha Proteins

Table 7.2 lists the known G-alpha proteins along with the effector systems they activate. They are divided into four families based on amino acid sequence homology [224].

G_s: Activation of Adenylyl Cyclase G_s is the term for the stimulatory G protein that activates adenylyl cyclase. The α subunit of this protein (called $G_{s\alpha}$) consists of a polypeptide of 394 amino acids. The G_s subunit is able to form associations with three separate molecules (Figure 7.8):

- First, in the inactive state (in the absence of ligand binding to receptor), the G_s binds a molecule of GDP.
- Second, after ligand has bound to the receptor, the third cytoplasmic loop of the receptor binds to the G_s–GDP complex. G_s replaces GDP with a new molecule of GTP and simultaneously reduces the affinity of the receptor for agonist.
- Third, the presence of GTP causes the G_s to activate the membrane-bound effector (adenylyl cyclase) to form cAMP. G_s is also a GTPase, and immediately after adenylyl cyclase activation, the active state of G_s is ended by the hydrolysis of GTP to GDP. Thus, the *inactive* $G_s = G_s$ bound to GDP, and the *active* $G_s = G_s$ bound to GTP.

G_i: Inhibition of Adenylyl Cyclase and Direct Action on Ion Channels In addition to hormonal mechanisms that increase cAMP, important regulatory processes are mediated by hormone-related inhibition of adenylyl cyclase. G_i refers to the inhibitory G protein that decreases adenylyl cyclase activity. The first such cyclase-inhibitory receptor to be studied was the α_2-adrenoceptor in blood vessels. When this receptor binds its agonist (NE), G_i is activated. The effect of G_i activation is a decrease in cAMP, which decreases the rate at which Ca^{2+} is being pumped out of smooth muscle cells. Ca^{2+} then increases within the cell, producing contraction.

G_t: Transducin in the Visual System The G_t protein is found only in association with the visual system; it is present in the intracellular disks of the outer segment of retinal photoreceptor cells (Figure 7.9).

Table 7.2 GPCR transducer and effector proteins.

Families	G transducers	Effector	Receptor types
G_s (stimulatory)	G_s	Adenylyl cyclase	Diverse
	G_{olf} (olfactory)	Adenylyl cyclase	Olfactory receptors
$G_{i/o}$ (inhibitory)	G_i	Adenylyl cyclase	Diverse
	G_o	Adenylyl cyclase	Diverse
		K^+, Na^+, Ca^{2+} channels	
	G_{gust}	"Gustducin" in insects	Taste receptors
	G_t (transducin)	Phosphodiesterase	Rhodopsin
$G_{q/11}$	G_q	Protein kinase C	Diverse
		Phospholipase C	
		Inositol triphosphate	
		Diacyl glycerol	
		K^+, Na^+, Ca^{2+} channels	
$G_{12/13}$	$G_{12/13}$	Na^+/H^+ exchange, transcription	Diverse

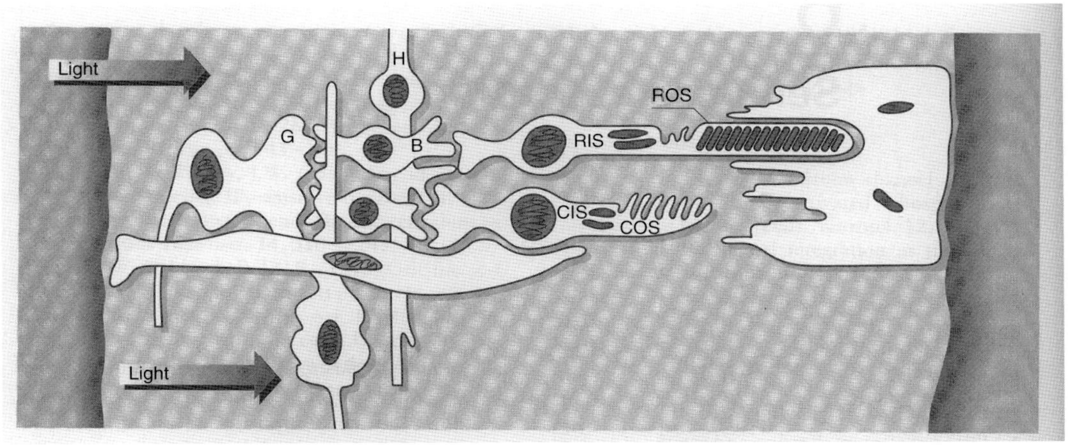

Figure 7.9 Schematic view of small region of the vertebrate retina with rod cell showing discs in the rod outer segment (ROS) that contain rhodopsin, as well as neurons that communicate the rod output to the optic nerve. Figure 6.1 in Ref. [91]. Used with permission.

Two similar α subunits are known from the bovine retina: the G_{t1} and G_{t2}. Based on information from antibody binding experiments, these two proteins seem to be specific to the rods and cones, respectively [227]. In the following discussion, the rods and cones act in similar ways, except that rods have higher sensitivity in low light than do cones.

Function of G_t in Vertebrate Vision In this description, the term "dark" refers either to conditions of zero ambient light or to conditions in which a small dark region, such as a black letter on a page, is viewed by the retina. In both cases, a given photoreceptor cell views a dark field. The cell has a high concentration of **cyclic GMP** (cGMP), the intracellular second messenger that binds to the cation channels of the plasma membrane to keep them open and allow Na^+ or Ca^{2+} into the cell. cGMP is regenerated by guanylyl cyclase within the rod cell; this enzyme is inhibited by Ca^{2+}. **Phosphodiesterase** (PDE) breaks cGMP down to GMP, but in the dark PDE is inactive so little or no cGMP is lost. Thus, the open channels keep the dark-exposed cell in a state of depolarization. The depolarized cell continuously releases its transmitter (glutamate), which activates inhibitory cells of the retina, preventing optic nerve activity (Figure 7.10a "dark").

For those rods or cones that receive light (Figure 7.10b "light"), the retinal bound to the opsin is altered from the 11-cis form to the all-trans form, and thereby changes the conformation of the opsin. This change activates the G_t α subunit, which then exchanges GDP for GTP and causes the dissociation of the α from the β/γ subunits. The GTP-activated α subunit then stimulates a cGMP-specific PDE that catalyzes the breakdown of cGMP. The loss of cGMP from the outer segments then allows the closure of Na^+ channels in the rod membrane. This event hyperpolarizes the membrane, and stops the firing of the cell. The decrease in output of transmitter ends the inhibition of optic nerve. The action of PDE is terminated when the GTP on the G_t α subunit is broken down to GDP and P_i, thus decreasing the affinity of the α subunit for PDE and increasing its affinity for the β/γ units. Further, the reduction in Ca^{2+} stimulates guanylyl cyclase to synthesize more cGMP.

For invertebrate mechanisms in vision, see the section below on G_q proteins.

G_o (Other Effects) This G protein was discovered in brain tissue by use of radiolabeled GTP analogs as a probe for G_i [224]; its function is not well understood, but one action involves inhibition of transmitter release in CNS neurons [200].

G_q: Activation of Phospholipase C and Protein Kinase C As discussed in Chapter 2, the plasma membrane contains a number of unusual phospholipids that have both structural and signaling roles. One such lipid is phosphatidyl inositol 4,5-bisphosphate (PIP_2), which is found in certain cell membranes in high concentration surrounding receptor proteins, in particular the muscarinic and α_1-adrenoceptors. The binding of transmitter (acetylcholine (ACh) to muscarinic receptor and norepinephrine to α_1-adrenoceptor) causes activation of the G_q form of G protein, which activates the membrane enzyme phospholipase C

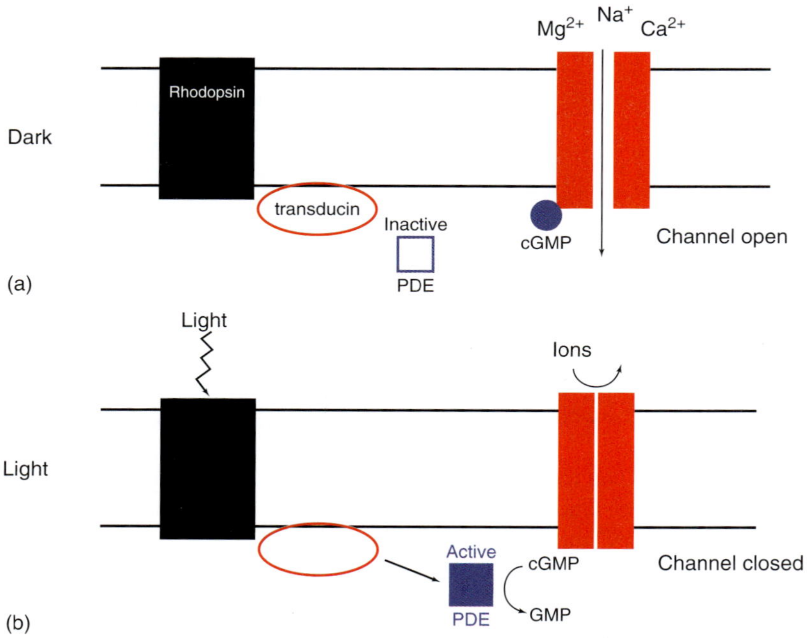

Figure 7.10 Vertebrates. The rod or cone disc membrane is the flattened membrane structure seen in Figure 7.9. (a) Absence of light (or for cells viewing a dark field), rhodopsin does not activate transducin; PDE is not activated; (cGMP) remains high and binds to ion channels, allowing ion current to keep the cell depolarized. (b) In the light (or for cells viewing a light field), rhodopsin activates transducin; PDE is activated to break cGMP down to GMP, which cannot bind to channels. Channels close, allowing outward K current to hyperpolarize rod or cone cell.

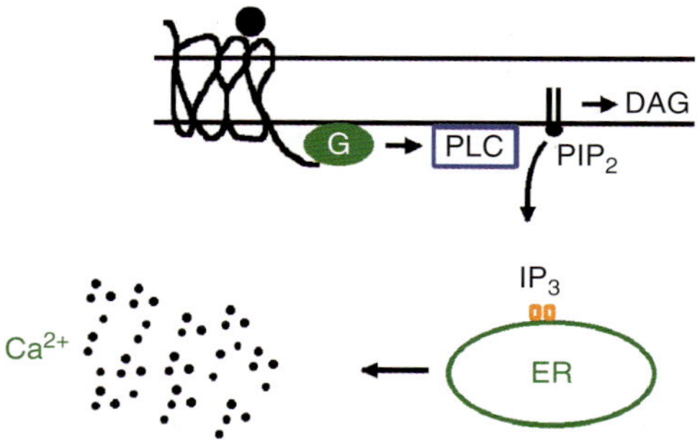

Figure 7.11 The phosphatidylinositol transduction cycle. Membrane-bound phospholipase Cβ (blue rectangle) is activated by GPCR agonist (black circle) binding via G_q (green oval). PIP_2 is split to IP_3 and DAG and IP_3 is a second messenger and binds to its receptor (orange) to open Ca^{2+} channels in ER. This releases Ca^{2+} for further signaling.

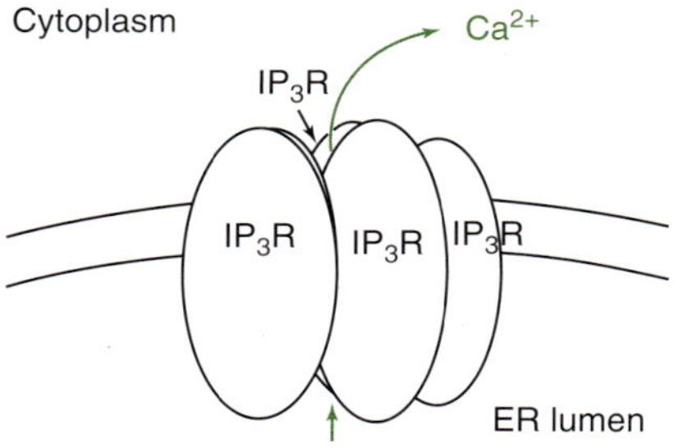

Figure 7.12 Model of the inositol-1,4,5-trisphosphate receptor on cellular Ca^{2+} store (e.g., lipid bilayer of the ER). The source of the IP_3 is plasma membrane PIP_2, and is generated by the action of phospholipase C (Figure 7.11). Calcium ion movement is shown as the green arrow.

(PLC). This enzyme cleaves the nearby PIP_2 into the two second messengers IP_3 and diacyl glycerol (DAG). The process [231] is shown in Figure 7.11.

Each of these messengers has a separate role to play in cell activation:

- IP_3 is soluble in the cytoplasm; it diffuses to the endoplasmic reticulum (ER) and binds to its receptor on the surface of the ER. This receptor is a member of the ion channel superfamily (Chapter 6) and is an ER channel for Ca^{2+} ions. Binding of IP_3 opens the channel to allow Ca^{2+} to flow out into the cytoplasm of the cell (Figure 7.12), where it activates cellular processes such as contraction of smooth muscle or glandular secretion [62].
- DAG remains associated with the membrane. It permits the enzyme protein kinase C (PKC) to bind to the inner surface of the cell membrane and phosphorylate cellular proteins such as enzymes

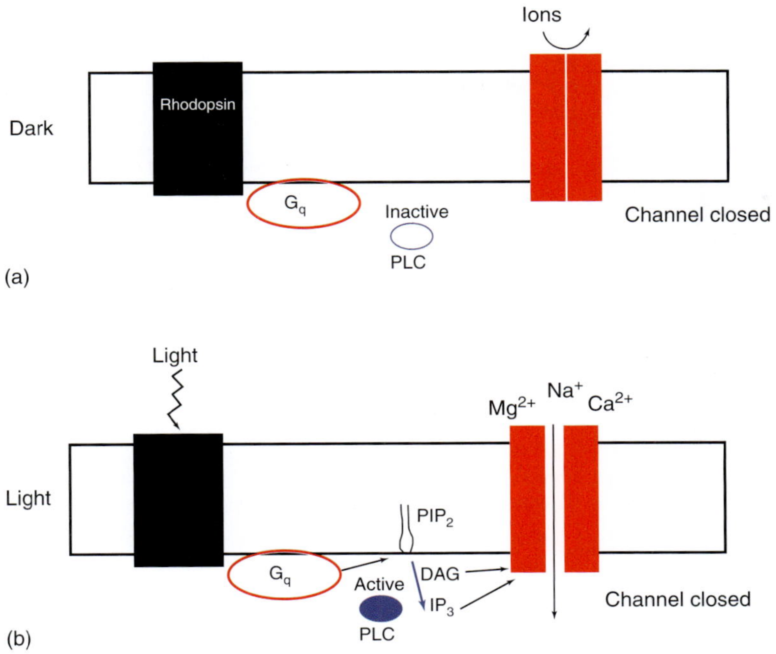

Figure 7.13 Invertebrates: Depolarization by light of light-sensitive cells. (a) In the dark, opsin is inactive and the ion channel is closed. (b) In the light, opsin activates a G_q protein, which allows phospholipase C (PLC) to liberate DAG and IP_3 from the disc membrane. One (or both) opens the light channel to allow in cations that depolarize the cell.

and ion channels. Phosphorylation is one way of sensitizing ion channels so they remain open longer in response to other transmitters. This process provides a mechanism for the potentiation of memory (an action to be discussed in more detail in Chapter 13).

The invertebrate eye uses the G_q protein to allow vision, but uses a reverse version of the vertebrate mechanism. In invertebrates such as squid and *Drosophila* [232], light activates an ion current that depolarizes the light-sensitive cell. The ion current is generally Ca^{2+}, and the ions enter through a transient receptor potential (TRP – see Chapter 6) channel in the eye membrane. The activation process is shown in Figure 7.13.

In this mechanism, light activates rhodopsin to couple to the G_q protein, which stimulates PLC to split membrane phospholipids to release DAG and PIP_3. These directly open cation channels that depolarize, and thereby activate, the light-sensitive cell in response to light. The photocell mechanisms of vision in invertebrates thus operate in a manner opposite to that of vertebrates.

G_{12}: Activation of *Ras* and *MAP Kinases* All transduction by GPCR occurs via exchange of GDP for GTP on the G_α protein for GTP. As GEFs are required for this process to be rapid, it is not surprising to see GEFs involved in signaling with the above G proteins (G_s, G_i, and G_q). However, the G_{12} family of G proteins directly activates GEFs, and thus *only* signals via GEF-induced activation of downstream processes such as Ras and MAP (mitogen-activated protein) kinase. The receptors involved in G_{12} signaling also activate other G proteins; for example, the vasoconstrictor peptides endothelin and angiotensin II operate via G_{12} proteins [233]; however, endothelin also activates G_q [234], and angiotensin II also activates G_s, G_i, and G_q proteins.

7.3.2.2 Roles of the Beta and Gamma Subunits

The β−γ subunits bind each other tightly but associate loosely with the α subunit via a helix at the N-terminus of the α subunit (Figure 7.8). β−γ subunits dissociate from the α subunit when the bound GDP is replaced by GTP. They then act either on membrane-bound effector proteins or on soluble proteins within the cytoplasm.

Wang *et al.* [230] showed that the β subunit could operate independently of the α subunit to transduce processes related to mating in fungi. The main functions involve responses to mating pheromones

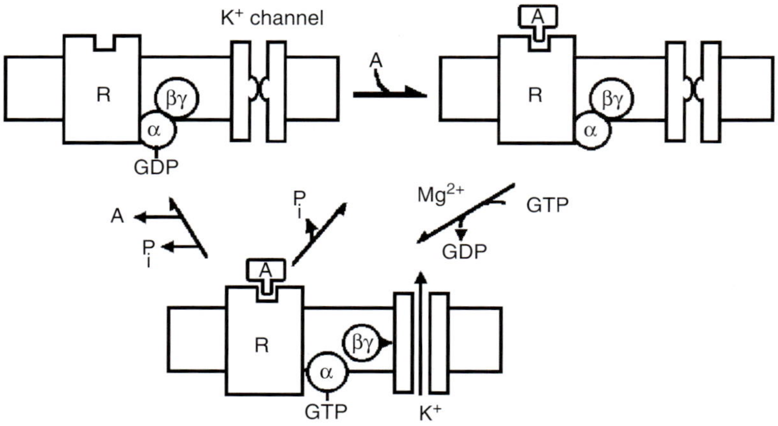

Figure 7.14 Activation of the inwardly rectifying K channel by the muscarinic action of ACh. R: receptor; α, β, γ are subunits of the G protein. ACh (A) binding activates receptor to exchange GDP for GTP. This step liberates β−γ, which opens the channel. Figure 6 of Ref. [229]. Used with permission.

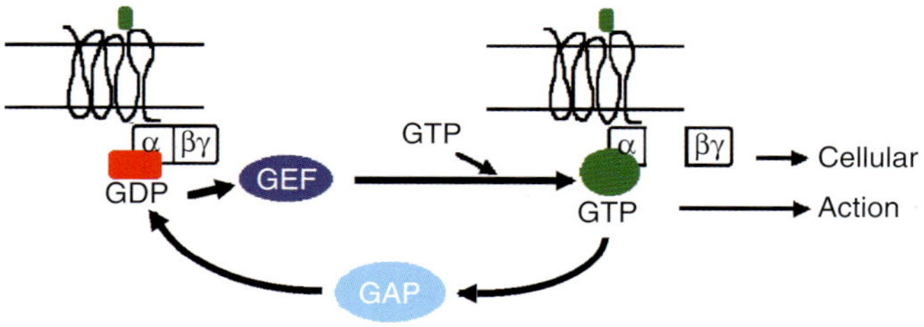

Figure 7.15 An example of modulation of G protein function by additional proteins. Here the protein GEF (guanine nucleotide exchange factor) promotes exchange of GTP for GDP to initiate response, and GAP (GTPase-activating protein) accelerates the hydrolysis of GTP (removal of phosphate P_i), causing return to GDP and rapid cessation of the response to ligand.

and the change of phenotype to one that produces spores for dissemination of fungal cells.

Ion Channel Activation β−γ subunit transduction can directly regulate ion channels for K^+ (Figure 7.14).

In this process, seen in atrial cells of the heart, ACh binding activates the G protein, and the βγ subunits open the K^+ channel [229] (Chapter 6). The efflux of K^+ hyperpolarizes the cell so it takes longer to depolarize during normal pacemaker activity. The direct nature of the relationship comes from the fact that the G protein subunits activate the channel without requirement for an intermediate second messenger.

Inhibition of Transmitter Release Several receptors in neurons (M2 muscarinic; α_2-adrenoceptors; opioid receptors) act as negative feedback mechanisms to decrease release of neurotransmitter. These receptors activate the β−γ subunits, and these bind to membrane calcium channels and prevent the inward movement of calcium that leads to transmitter release [200].

7.3.3 Proteins That Enhance (GEF) or Inhibit (GAP) GTP Binding

7.3.3.1 GEF Protein

In the inactive state, GDP is bound to the G protein α subunit. Upon agonist binding, GEF ([235] – Figure 7.15) promotes the replacement of GDP by GTP.

The binding of GEF to the α subunit lowers α affinity for GDP, and the nucleotide dissociates. It is replaced by a molecule of GTP because GTP concentration in the cell is higher than that of GDP [235]. This step activates the G protein because the α-GTP combination has low affinity for the β−γ subunits. GTP-bound α thus becomes free to associate with effector molecules such as adenylyl cyclase.

7.3.3.2 GAP Protein

GAP (Figure 7.15) promotes hydrolysis of the GTP attached to the α subunit of the G protein, thus accelerating dissociation of the α subunit from the cellular effector (e.g., adenylyl cyclase) and preventing

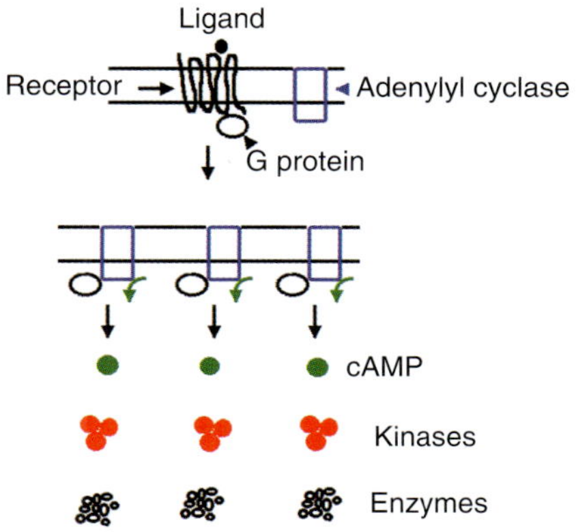

Figure 7.16 Amplification of the ligand-activated catalytic cascade. A single ligand binds to its receptor; the receptor generates several G_s, each of which activates adenylyl cyclase for a prolonged period. Each adenylyl cyclase thus produces many cAMP, which activate multiple kinases and other enzymes.

prolonged response to agonist [236]. GAP belongs to a group of RGS (regulators of G protein signaling) proteins important for their ability to modulate GPCR signaling.

7.3.4 Signal Amplification

One of the important features of signal transduction in general, and the G protein system in particular, is the amplification of the original single hormone-binding event into hundreds or thousands of final biochemical events within the cell. The amplification cascade extends from receptor to G protein to adenylyl cyclase to cAMP to kinases and other enzymes (Figure 7.16).

7.3.5 Signal Cessation – Several Processes Decrease Receptor Activity

When environmental stress causes an organism continuously to secrete a large amount of a particular hormone, two types of events follow: in the short term, the tissue responds; in the long term, the response decreases even in the presence of hormone. This long-term **desensitization** (also called tachyphylaxis, refractoriness, or **tolerance**) occurs in clinical situations when a drug is administered for a long period of time. Several mechanisms contribute to this decrease, and are outlined in the next section.

Desensitization is accomplished by two main mechanisms [237], shown in Figure 7.17.

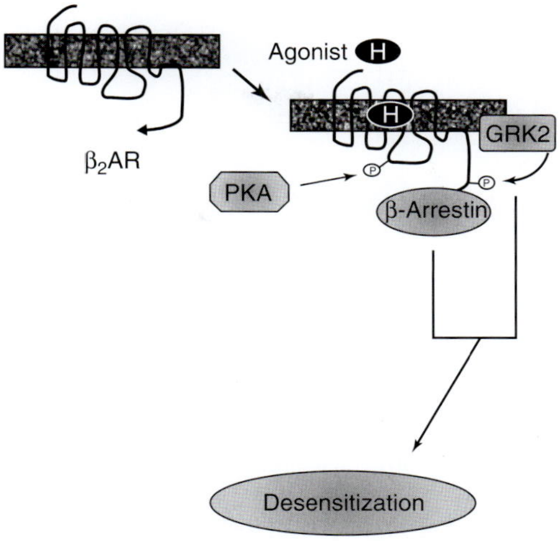

Figure 7.17 Desensitization of β-adrenoceptors by covalent modification. GRK2 phosphorylates several amino acids of the C-terminus of the receptor; this creates a recognition site for β-arrestin, whose binding prevents association of receptor and G protein. PKA phosphorylates another site of the receptor, altering it so it does not associate with G protein. Modified from Figure 11 of Ref. [237]. Used with permission.

Inactivation occurs by the actions of three intracellular enzymes, PKA, GRK, and β-arrestin. The kinase A phosphorylates the receptor directly and thus inactivates it [237]. GRK promotes phosphorylation on several threonine and serine residues at the carboxy terminal end of the receptor, and makes the receptor available for attachment to β-arrestin, which binds to phosphorylated receptors and prevents their association with G proteins. A phosphorylation process also stops action of opsin in retinal cells: the photoactivated form of opsin (photorhodopsin) is phosphorylated by a kinase, and the modified receptor is bound by the protein arrestin (similar to β-arrestin) preventing further association of opsin with G proteins of the rod disk [237].

The arrestin–GRK system is actually more complex than shown above; following desensitization of a GPCR, the arrestin goes on to activate additional signaling cascades such as the MAP kinase system (this connection is further developed in Chapters 8 and 10).

7.3.6 Interactions between Receptors and G Proteins

The various GPCRs have similar overall shapes and similar primary amino acid sequences, yet they control very different cellular events. Thus, the small differences in structure must be important in determining activity.

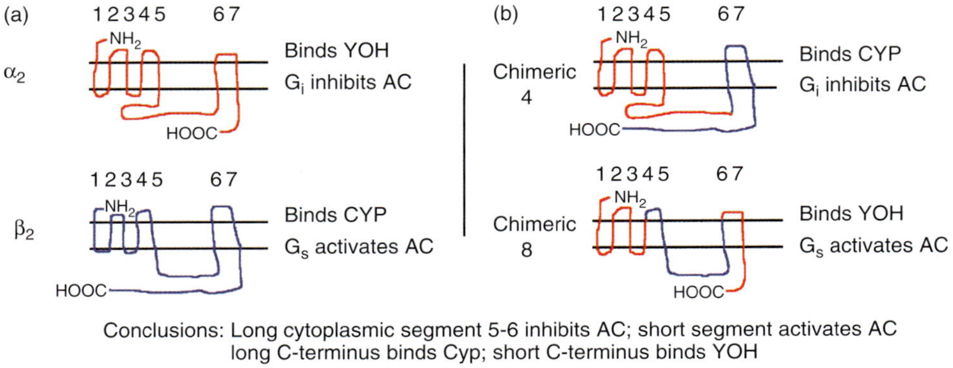

Figure 7.18 Functional domains in G-protein-coupled receptors. Genetic chimeras were expressed in *Xenopus* oocytes and used to show locations of ligand-binding and G protein activation. (a) Structure of α_2- and β_2-adrenoceptors showing different long loop between helices 5 and 6 in the α_2 and the long C-terminus in the β_2. The α_2 receptor binds yohimbine (YOH) and the β_2 receptor binds Cyanopindolol (CYP). (b) Chimeric proteins and their characteristics.

The best example is the relationship between the α_2- and β_2-adrenoceptors [238]. The α_2-receptor activates G_i and the β_2-receptor activates G_s. In addition, though both receptors can be activated by epinephrine, they also are bound with high specificity by different antagonists – the α_2-receptor by yohimbine and the β_2-receptor by cyanopindolol. Something about the structures of these receptors must control these differences in activity and ligand-binding specificity.

Looking at the details of structural differences in some receptors, one sees patterns that might explain differences in ligand binding and activation of different G proteins [238] (Figure 7.18a).

For example, the α_2- and β_2-receptors differ in the length of the cytoplasmic segment between the fifth and sixth membrane-spanning regions, as well as in the length of the carboxy-terminal end. Thus, the "α_2-"nature of that receptor may be established either by the longer cytoplasmic region or by the shorter carboxy-terminal end. Likewise, the "β_2-"nature of that receptor may be determined either by the short cytoplasmic segment or by the long carboxy-terminal end.

Kobilka *et al.* [238] performed an elegant set of experiments to determine where in the receptors these ligand-binding and transduction roles were controlled. They constructed 10 chimeric genes for the two receptors by first cloning the genes separately, then ligating different portions of each together in several combinations. They then expressed the genes by injecting the messenger RNAs for the chimeras into cells that normally do not express these receptors, for example, frog oocytes or COS-7 cells, a line of cultured monkey kidney cells. They then isolated the membranes and determined the ligand-binding properties and G-protein-activating properties of the chimeric receptor proteins (two of their chimeras are shown in Figure 7.18b).

G_s Activation: In each case, the chimeric receptor that contained the short cytoplasmic loop was able to activate G_s rather than inhibit it. The receptor bearing the long loop always inhibited G_s. Thus, the specificity of G protein coupling lies in the fifth to sixth intermembrane loop.

Ligand Binding: In each case, the chimeric receptor that contained the long C-terminal end bound cyanopindolol with greater affinity than yohimbine. The receptor with the short C-terminal end bound yohimbine better. Thus, the specificity of ligand binding lies in the carboxy-terminal end of the receptor.

We can thus conclude that the properties of the β_2-receptor are indeed determined by the short cytoplasmic loop (activates G_s) and the long carboxy terminus (binds cyanopindolol), while the properties of the α_2-receptor are determined by the long cytoplasmic loop (inhibits G_s) and the short carboxy terminus (binds yohimbine).

7.3.7 Summary of Actions of GPCRs: Agonists, Receptors, G Proteins, and Signaling Cascades

As indicated in Figure 7.19, three sets of proteins are directly involved in the initial steps from agonist binding to cellular action: the receptor, the G protein, and the effector (enzymes that produce cAMP, IP_3, and DAG).

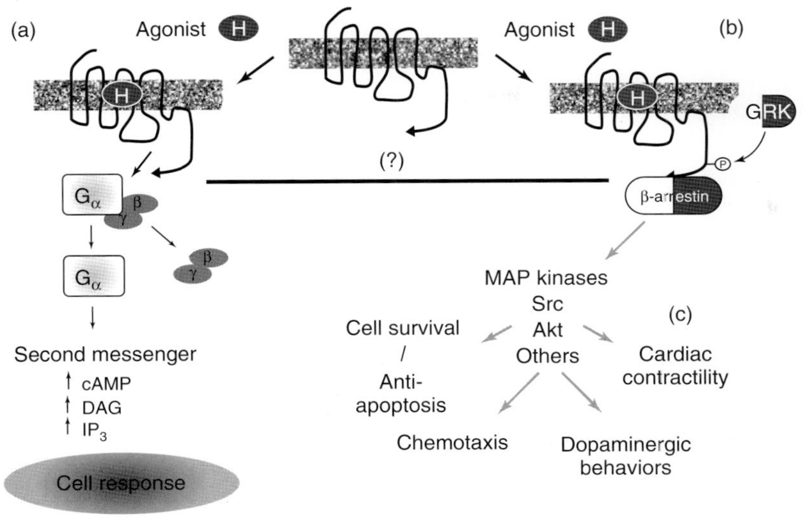

Figure 7.19 Three actions of GPCRs: (a): stimulation of cell processes by agonist binding. (b): After stimulation, the activated receptors are recognized and phosphorylated by GRKs, leading to the binding of a second protein, β-arrestin, which then activates alternate cascades (c) and desensitizes receptors, preventing further action. Modified from Figure 12 of Ref. [237]. Used with permission.

The next step, activation of GRK and β-arrestin, serves two functions – to initiate additional cascades via activation of MAP kinases, and to attenuate or terminate receptor action.

7.4
The Major Families of G Protein-Coupled Receptors

These receptors as found in eukaryotes belong to six families, but the ancestral sequence had such an early origin that sequence comparisons do not allow clear determination of the relationships among the families [213].

- *Family A:* Rhodopsin-like: the largest family, containing the rhodopsins, the amine receptors, odorant receptors, bitter taste receptors, nucleotide receptors, and receptors for some peptides such as the opioids.
- *Family B:* Secretin-like, containing receptors for secretin, glucagon, vasoactive intestinal peptide (VIP), gastric inhibitory peptide (GIP), and the hypothalamic releasing hormones.
- *Family C:* The metabotropic glutamate receptors (mGluRs).
- *Family D:* Fungal pheromone receptors.
- *Family E:* cAMP receptors.
- *Family F:* Frizzled/smoothened receptors.

Not all families occur in all eukaryotes: the five families in mammals are often described by the acronym GRAFS [206], for glutamate (G), rhodopsin (R), adhesion (A), frizzled/smoothened (F), and secretin (S) families. We provide information about all six families, but emphasize the mammalian receptors.

7.4.1
Family A – Rhodopsin-Like

These receptors have been divided into four classes, α, β, γ, and δ.

7.4.1.1 The α Subfamily

These receptors consist of the visual protein rhodopsin, the monoamine receptors, and other non-peptide transmitter receptors.

The best-studied ligands for amine receptors are epinephrine and norepinephrine (Chapter 4) secreted from the sympathetic nerves and the adrenal medulla. Extracts of the adrenal gland had formed the basis for Langley's early studies of the sympathetic nervous system [239], and modern concepts of receptor action developed in parallel with our understanding of the ways that catecholamines work in the body.

Beta-adrenoceptors These are considered the prototypes of GPCRs, mainly because their role in activating adenylyl cylase and cAMP formed the basis for much of our current understanding of metabolic control mechanisms. Their actions concerning adenylyl cyclase and cAMP were covered earlier *in* this chapter.

Three subtypes of β-adrenoceptors have been described: β_1, β_2, and β_3. The three have slightly different molecular structures, are found in different tissues, and control different physiological processes. For example, the β_1 receptors are found mainly in the heart, where they control rate and strength of cardiac muscle contraction, and in the gut, where they inhibit contraction and movement of intestinal smooth muscle. The β_2 receptors are found mainly in the peripheral circulation and in the airways, where they cause relaxation of smooth muscle, allowing vasodilation and bronchodilation. The β_3 receptors have only recently been identified, and their functions, while not yet completely understood, involve relaxation of vascular smooth muscle and stimulation of lipolysis [240]. Increased cardiac contractility in response to β_3 stimulation has been reported by some investigators [240], while others [241] have reported decreased cardiac contractility. β_1- and β_2-adrenoceptors activate G_s proteins and increase cAMP levels, while β_3-receptors activate G_i proteins and inhibit cAMP production.

Alpha-adrenoceptors These were among the first receptors to be studied in the nineteenth century because early medical researchers recognized the importance of the sympathetic nervous system to human health. α-Adrenoceptors are bound by the neurotransmitter norepinephrine (NE), which is released by sympathetic nerves in the CNS and in the smooth muscle of blood vessels, gut, airways, and iris. The main systemic effect of α-adrenoceptor action is increase in intracellular Ca^{2+} and contraction of vascular smooth muscle, leading to constriction of blood vessels and an increase in blood pressure.

The α-adrenoceptors have been divided into two types according to the source of the activator Ca^{2+}: in general, α_1-adrenoceptors activate a G_q protein to cause release of IP_3, which causes release of Ca^{2+} from intracellular stores [197]. The α_2-adrenoceptors activate G_i proteins and have two effects: inhibition of adenylyl cyclase and opening of L-type Ca^{2+} channels (Chapter 6) to produce an increase in Ca^{2+} movement into cells.

One type of blood vessel may use several subtypes of both α_1- and α_2-adrenoceptor for contraction [149]. A major part of *blood pressure control* occurs by constriction/dilation of peripheral vessels influenced by α_1-adrenoceptors [242] while *thermoregulatory control* of blood vessel constriction resides primarily in the α_2-adrenoceptors [243, 244].

Dopamine Receptors The CNS contains several types of dopamine (DA) receptors, each operating in different areas to affect movement, motivation, cognition, and emotion. The anterior pituitary also contains DA receptors that respond to DA from the hypothalamus to inhibit release of the peptide hormone prolactin.

Five types of DA receptors are currently recognized. They fall into two functional groups, the D_1-like and the D_2-like. The D_1-like receptors are the D_1 and D_5 subtypes, which activate adenylyl cyclase and PLC. The D_2-like receptors are the D_2, D_3, and D_4 subtypes, which inhibit adenylyl cyclase, activate K channels, and inhibit Ca^{2+} channels.

Histamine Receptors The receptors for histamine within the GPCR belong to four subtypes (H_{1-4}), though only the H_{1-3} subtypes have been pharmacologically characterized. H_1 receptors activate G_q proteins to increase IP_3 production; they occur in the CNS, the heart, the adrenal medulla, and on most smooth muscle, cells; they cause neuronal firing, vasoconstriction (but cardiac inhibition), and increased vascular permeability. H_2 receptors activate G_s proteins to increase cAMP production. They occur in the CNS, the heart, vascular smooth muscle, and on gastric parietal cells. Histamine acts on H_2 receptors to decrease neuronal firing and cause vasodilation (along with cardiac excitation). H_3 receptors appear to activate G_i proteins and decrease cAMP production; they are found in the CNS and on peripheral nerves, and are autoreceptors that control the release of histamine from neurons [133].

Serotonin Receptors Two main types of serotonin (5-Hydroxy tryptamine or 5HT) receptors are recognized today – the $5HT_{1, 2, \text{and } 4}$ receptors and the $5HT_3$ receptors. The $5HT_{1, 2, \text{and } 4}$ receptors are GPCRs and the $5HT_3$ receptors are ligand-gated ion channels (Chapter 6).

The $5HT_{1, 2, \text{and } 4}$ receptors contain five subtypes placed into two functional classes: the $5HT_1$ receptors cause hyperpolarization of cell membranes, while the $5HT_{2 \text{ and } 4}$ receptors cause depolarization. Quite different effector mechanisms bring these membrane changes about: $5HT_{1A, 1B, \text{and } 1D}$ receptors inhibit adenylyl cyclase, the $5HT_{2 \text{ and } 1C}$ receptors activate PLC, and the $5HT_4$ receptor activates adenylyl cyclase. Finally, the hyperpolarizations caused by $5HT_{1A}$ receptors and $5HT_{1C}$ receptors occur by different ion flows: $5HT_{1A}$ receptors increase K^+ conductance and inhibit Ca^{2+} conductance, while $5HT_{1C}$ receptors increase Cl^- conductance [197].

Muscarinic Receptors Because of the numerous effects of ACh on visceral organ function, this was also one of the first receptor classes to be studied. The name comes from that fact that muscarine (a compound from mushrooms) is an agonist and mimics the actions of ACh in the heart, gut, airways, and blood vessels. ACh also activates nerves and skeletal muscles, where muscarine has no effect. Because nicotine acts like ACh on nerves and skeletal muscle, those actions were called *nicotinic*. It was a great mystery [245] for decades why ACh should have such different effects; the mystery was solved only when it was discovered that muscarinic receptors are G-protein-coupled and nicotinic receptors are ion channels (Chapter 6).

The muscarinic receptors are coupled to G proteins, and act on intracellular effectors in ways similar to that of the other GPCRs. They are divided into five types, named M1 through M5. The M1, M3, and M5 types all stimulate PLC and increase intracellular Ca^{2+} by an IP_3 mechanism (see above). This Ca^{2+} then may cause contraction of smooth muscle, for example, in the airways [142]. The M2 and M4 types, in contrast, inhibit adenylyl cyclase by activating an inward-rectifying K channel (Chapter 6), causing hyperpolarization of the cardiac cell membrane [229]. This action was discussed above in the section on $\beta-\gamma$ subunit function.

Trace-Amine-Associated Receptors (TAARs) This family was only recently discovered, and is not yet fully characterized. The receptors bind cellular amine substances (e.g., tryptamine, tyramine, and phenylethylamine), and are found in several cell types, most prominently in the brain regions responsible for actions of the amine transmitters (dopamine, epinephrine, and norepinephrine) and olfactory tissues.

In the brain, trace amines can affect the synthesis, release, and uptake of amine transmitters, thus altering neuronal responsiveness and also affecting behavior. In the olfactory system, trace amines serve as olfactory signals. Certain amino acids are broken down to **diamines**, compounds that are found in rotting meat and that often have foul odors (putrescine and cadaverine are examples whose names are descriptive). Scavenging animals are attracted by these diamines, while non-scavengers avoid them. Both attraction and avoidance imply the existence of specific receptors for the substances, and the olfactory trace-amine-associated receptors (TAARs) are those receptors [246].

Receptors for Purine and Pyrimidine Nucleotides Nucleotides (base plus ribose plus phosphate) and nucleosides (base plus ribose alone) have long been known as crucial intracellular energy sources and components of nucleic acids and metabolic cofactors. Our understanding of their roles as extracellular signals is more recent, and began with a 1929 report by Drury and Szent-Gyorgyi of the effect of tissue extracts on heart function [247]. The extracts were shown to slow heart rate and reduce arterial pressure through a vascular dilation and direct effect on cardiac cells. They showed that the active substance was the nucleoside adenosine or the nucleotide adenosine monophosphate (AMP). Later, these actions of AMP were found to be mediated by cell-surface receptors, termed by Burnstock [248] **purinoceptors** or purinergic receptors.

Three major families of receptor are sensitive to nucleotide and related compounds. Two, belonging to the GPCR superfamily, are covered in this section; a third, belonging to the ligand-gated ion channel family, is covered in Chapter 6.

P1 (Adenosine Receptors) These receptors are most closely related to the monoamine receptors within Family A of the GPCRs. The early report [247] suggesting the cardiovascular actions of nucleosides probably described the actions of adenosine on P1 receptors; these receptors have also been shown to affect immune system and nervous system responses [249]. P1 receptors are selective for adenosine, a metabolite of ATP that is produced in extracellular fluid by enzymatic degradation. Four subtypes have been identified, A_1, A_{2A}, A_{2B}, and A_3. The receptors couple to adenylyl cyclase (the A_1 and A_3 subtypes interact with $G_{i/o}$), and the A_{2A} and A_{2B} subtypes activate G_s.

P2Y (ATP/ADP/UTP Receptors) The P2Y receptors are not bound by nucleosides such as adenosine. Instead, they are bound by nucleotides, both adenine- and uracil-based. They are thus not strictly purinergic, but are also pyrimidinergic [249]. Eight subtypes have been described ($P2Y_{1,2,4,6,11}$ belong to one clade; $P2Y_{12,13,14}$ belong to another clade). The two clades are closely related to each other, but are only distantly related to the P1 (adenosine) receptors. Their closest relatives among the rhodopsin-like receptors are the platelet-activating-factor receptors (see Chapter 14).

Subtypes $P2Y_{1,2,4,6,11}$ couple to G_q, and therefore increase levels of the second messenger IP_3 within cells. This causes an increase in intracellular Ca^{2+} that activates enzymes or cell membrane ion channels. Subtype $P2Y_1$ also affects gene transcription via activation of the MAP kinase pathway (Chapter 10). Subtypes $P2Y_{12,13,14}$ couple to G_i and inhibit adenylyl cyclase.

Table 7.3 Prostaglandin receptors, effectors, and actions.

Agonist	Receptor	Main G protein	Cardiovascular	Uterus	Airways	GI	Inflammation
PGD_2	DP	G_s	Dilate	—	—	—	—
PGE_2	EP1	G_q	—	Constrict	—	Constrict	—
PGE_2	EP2	G_s	Dilate	—	Dilate	Relax	Pro-
PGE_2	EP3	G_i	Dilate	Constrict	—	Constrict	Fever
PGE_2	EP4	G_s	—	—	—	—	—
PGI_2	IP	G_s	Dilate	—	Dilate	—	—
PGF2α	FP	G_q	—	Constrict	Constrict	—	—

The two families of nucleotide/nucleoside receptors are functionally connected by the complexity of structure and function of ATP itself. It is released from neurons in the CNS, and acts on P2Y receptors and on P2X ion channel receptors (Chapter 6). While in the extracellular space, ATP may be degraded, both to ADP and to adenosine. ADP acts on P2Y receptors on postsynaptic cells, and adenosine acts on presynaptic P1 receptors [250].

Prostaglandin (PG) Receptors Prostaglandins (PGs) activate several subtypes of GPCRs; these are termed DP1-2, EP1-4, FP, IP1-2, and TP, and each subtype binds a different PG (Table 7.3) [251].

As discussed further in Chapter 14, the letter designations do not indicate true subfamily relationships; for example, the E_{1-4} receptors all bind PGE_2, but are not closely related, and in some cases activate different G proteins, leading either to increased or decreased cAMP. Indeed, the EP_2 and EP_4, along with IP and DP, all activate G_s and increase cAMP. The EP_3 alone activates G_i, resulting in decreased cAMP; the EP_1, FP, and TP all activate G_q, which increases intracellular Ca^{2+}.

Each of these intracellular events produces a predictable physiological response: G_s signaling and increased cAMP tends to cause relaxation of smooth muscle; G_i signaling decreases cAMP and causes constriction of smooth muscle; and G_q signaling increases activity of PLC, promoting increased intracellular Ca^{2+} and leading to smooth muscle constriction.

Cannabinoid Receptors (CBRs) These receptors are best known for their ability to bind cannabis (Δ^9-tetrahydrocannabinol (THC)) and mediate its psychoactive effects. They actually evolved to promote the effects of natural brain substances, the endocannabinoids (eCBs) [252]. Two types of receptors are known in mammals for the eCB. These are the CB_1R (in brain tissue and peripheral nerves) and the CB_2R (only in peripheral nerves). The CB_1R are the most abundant GPCR found in the brain, and are primarily involved in inhibition of neurotransmitter release. The cannabis-derived compounds such as Δ^9-THC bind the two receptors equally well.

The eCB compounds anandamide and 2-arachidonoylglycerol are produced in postsynaptic cells; they then diffuse out and to the presynaptic cell, where they bind a CB_1R to inhibit release of the transmitter from the presynaptic cell. This action of eCB is known as *retrograde*. Until recently, only a retrograde action of eCB was known, but recent research has added two additional mechanisms of eCB action – autocrine signaling within postsynaptic cells and astrocyte signaling (Figure 7.20).

The cannabinoid receptors (CBR) shown in Figure 7.20 are on the presynaptic cell (A), the postsynaptic cell (B) or nearby astrocytes (glial cells that have the unusual ability to release glutamate as transmitter). When glia release glutamate, it binds to mGluRs (see below) on presynaptic cells to inhibit release of transmitter. Thus, eCB seem not to be used as standard transmitters, but rather as inhibitors of transmission within the pathways shown. Both CB1R and CB2R activate the $G_{i/o}$ G proteins, leading to inhibition of cAMP.

7.4.1.2 The β Subfamily

This group in the rhodopsin family contains receptors for neuropeptides (most of the hypothalamic releasing peptides, endothelin, neuropeptide Y, and cholecystokinin.

7.4.1.3 The γ Subfamily

These are also peptide receptors, though different from those in the β class, and include receptors for somatostatin, the opioids, chemokines, angiotensin II, bradykinin, and the non-genomic actions of estrogen (Chapter 9).

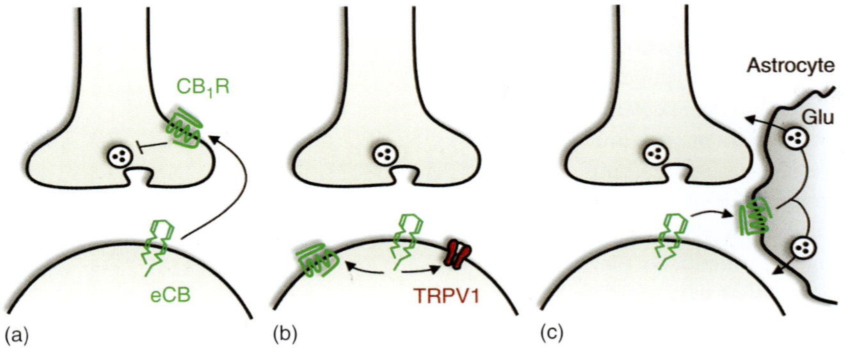

Figure 7.20 Endocannabinoid (eCB) signaling. (a) Retrograde signaling, in which eCBs are released from postsynaptic cells and bind CB receptors on neurons to inhibit eCB release. (b) Standard signaling, in which eCBs from neurons target CB receptors or TRPV1 channels (Chapter 6 – modulate pain transmission). (c) Astrocyte signaling, in which released eCBs initiate transmission in glial cells. Figure 1 of Ref. [252]. Used with permission.

Opioid Receptors The existence of these receptors was demonstrated by Pert and Snyder [253] and by Simon *et al.* [254] in 1973, based on high-affinity binding of tritium-labeled opiate drugs to brain tissues. Hughes *et al.* [255] later determined that the natural ligands of these receptors were proteins, and called them enkephalins (in-brain molecules). The receptors for the opiates and natural enkephalins became known as *opioid receptors*. **Opioids** are natural vertebrate peptide transmitters that activate brain receptors; **opiates** are plant derivatives that happen to bind the same receptors, mimicking the effects of opioid peptides.

The contributions of these receptors to drug **addiction** are discussed in Chapter 13, but a quick overview of receptor actions follows. The opiate drugs act by inhibiting pain reception. As such, they are similar to the local anesthetics such as Novocain. However, the mechanisms of local anesthesia are quite different from those of opiates: Novocain blocks axonal transmission of a nerve impulse by closing voltage-gated Na channels (Chapter 6). The opioid peptides, in contrast, prevent release of excitatory neurotransmitters in these pathways. The receptors for opioid peptides are on the terminals of axons involved in pain transmission. When an opioid peptide binds such a receptor, it blocks release of the transmitter by reducing the size of axonal action potentials or by preventing Ca^{2+} influx that causes exocytosis of transmitter.

Four types of opioid receptors are recognized: three designated by Greek letters (δ, κ, and μ) and one designated ORL (opioid receptor-like) or NOP (nociceptor, or "pain" receptor). Each receptor is acted on by one or more of the naturally occurring endorphin-like peptides (Chapter 4). The delta opioid receptors (DORs) are bound best by enkephalins; the kappa opioid receptors (KORs) by dynorphin, and the mu opioid receptors (MORs) by the endomorphins, as well as by enkephalin and β-endorphin [94]. Evolutionary relationships of the four receptors are shown in Chapter 14.

In addition to their binding of opioid peptides, these receptors also bind opiate drugs. MORs have especially high affinities for morphine, explaining the analgesic effects of this drug. All three receptors inhibit adenylyl cyclase, resulting in opening of K^+ channels that cause cell hyperpolarization and thus inhibition of action potentials. This is their main mechanism of action to inhibit release of excitatory transmitters in pain pathways, but they also inhibit Ca^{2+} channel opening, thus directly limiting transmitter release [256]. The MORs affect neurons releasing epinephrine, dopamine, ACh, and substance P, while the δ receptors affect glutamate-releasing neurons. The endomorphins are more recently discovered [94] tetrapeptides (YPFF and YPWF) that appear to be the natural agonists of the MOR.

Peptide Venom Receptors Venomous lizards release protein-based poisons that affect different receptors from those of the venomous snakes and cone snails described in Chapter 6. Poisonous lizards are the Indonesian monitor lizard (*Varanus* sp.) and the American gila monster (*Heloderma* sp.). In both these animals, the venom glands are located in the lower jaw, whereas snake venom glands are in the upper jaw.

The venom of the monitor was once thought to consist mainly of bacteria that cause infection in the victim, leading to eventual weakness and allowing the lizard to subdue the prey long after biting it.

Recent research has produced a more conventional concept: *Varanus* venom does contain bacteria, but the primary effects of a bite come from peptides that cause two debilitating effects: intense intestinal smooth muscle contraction via action on a family of GPCR – the prokineticin peptide receptors; and profound vasodilation, producing decreased blood pressure and loss of consciousness [257]. Both these effects disable the prey, though less immediately than in the case of prey bitten by snakes that release ACh receptor antagonists.

The venom of the gila monster consists of peptides that also target GPCR. The primary effect of a gila monster bite is a massive vasodilation caused by peptides of the VIP-glucagon family, which target VIP receptors. These peptides are more effective agonists at the VIP receptors than is VIP itself, explaining the intense drop in blood pressure and loss of consciousness [258].

7.4.1.4 The δ Subfamily

These are the receptors for olfactory molecules and three anterior pituitary hormones. Odorant molecules bind their receptors in the same manner as hormone and transmitter ligands. The binding event is transduced by G_{olf} proteins (similar in structure and function to G_s), and the effector is an unusual form of adenylyl cyclase that activates an ion channel to cause depolarization of the sensory neuron. This system is reminiscent of, but opposite to, the visual system (above), in which opsin and light work through a G_t protein to *de*activate an ion channel in the rod outer membrane.

Genes encoding the G-protein-coupled olfactory receptors have been cloned; the surprising result of this work is that about 1000 genes have been described, each encoding a different odorant receptor molecule [215]. This is equivalent to having 1000 subtypes of β-adrenoceptors, for example, rather than just three. The odorant receptor class of genes thus is the largest known family of GPCRs [259]. It is possible that these receptors are encoded by gene rearrangement, as in the genes that generate antibody diversity, but if they are not, then the genes for odorant receptors may be the largest constant-structure gene family in vertebrates.

Taste 2 receptors (T2Rs) for bitter taste: The receptors for bitter taste are members of the rhodopsin family, while receptors for sweet and umami tastes are members of Family C (below). About 30 different related receptors occur as dimers and confer ability to taste a wide variety of bitter substances (Figure 7.21).

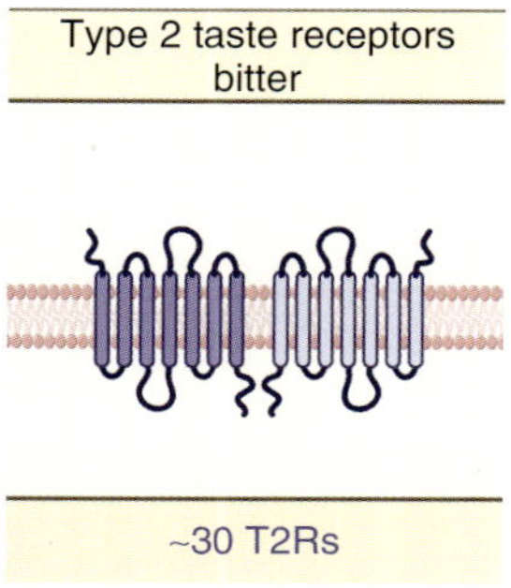

Figure 7.21 Receptors for bitter taste. Dimers form between most or all of the 30 different T2R receptors. Modified from Figure 2 of Ref. [214]. Used with permission.

Bitter receptors bind with high affinity, an adaptive situation for detection of toxic or noxious (and generally non-nutritional) substances. Some of these substances, found to be bitter by large numbers of people, are PTC (phenylthiocarbamide), saccharine, and quinine [214]. T2R (as well as the taste 1 receptor (T1R) discussed below) couple to an unusual form of G protein called **gustducin** [260].

7.4.2 Family B – Secretin-Like

These receptors are activated only by peptides, and, of those, only the peptides of the secretin group, which also includes calcitonin, calcitonin-gene-related peptide (CGRP), corticotropin-releasing hormone (CRH), growth-hormone-releasing hormone, VIP, and glucagon.

Receptors in this class show two methods of binding and transduction. The receptors for CRH (Chapter 4), bind ligand in the standard way and activate cAMP via G_s protein. Other receptors, such as the CGRP receptor, only bind ligand in the presence of additional membrane proteins called **receptor-activity-modifying-protein s** (RAMPs) [261]. The receptor and the RAMPs form **heterodimers** in the membrane. These heterodimers do not form through covalent bonds, and their functional significance is not yet understood, though they are important for proper ligand binding.

7.4.3 Family C – Metabotropic Glutamate and Sweet/Umami Taste Receptors

The Family C GPCRs are the mGluRs, $GABA_B$Rs, Taste1 receptors (for sweet and umami sensation), the calcium-sensing receptor, and some of the receptors

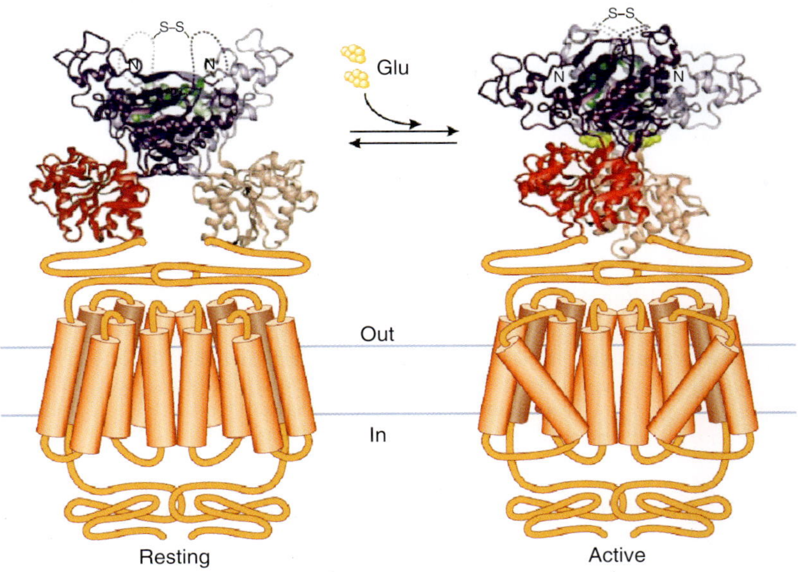

Figure 7.22 Schematic representation of the membrane topology and dimeric structure of metabotropic glutamate receptors (unbound at left and bound by ligand at right). Two identical receptors combine to bind two molecules of glutamate (in the space between blue and red domains), and both receptors associate with G proteins (not shown). Modified from Figure 1 of Ref. [209]. Used with permission.

for pheromones. These receptors all have large extracellular N-termini that bind their agonists. mGluRs and $GABA_B$Rs are unusual in that their agonists (the amino acids glutamate and gamma-amino butyric acid (GABA)) are primarily known as ligands for ion channel receptors (Chapter 6). These two receptors form dimers when they bind the ligand; the mGluR is shown in Figure 7.22.

The type of dimerization seen here is different from that seen in the Class B receptors, and it seems to be crucial for proper activation of G proteins and effectors. The mGluRs are important in learning and memory, as discussed in detail in Chapter 13. The discovery of dimerization and the association between receptors and other membrane proteins such as RAMPs represent exciting new developments in GPCR research, and may well be important for the function of many types of receptors in this large family [262].

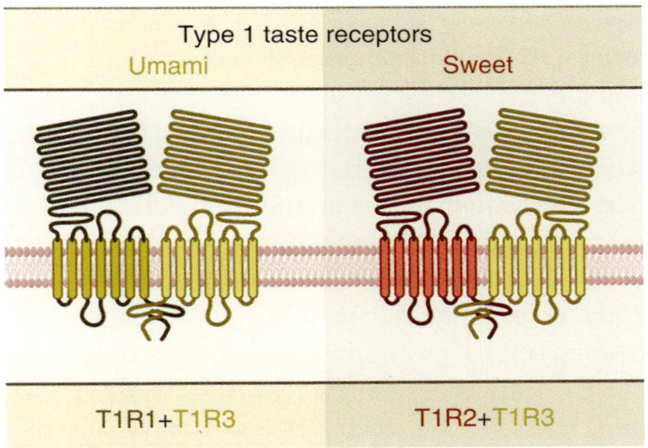

Figure 7.23 Schematic representation of the membrane topology and dimeric structure of human taste receptors. The sweet and umami taste receptors are based on the T1 structure with long N-termini and share one receptor monomer (T1R3). Modified from Figure 2 of Ref. [214]. Used with permission.

7.4.3.1 Taste 1 Receptors (T1Rs)

Sweet and umami (the attractive tastes) are sensed by members of Family C GPCRs: three different receptors are found in mammals, and they occur as dimers. Umami tastes are detected by T1R1–T1R3 dimers and sweet tastes are detected by T1R2–T1R3 dimers (Figure 7.23).

Umami tastes are generally associated with L-amino acids (e.g., L-aspartate or L-glutamate), or with the purine nucleotide monophosphates (GMP and AMP). Both receptor dimers show low-affinity binding of ligands; this is adaptive as umami or sweet molecules are rarely toxic, and receptors will not saturate below nutritionally relevant concentrations. Some examples of substances that activate sweet receptors are mono- and disaccharides, saccharin (as well as other artificial sweeteners), and the D-amino acids. Not all mammals have sweet sensation: carnivores of the cat family (Felidae) have a mutated T1R2 protein that does not bind sugars [214]. T1Rs also couple to the G proteins G_s and G_q, as well as to the gustducin to which T2Rs (above) couple [260].

7.4.3.2 Calcium-Sensing Receptors

These receptors are located on parathyroid gland membranes; they are unusual GPCRs in that their agonist is a calcium ion. Binding of Ca^{2+} to the receptor activates G_q to increase activity of PLC, thus leading to production of IP_3 and inhibition of release of parathyroid hormone (PTH) from bone [255]. Because the PTH causes increased Ca^{2+} levels in the blood, the calcium-sensing receptor acts as a negative feedbac k control for blood Ca^{2+}.

7.4.4 Family D – Adhesion Receptors

These receptors are poorly understood, but are seven-membrane-spanning proteins with very long extracellular (N-terminus) peptide sequences. These sequences seem to be able to bind cell-surface proteins on other cells and hold the two cells together in tissues [206].

7.4.5 Family F – Frizzled-Smoothened Receptors

This small family is important in embryonic development of animals, where it is involved in signaling processes that control cell division and differentiation [263]. Its signaling mechanisms involve kinases that ultimately activate transcription factors in the nucleus. Details of the pathway are considered in Chapter 11.

7.4.6 Family E – Cyclic AMP Receptors

This family is found only in slime molds that use cAMP as an extracellular (chemotaxic) signaling molecule for movement and association of single cells into multicellular aggregates.

7.4.7 Other G-Protein-Coupled Receptor Types in Eukaryotes

7.4.7.1 Yeast Mating Pheromone Receptors

Many unicellular fungi have two "mating types" (MAT**a** and MAT**α** (monoamine neurotransmitter) – equivalent to "sexes") that communicate a mating signal from cells of one mating type to the other, the equivalent of gamete binding among multicellular eukaryotes. MAT**α** cells secrete **α** peptide, and MAT**a** cells secrete the **a** peptide. The **a** peptide binds to a GPCR (STE3) on **α** cells, and the α peptide binds to another GPCR (STE2) on **a** cells. Binding of both pheromones to receptors activates G proteins to promote processes involved in cell growth, protein transcription, cell fusion, and division (Figure 7.24).

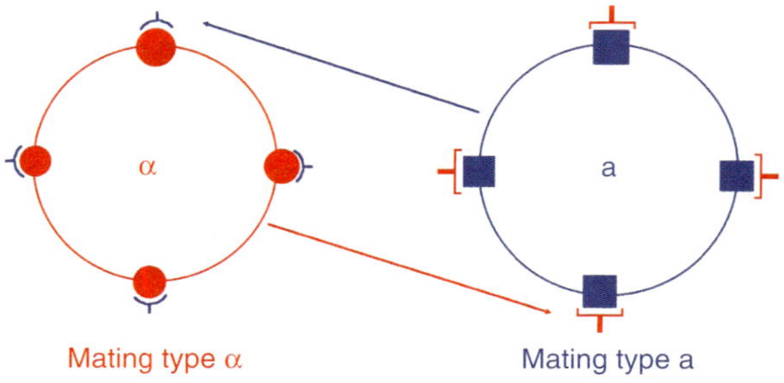

Figure 7.24 Mating types of haploid yeast cells are determined by G-protein-coupled receptors. α (red) and **a** (blue) are two mating types. Cells of type **a** have receptors (blue squares) for α mating peptide (red ligand) and cells of type α have receptors (red circles) for **a** peptide (blue ligand).

7.4.7.2 Insect Taste Receptors

The arthropods express on their antennae, proboscis, and even ovipositors several dozen different GPCRs that are sensitive to specific "taste" molecules. The family of receptors is diverse and shows no clear sequence homology with vertebrate GPCRs [206]. The receptors are able to recognize pheromones and to discriminate among sweet and bitter tastes. The transduction mechanisms are not known [264].

7.4.7.3 Nematode Chemoreceptors

Nematodes (roundworms) express a large number of GPCRs that transduce chemical signals from their soil environment. Though they have a seven-membrane-spanning structure, the amino acid sequences are not homologous to insect or vertebrate taste or odorant receptors, and their G protein mechanisms of sensory transduction are not known [264].

8 Transduction III: Receptor Kinases and Immunoglobulins

> Reversible phosphorylation is the most widespread posttranslational protein modification, playing (a) regulatory role in almost every aspect of cell life.
>
> *Ciesla, Fraczyk, and Rode [265]*

8.1 Protein Kinases

Kinases are enzymes that transfer phosphate groups from ATP to amino acid residues on cellular proteins. Kinase activity is an important intracellular process, and 30% of human proteins are phosphorylated at one time or another [266]. The added phosphate alters the activity of the receiving enzyme, marks it for destruction, or facilitates its intracellular movement. This chapter focuses on the receptors that are responsible for phosphorylation of tyrosine, serine, or threonine residues, though several other amino acids are also sites of phosphorylation in prokaryotic and eukaryotic organisms [265]. Figure 8.1 shows the hydroxyl that serves as the phosphorylation site for the kinases we discuss.

The first kinases were discovered in the 1940s and 1950s as intracellular enzymes – activators of processes involved in carbohydrate metabolism [26]. These enzymes, such as phosphorylase kinase and protein kinase A, phosphorylate serine and/or threonine residues. Additional kinases involved in cell division were discovered in the 1970s and 1980s; these, such as Src (sarcoma), were shown to phosphorylate tyrosine residues.

At the time of their original discovery, all the tyrosine kinases and serine/threonine kinases were thought to be exclusively intracellular enzymes; then a group of cell-membrane-bound receptor kinases were discovered in the 1970s. These were the **receptor tyrosine kinases** (RTKs). Large numbers of RTKs have been discovered in the years since, and the number of currently recognized human receptor kinases (58) is greater than the number of intracellular kinases (32) [267]. A final class of receptor kinases – the **receptor serine/threonine kinases** – was described in 1987 [268]: this small group consists of receptors for the transforming growth factor-β signaling peptides [269].

The receptor kinases have a distinctive extracellular domain (the immunoglobulin domain or Ig domain) of several antiparallel β strands (Figure 8.2) that is involved either in ligand-binding or inter-receptor interactions.

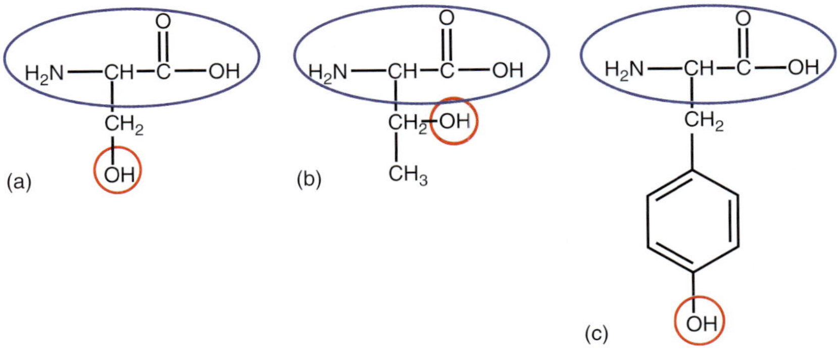

Figure 8.1 Molecular structures of the amino acids ((a) serine, (b) threonine, (c) tyrosine) phosphorylated by kinases. Blue oval encloses the amino acid core of each; red circle indicates the hydroxyl on the side chain that is the substrate for kinase addition of the phosphate. These are the only three amino acids with hydroxyl groups on their side chains.

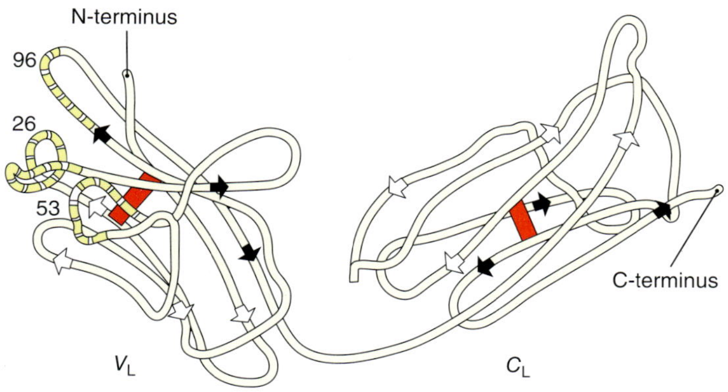

Figure 8.2 Structure of two Ig domains of the immunoglobulin G antibody light chain. V_L: variable amino acid sequence; banded region: variable amino acids. C_L: constant sequence; orange bars: disulfide bonds. Arrows indicate directions of β sheets. Figure 4.5 of Ref. [270]. Used with permission.

The Ig domain is found in antibodies, immune cell receptors, and receptors of the tyrosine kinase family. The discovery of this domain led to the concept of a "superfamily" of proteins, and the Ig superfamily was the first receptor superfamily to be recognized [271]. Most members of this superfamily are involved in promotion of cell division, metabolism, and individual identity [267]. The families within it include the receptor kinases, **cytokine** receptors, immune system proteins (immunoglobulins, **major histocompatibility antigens** (MHC), and T-cell receptors). These apparently disparate groups have been placed together because

- their structures indicate derivation from a common ancestral sequence [272];
- they share structures, such as the Ig domain, a single transmembrane domain, and cysteine- or leucine-rich regions of the extracellular domain (Figure 8.3);
- they either are kinases or link to intracellular kinases.

8.2 Receptors for Cell Division and Metabolism

8.2.1 Overview of Family Members

Figure 8.3 illustrates the subfamilies that comprise the family of RTKs. The common ancestry of all these receptors is demonstrated by the sequence similarities of the intracellular domains of the RTKs; even so, their different binding properties and cellular functions are associated with extensive differences in their extracellular domains.

Two features are common to all RTKs, regardless of specific structure or mechanism of activation: (i) the receptors operate as dimers or oligomers and (ii) the receptor dimerization process invokes predictable interactions between the intracellular kinase domains of the associated receptors.

8.2.2 Overall Functions of RTK

The various receptor types affect a multitude of cellular processes but act through a more limited number of core processes (Figure 8.4). The core processes (e.g., MAP (mitogen-activated protein) kinase, Ras, JAK (Janus kinase), JNK (c-Jun N-terminal kinase), or Akt) are covered in more detail in Chapter 10, but are briefly mentioned here in relation to their primary signaling receptors.

8.2.2.1 Extracellular Domains

The primary extracellular domains consist of regions that are rich in cysteine and those that are folded into the immunoglobulin domain. Though the receptors require ligand binding to the external domain in order to fully activate the internal kinase domain, the mechanism by which this external activation occurs differs between receptors. Figure 8.5 shows four receptor types, each with its own mode of interaction with ligand and with another receptor.

This figure shows the spectrum of ligand-binding-dimerization patterns in the RTK family: the ErbB (EGF, epidermal growth factor) receptors (a) show maximal receptor interaction with minimal ligand interaction; the Trk (NGF, nerve growth

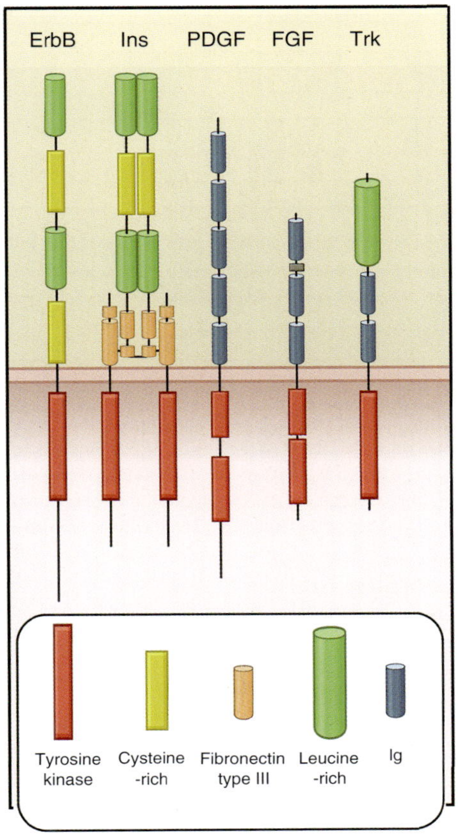

Figure 8.3 The subfamilies of receptor tyrosine kinases that are treated in this chapter. Cell membrane is the horizontal double line; intracellular domain (red) is similar in all the RTKs; the different receptor subfamilies have different extracellular domains (legend at bottom). The Ig domain of the PDGF is shown in detail in Figure 8.2. Modified from Figure 1 of Ref. [273]. Used with permission.

factor) receptors show maximal ligand interactions with minimal inter-receptor association; the FGF (fibroblast growth factor) receptors (b) and KIT (PDGF, platelet-derived growth factor) receptors (c) show intermediate degrees of ligand and receptor interactions.

The differences among the receptor types reflect functional differences in the various extracellular domains, and can be important when scientists consider how to design drugs to modify receptor activation. For example, if ligand is required for dimerization of a receptor, molecular blocking of the ligand-binding site will potentially prohibit receptor dimerization. Mutations affecting ligand-binding sites can also be important in developing receptor-based diseases, as described in Chapter 12.

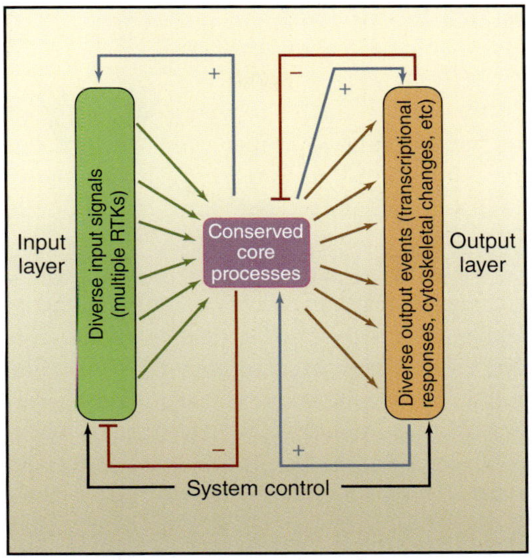

Figure 8.4 Overall aspects of RTK signaling. At left is the "input layer," or signaling from many types of RTKs. At right is the "output layer," or the diverse set of resultant cellular effects. In the center is the "core," or limited number of mechanisms that transduce the receptor signals to their downstream targets. Figure 4B of Ref. [273]. Used with permission.

8.2.2.2 Intracellular Domains

The receptors all have similar tyrosine kinase domains that consist of "lobes" termed N and C. The C lobe is located toward the carboxyl end of the intracellular domain. In the absence of ligand binding, each lobe is oriented in such a way that it inhibits its own activation; ligand binding allows activation by altering the inhibitory arrangement, promoting catalytic interactions between the N lobe of one receptor of the dimer and the C lobe of another.

The mechanisms by which these conformational changes occur differ between receptor subfamilies, in part depending on whether the receptors dimerize or oligomerize. This process allows interactions that stabilize the active form of the intracellular kinase, thus allowing it to function (Figures 8.6 and 8.7).

A key characteristic of these receptors is that their regulation results in activation and deactivation of major signal transduction pathways leading to mitogenic (cell division) processes. The phosphorylated tyrosines are binding sites for intracellular signaling proteins such as those with Src homology-2 (SH2) or phosphotyrosine-binding (PTB) domains [267], endocytic **downregulation** through **clathrin**-coated pits and caveolin vesicles, and inhibition of dimerization mechanisms (discussed below).

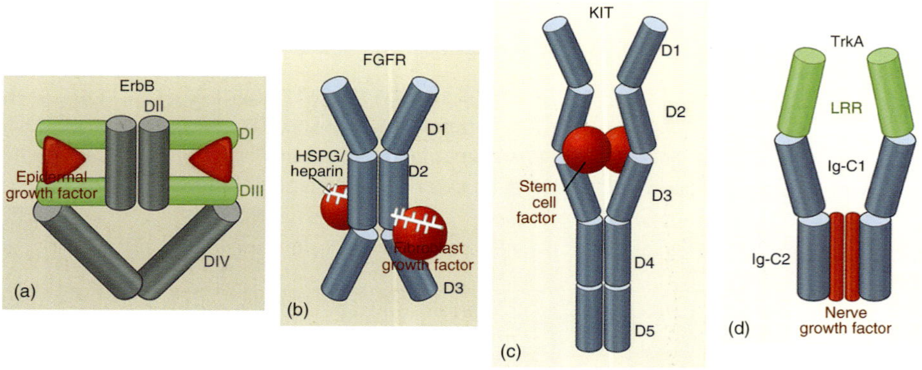

Figure 8.5 Extracellular ligand-binding mechanisms for four of the subfamilies shown in Figure 8.3. Ligands are depicted in red in each panel. (a) ErbB (FGF) receptor: Dimerization of the receptor (in this case, via DII domains) is independent of ligand binding. (b) FGF receptor: Dimerization of external domains and binding of heparin can facilitate the interaction of ligand. (c) KIT (PDGF) receptor: Ligands and external domains can simultaneously interact and dimerize. (d) Trk (NGF) receptor: Ligands facilitate the interaction of extracellular domains but receptors themselves do not physically interact. Modified from figure 2 of Ref. [273]. Used with permission.

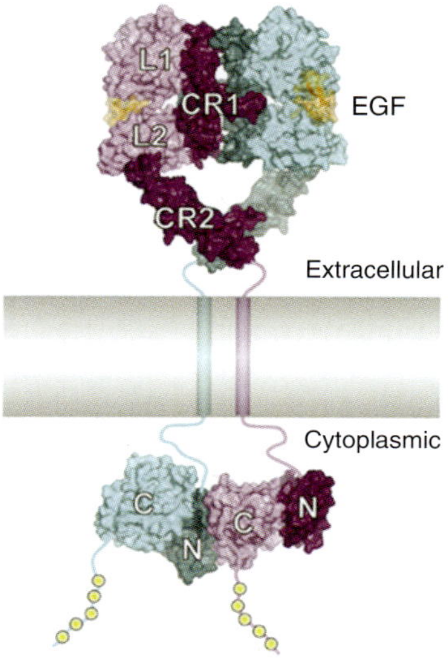

Figure 8.6 Events in EGF receptor binding, dimerization, and intracellular activation. The ligands are shown as orange structures bound to the extracellular domains. The C lobe of one receptor binds to the N lobe of the partner, activating it following ligand binding. Figure 1 in Ref. [267]. Used with permission.

In the following sections, each receptor subfamily is described, including the mechanisms of aggregation, ligand activation, and kinase domain inhibition and activation.

8.2.3
Receptor Tyrosine Kinase Subfamilies

The RTKs covered in this chapter are those for the peptide ligands EGF, insulin (INS), FGF, PDGF, and NGF.

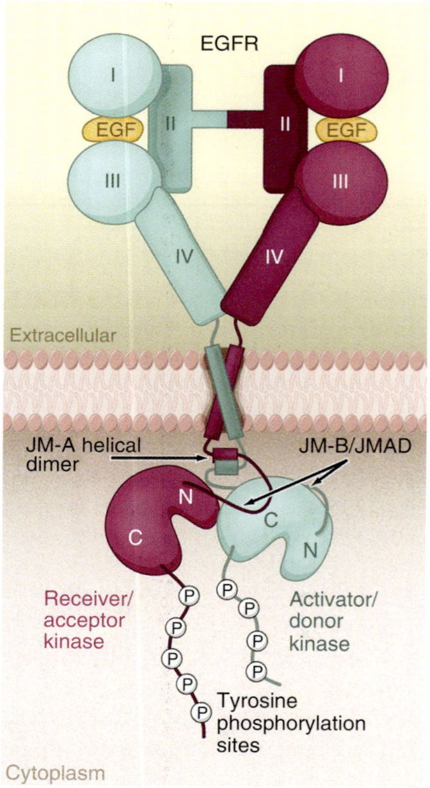

Figure 8.7 Mechanisms for activation of the intracellular tyrosine kinase domain, here shown for the EGFR. Ligand-mediated receptor dimerization positions the two cytoplasmic kinase domains for allosteric phosphorylation of tyrosine residues (circled Ps). The C-terminus of the activator/donor kinase phosphorylates the N-terminus of the receiver/acceptor kinase. Figure 1 of Ref. [274]. Used with permission.

8.2.3.1 EGF Receptor Subfamily

EGF receptors exist in the unbound state as monomers and are dimerized by ligand binding; this contrasts with INS receptors, which exist in the dimerized state even before ligand binding (Figure 8.2). Four subtypes of ErbB receptors are found in humans, and they can associate in either homodimer or heterodimer pairs. In the EGF receptor, the intracellular kinase domain of one receptor activates the corresponding domain of its partner by stabilizing the active state of the kinase so it can phosphorylate and thereby alter the behavior of targeted cellular proteins [267] (Figure 8.6).

This process is shown diagrammatically for the epidermal growth factor receptor (EGFR) in Figure 8.7.

Once activated, the kinase domain serves as a docking site for intracellular proteins such as SH2 or PTB proteins that regulate transcription of new proteins controlling metabolic or mitotic processes [267].

8.2.3.2 Insulin Receptor Subfamily

The INS receptor (Figure 8.8) consists, even in the unliganded state, of two polypeptide chains attached by a **disulfide bridge**.

Insulin Receptor Function Binding of INS causes a conformational change in the kinase domain, which then phosphorylates itself **(autophosphorylation)** at any of several tyrosine residues on the receptor molecule [273]. Phosphorylation activates processes

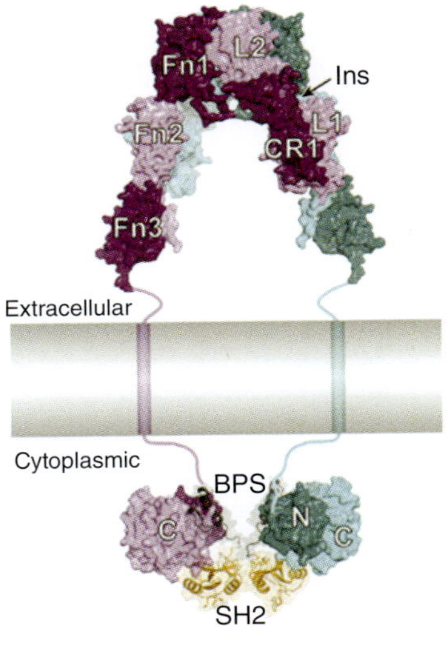

Figure 8.8 Architecture of the insulin receptor. One molecule of insulin binds to the already dimerized receptor; the intracellular kinase domains are activated and the SH2 domain serves as the activation site for similar domains of intracellular kinases. Figure 1 in Ref. [267]. Used with permission.

for glucose transport into the cell, in addition to activating enzymes involved in storage of glucose as glycogen or fat. The SH2 domain on the receptor (Figure 8.8) interacts with several intracellular proteins, some activating vesicular transport of GLUT4 molecules from microsomes within the cytoplasm to the plasma membrane (Figure 8.9) and others activating enzymes to promote lipid and glycogen synthesis [275].

8.2.3.3 FGF and PDGF Receptor Subfamilies

FGFR These receptors bind the peptide growth factors FGF (fibroblast growth factor), and PDGF (platelet-derived growth factor) and activate **mitogenesis**. They are especially important as regulators of embryonic development (Chapter 11), and mutated versions of the receptors play roles in the origins of many types of cancer (Chapter 12).

In humans, 22 genes have been described for FGF ligand production. FGF peptides are secreted from signaling cells during development and bind to one of five different FGF receptors on nearby cells to promote limb development and often facial and cranial development (Chapter 11). In adults, FGFs promote tissue repair and response to injury, and, when abnormally expressed, can also lead to cancer phenotypes.

PDGFR The PDGF ligand family is comprised of four peptides (A, B, C, and D). The four ligands themselves form homodimers, and A and B form a heterodimer, producing in all five different dimer ligands. The receptor exists in two forms, α and β. Ligand dimerization occurs first, and then binding leads to receptor dimerization and activation (Figure 8.10). Kinase activity leads to activation of the receptor's SH2 and PTB domains and the activation of several types of signaling proteins [276].

8.2.3.4 NGF Receptor Subfamily

These receptors are sensitive to four different peptides – NGF, brain-derived neurotrophic factor (BDNF), and two neurotrophins (NTs). Activation leads to receptor dimerization and activation of the intracellular tyrosine kinase domain; signaling induces neural cell migration and proliferation in other cell types. NGF receptors are also called tropomyosin receptor kinase or Trk receptors. Binding of NGF to TrkA couples to phospholipase C and MAP kinase signaling pathways, expressing genes for cell proliferation and outgrowth of cellular membranes during neural cell migration [278].

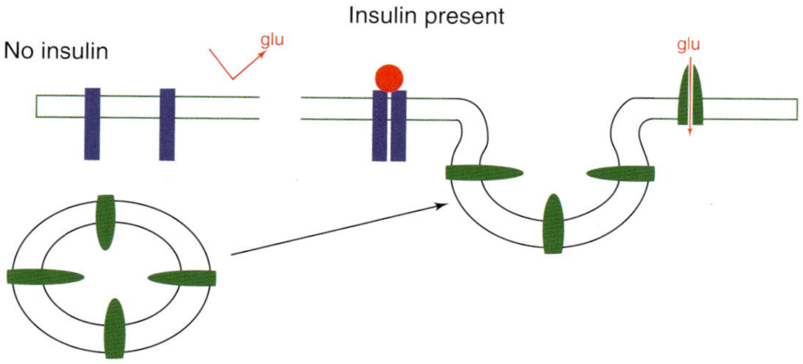

Figure 8.9 Actions of insulin receptor in glucose transport into cell. Left: In the unstimulated state, transporter (green) is in inactive form in cytoplasmic vesicles. Right: Insulin (red circle) binding promotes delivery of transporter to cell surface, where it binds extracellular glucose and transports it into the cell.

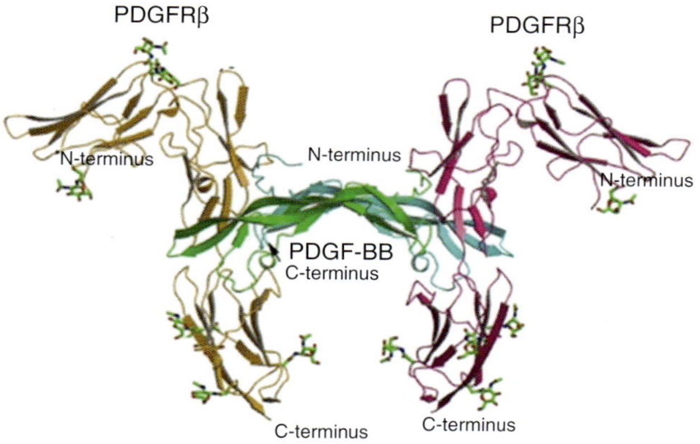

Figure 8.10 PDGF ligand peptides (green) dimerize in advance of receptor binding, then bind to and dimerize their PDGF receptors. Approximate membrane position is shown. Modified from Figure 2 A of Ref. [277]. Used with permission.

8.3 Receptor Serine/Threonine Kinases

8.3.1 Transforming Growth Factor-Beta (TGF-β) Receptor

The TGFR family of receptors is involved in embryonic development and tissue repair. Its receptors are divided into three classes: TGF-β, activin, and a large group of receptors that promote mesoderm formation in embryos [272]. The receptors are paired with ligands belonging to the TGF-β family of peptides (activin, inhibin, TGF-β, and BMP (bone morphogenetic protein) – Chapters 4 and 11).

These receptors have a cysteine-rich extracellular domain, a single transmembrane domain, and an intracellular serine/threonine kinase domain. They exist in the membrane as monomers, but on ligand binding form heteromeric receptor–ligand complexes; the TβR-II recruits the TβR-I, and the TβR-I activates **Smad** proteins (proteins specific to the TGF-β pathway) that move to the nucleus and act as transcription factors for developmental processes [279]. Figure 8.11 shows the two TGFβ receptor subtypes in their membrane location [269].

8.4 The Guanylyl Cyclase Receptor Subfamily – Natriuretic Peptide Receptors

The natriuretic peptide receptors (NPRs) are unlike the other receptors of this chapter in having a guanylyl cyclase (GC) as their intracellular signaling domain. Their homology with RTK is shown by their homodimerization on ligand binding, their single transmembrane (TM) domain, and their protein kinase intracellular domain [280].

These receptors are involved in blood pressure and volume regulation by the cardiovascular system and kidney tubules. The tubules contain fluid that has been filtered out of the blood, and is destined to

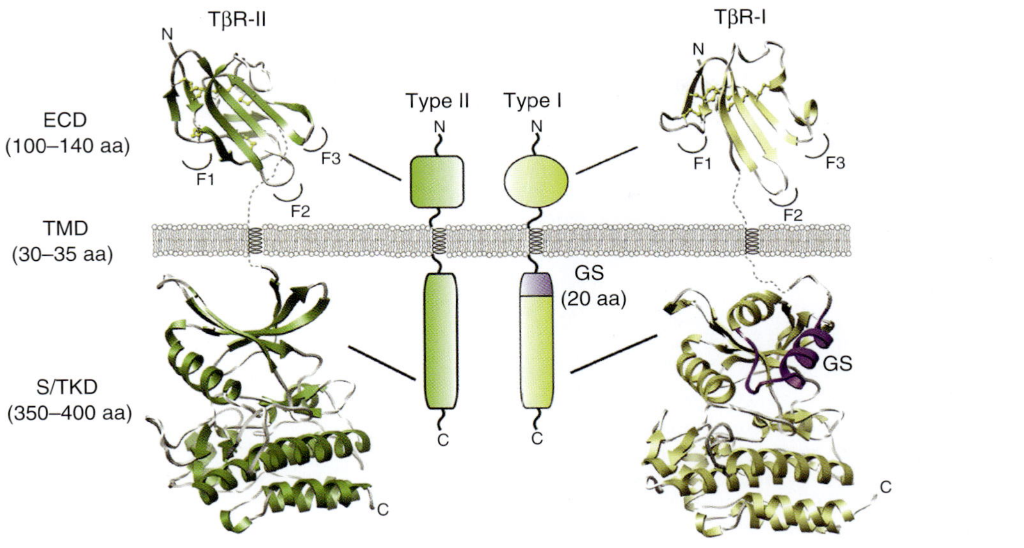

Figure 8.11 TGFβ receptors (TβR-II and TβR-I) shown in center as simple shapes; to left and right of each are shown their ribbon structures. ECD, extracellular domain; TMD, transmembrane domain; S/TKD, serine/threonine kinase domain. GS is a regulatory domain rich in glycines and serines found in type I receptors. Figure 3 in Ref. [269]. Used with permission.

be excreted. Because blood Na^+ (natrium in Latin) is osmotically active, any substance that causes a decrease in Na^+ uptake from the renal tubules would have a moderating influence on blood pressure.

The various natriuretic peptides are secreted in response to increased blood pressure. They bind to three subtypes of NPR, NPRA, -B, and -C. NPRA and -B have similar physiological effects; NPRC is a distinct receptor of uncertain structure that is responsible for clearance of the natriuretic peptides from the circulation [280].

Figure 8.12 shows the membrane orientation and architecture of the NPRA.

Binding activates the GC domain to produce cyclic GMP (cGMP), which activates phosphodiesterase and protein kinase 1. These proteins go on to cause vasodilation and increased vascular permeability (leading to decreased blood pressure) and reduction in uptake of Na^+ from kidney tubules [281] (leading to increased excretion of Na^+ and reduction in blood volume).

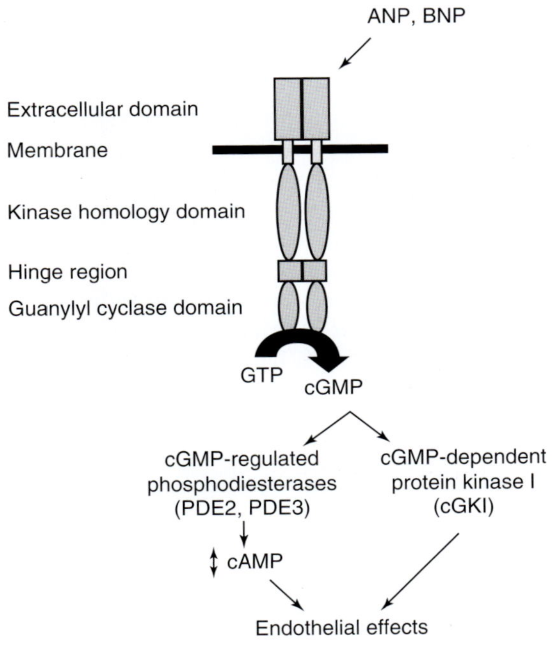

Figure 8.12 The atrial natriuretic receptor exists as a homodimer in the cell membrane; its agonists are atrial natriuretic peptide (ANP) and brain natriuretic peptide (BNP). The extracellular domain binds agonists, and the intracellular domain has both kinase and guanylyl cyclase (GC) domains. Binding activates the GC domain to produce cGMP, which increases cellular phosphodiesterases and protein kinase 1. These exert the vasodilatory and Na^+ secretory effects of the receptors. Figure 1 in Ref. [281]. Used with permission.

8.5 Non-Kinase Molecules – LDL Receptors

8.5.1 Cholesterol Transport

Large molecules generally do not diffuse through membranes or pass through ion channels, but are

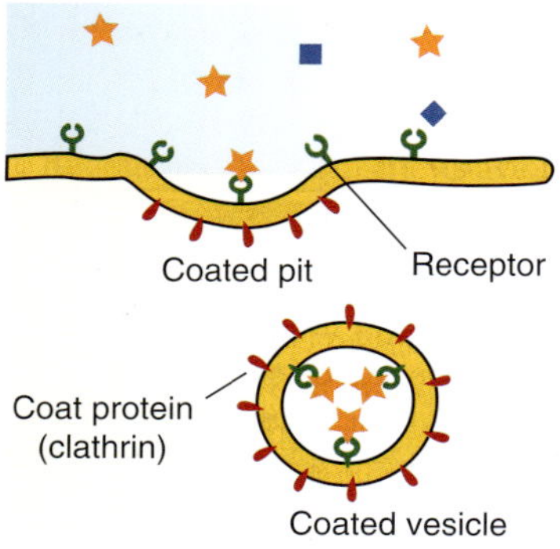

Figure 8.13 Receptor-mediated endocytosis promotes selective uptake of large molecules and particles. Ligand (LDL – star) binds its receptor, triggering the internalization of the receptor–ligand complex. The vesicle thus formed buds off and carries ligand to the interior of the cell. Clathrin, an intracellular membrane protein, (red shapes) lines the pit and the vesicle. http://en.wikibooks.org/wiki/Structural_Biochemistry/Protein_function/Ligand. Accessed 27 February 2015.

brought into cells via specific processes involving movement of extensive pieces of plasma membrane [282]. Two non-specific methods for movement of large particles and fluid into the cell are **phagocytosis** and **pinocytosis**. In these processes, a substance approaches the membrane and is engulfed by movement of the membrane around the boundary of the substance.

Receptor-mediated **endocytosis** is a related but different mechanism used to bring substances – for which specific cell-surface receptors exist – into the cell. The types of substances translocated are generally large proteins or proteins complexed with another substance. The best example of endocytotic receptors is that for cholesterol via binding and absorption of **low-density lipoproteins** (LDLs – Figure 8.13):

8.5.2 The Low-Density Lipoprotein (LDL) Receptor

Cholesterol is a fatty substance that makes up a large part of the cell membranes of eukaryotes, and is the compound from which many steroids and steroid-like substances are synthesized (Chapter 4). It is insoluble in blood, and so must be transported within the bloodstream bound to protein. Several different cholesterol–protein complexes, termed lipoproteins, are involved in cholesterol transport. One of these, the LDLs, transport cholesterol to cells for use. LDLs are large spherical droplets of single-layered phospholipid membrane surrounding a core of 1500 cholesterol molecules esterified to triacylglycerols, with a single molecule of the protein apolipoprotein B-100 (apo B), embedded in the membrane [283] (Figure 8.14).

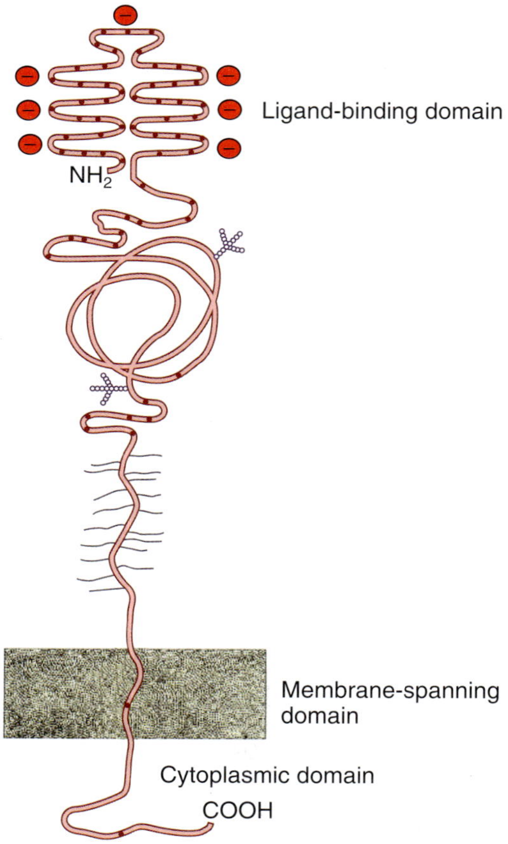

Figure 8.14 LDL receptor showing long extracellular region with ligand-binding domain; membrane domain with single membrane-spanning helix; cytoplasmic domain. http://en.wikiversity.org/wiki/Cholesterol_Regulation. Accessed 27 February 2015.

The apo B of the LDL particles binds specifically and with high affinity to LDL receptors on cell membranes. These receptors have three domains (Figure 8.14). The first domain is the binding site for LDL, near the amino-terminal end on the extracellular side of the plasma membrane; this domain contains EGF and fibronectin domains [284] that indicate the homology of this receptor with the RTKs. The second domain is the transmembrane domain that anchors the receptor in the membrane; the third is the intracellular domain. This domain is not catalytic; rather, it is responsible for aggregation of ligand-bound receptors and their internalization at clathrin-coated pits.

8.5.2.1 Clathrin-Coated Pits

These are regions where the inner surface of the membrane is lined by the protein **clathrin.**

The binding of LDL to its receptor begins a process in which the receptors aggregate and the coated pits are internalized, forming **endosomes**. The endosomes have a low pH (about 5) that reduces the affinity between the LDL and its receptor so the two dissociate. The endosome then splits into two portions, one that returns the receptors to the cell surface and the other that delivers the LDL to the lysosomes, membrane-bound vesicles containing enzymes. The lysosomes process the LDL to remove the cholesterol and direct it to the parts of the cell where it is needed [283].

The cholesterol brought into the cell serves as the basis for steroid synthesis and membrane construction. If the cellular level of cholesterol is excessive, it reduces the activity of the enzymes needed for cholesterol synthesis, increases the intracellular storage of the LDL form of cholesterol, and stops the synthesis of new LDL receptor [283].

Caveolins are transmembrane proteins that drive the formation of caveolae, small, flask-shaped invaginations along the interior of the plasma membrane. In contrast to clathrin-coated pits, caveolae are smooth vesicles and are not clipped via a GTPase-driven mechanism. Caveolar proteins can be regulated through phosphorylation and are linked to focal adhesions and **integrin** molecules. Caveolae have been found to form in the membrane and cause the endocytosis of growth factor receptors, notably the EGFR.

8.6 Cell–Cell Contact Signaling

Nearly all signaling of the Ig superfamily occurs via diffusible ligands interacting with receptors on cells that are not in contact with the signaling cell. Two exceptions conclude this chapter: first, the Notch–Delta developmental signaling pathway and, second, the signaling processes used by the immune system to generate antibodies.

8.6.1 Notch–Delta Signaling

The Notch protein is a cell-surface receptor for the Delta protein on a nearby cell; because Delta remains part of the cell membrane, the cells must make contact for Notch–Delta signaling to occur (Figure 8.15).

Two cells are involved in the process; the signaling cell has the "ligand" protein Delta in its membrane and the receiving cell has the receptor Notch. In the unstimulated state, gene transcription of the receiving cell is repressed by corepressor proteins bound to transcription factors (Figure 8.15a); to activate transcription, the signaling cell contacts the receiving cell so Notch–Delta interaction takes place (Figure 8.15b). This binding activates enzymes that cleave Notch (Figure 8.15c) so its ICD (intracellular domain) is released from the membrane and translocates to the nucleus (Figure 8.15d); binding of the ICD releases corepressors and activates transcription.

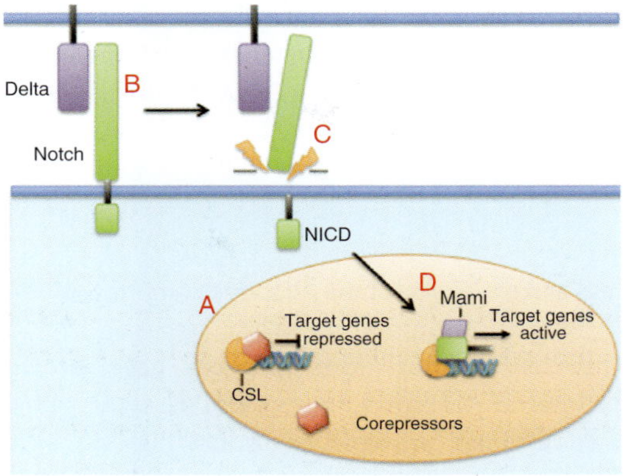

Figure 8.15 Notch–Delta signaling. Signaling cell membrane shown at top, with Delta "ligand" expressed on the cell-surface membrane. Receiving cell at bottom, with Notch receptor in membrane. A: In receiving cell before signaling, target genes are repressed by corepressors (red shape); B: Delta and Notch bind; C: two enzymes cleave the ECD of Notch from its TM and ICD domains; and D: the ICD of Notch translocates to the nucleus, where it replaces corepressors and initiates transcription. CSL and Mami are transcription cofactors. Figure 1 of Ref. [285]. Used with permission.

8.7 Immune System Receptors, Antibodies, and Cytokines

8.7.1 The Innate Immune Responses

All organisms have several "lines of defense" against disease-causing invaders: a surface that impedes entry, surface secretions that kill adsorbed bacteria, and non-specific chemical means of killing invading pathogens that have somehow gotten inside the body. The response to invading organisms is the double process of detection and attack: an organism must first be able to distinguish its own molecules from those of other objects or organisms in the environment. This ability to distinguish "self" is present in bacteria, unicellular eukaryotes, and animals, though the mechanisms are different in each group. Once the invader is recognized as foreign, it must be removed by a complex biochemical process of enzymatic destruction of invading cells [286]. The above processes make up the **innate immune response** common to all organisms. Vertebrates, however, are the only organisms with a well-developed specific cell-based **adaptive immune response**.

8.7.2
The Cells and Molecules of the Adaptive Immune System

For the past century, immunologists have known that cell-surface antigens on invaders were the signal for the response against them, leading to rapid proliferation of **lymphocytes**, immune response cells (indeed, as discussed in Chapter 1, the immune system provided early evidence for the receptor concept in general). Until the 1970s, cell-surface recognition processes between invader and host were thought to be the only mechanism of recognition. However, reports in the early 1970s suggested that various cell types in the immune system released substances having mitogenic signaling properties. The cells are known as **B(one)** and **T(hymic) cells** after their sites of origin. The substances they release were originally called **lymphokines**; the first lymphokines discovered were found to be produced by T cells and to act on T cells. They soon came to be seen as required for proliferation of immune system cells, and the cellular interactions that had earlier been thought to be the only basis of the response were newly reinterpreted as being needed mainly for setting the responsiveness of the T cells to the lymphokines.

The past 40 years have seen many changes in the understanding of the roles of signaling molecules and receptor molecules of the immune system. For example, many more lymphokines have been discovered (and they have been renamed **cytokines** to indicate that they may be produced by and act on cells other than lymphocytes), and the cellular interactions among immune system cells have been expanded.

8.7.3
T-Cell Receptors and Immunoglobulins

The structures of immune system antibodies and receptors are shown in Figure 8.16.

Box 8.1 A Brief History of Signaling in the Adaptive Immune System

Each lymphocyte can produce a different receptor type, even though the **antigen receptor**s of all lymphocytes have approximately the same structure. Thus, one lymphocyte can bind only one particular macromolecule, and a different lymphocyte can bind a different macromolecule. The differences in ability to bind antigen lie in details of the combining site of the receptor; these, in turn, depend on the way that genes encoding the antigen receptor proteins are processed.

Early history of the theory of antigen/antibody reactions:

Paul Ehrlich (Chapter 1) used the immune system as a model for the hypothesis that would later account for the actions of all receptors. In the "side-chain theory" he proposed that white blood cells possess specific receptors for antigenic substances, and that binding of antigen prompted the cell to synthesize more receptor. As the number of receptors increased, some were liberated into the blood and became antibodies. There was opposition to Ehrlich's views because it was difficult to see how blood cells could inherently possess the ability to produce the bewildering array of antigen receptors and antibodies for natural and artificial substances. Ehrlich's hypothesis was thus for a while replaced by the "template hypothesis," in which antigens were thought to direct the modification of existing receptors by forcing the receptors to "mold" themselves to the shape of the antigen [26].

The template hypothesis lasted until the 1950s, when it became apparent that it could not account for two well-known features of the immune response: the rapid rise in antibodies in the early stages of the response and the increased response of the body to the second challenge with the antigen. If an antigen were simply acting as template, then the number of antibodies could not grow much faster than the number of templates, and the second challenge with the antigen should produce the same intensity of response [26].

The clonal selection theory of antibody diversity:

Jerne [287] returned to concepts closer to Ehrlich's original view. He noted that the blood contains antibody-like molecules even in the absence of a specific immune response, and agreed with Ehrlich that there were a large number of cells, each with specificity to a single antigen, even in the absence of a foreign antigen. Talmage [288] proposed that the response to foreign antigen resulted in the production of a large number of new cells, each with the receptor for the antigen that initiated the response. The final form of the idea was developed by Burnet [289], who proposed that the antigen binding to the cell-surface receptor caused the synthesis of new receptors and the production of a clone of identical cells, each able to produce only one type of receptor.

The new idea explains all the data the template hypothesis could not: the rapid rise in antibody titer after exposure to antigen occurs because of new cell division, increasing the number of antibody-producing cells; the increased response to the second antigen exposure occurs because the first exposure increases the number of cells, and these form the basis for rapid production of new clones.

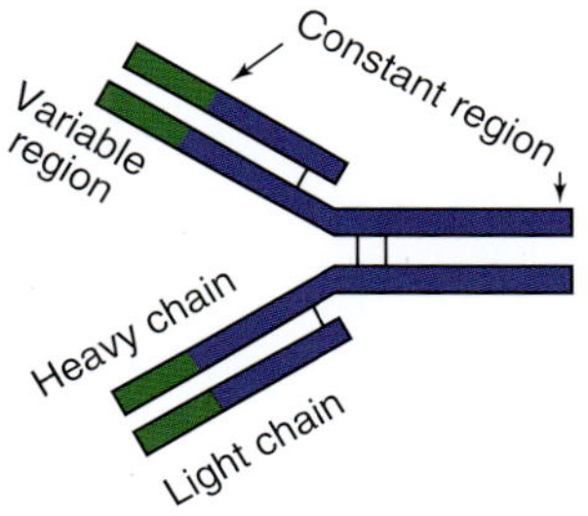

Figure 8.16 Structures of antibodies. The antibody consists of four proteins: two long chains (called "heavy chains") and two short chains (called "light chains"). The four N-termini are at the left (*variable* regions – green). Disulfide bonds (bars) hold the *constant* regions (blue) of the two heavy chains together and also link the light and heavy chains.

8.7.4 Cell-Surface Molecules

This group of molecules consists of the **MHC proteins**, antibodies, antigen receptors, and cytokine receptors.

- MHC proteins are cell-surface proteins that are the marks of "self" when in the producing animal and "nonself" when a cell is transplanted to or invades a foreign host.
- Antibodies are the immunoglobulins that are found on lymphocyte surfaces and in the blood.
- Antigen receptors are proteins on the lymphocyte cell surface that bind foreign antigens.
- Cytokine receptors bind cytokines to initiate immune cell responses such as proliferation and differentiation.

8.7.4.1 The MHC Proteins

This group of proteins is encoded by a set of closely linked genes in vertebrates. Their actions were first seen as mediators of tissue incompatibility or graft rejection reactions, but they can be more accurately described as binding sites for foreign antigens, allowing the antigens to be recognized and attacked by T cells (see below). The MHC proteins fall into three classes: MHC I, MHC II, and MHC III. The first two groups are cell-surface receptors with similar structures; MHC III are soluble proteins of the **complement** group, and are not considered here.

MHC I These molecules are found on the surfaces of almost all body cells. Because each different MHC I protein is encoded by a separate gene, this class constitutes one of the most polymorphic sets of cell-surface proteins known [286]. The three-dimensional structures of the MHC I proteins are all similar, but small differences in sequence create distinct antigen-binding domains with unique antigen-binding properties.

The MHC molecule is a peptide with a molecular weight of 43 kDa; it has a short intracellular domain, a single transmembrane domain, and three extracellular loops formed by disulfide bonds that serve as the ligand-binding domain (Figure 8.17a).

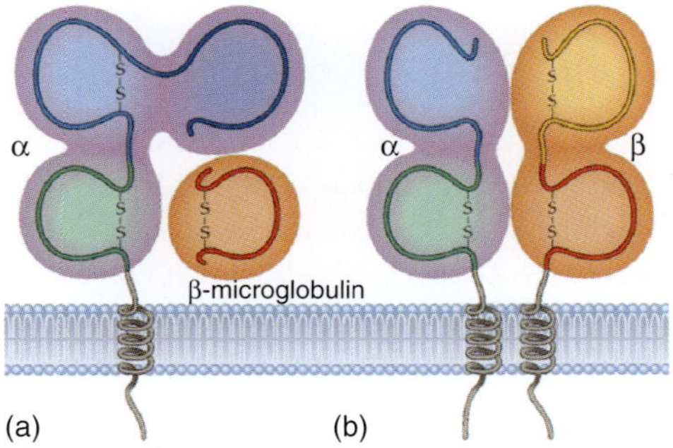

Figure 8.17 (a) MHC I molecule and (b) MHC II molecule. Source: http://what-when-how.com/acp-medicine/adaptive-immunity-histocompatibility-antigens-and-immune-response-genes-part-1/. Accessed 1 March 2015.

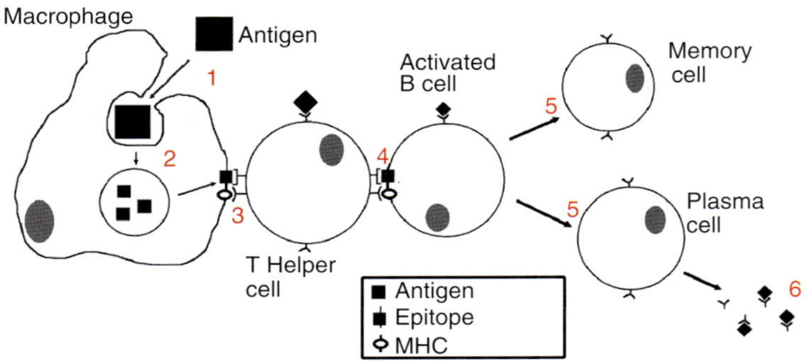

Figure 8.18 Steps in the activation of T and B cells for production of antibodies and generation of the immune response. Steps detailed in text.

In addition to the single transmembrane portion of the protein, the receptor is associated as well with a short peptide of 12 kDA (the β-2 microglobulin) that binds loosely to the larger portion. The microglobulin is encoded by a separate gene, and is identical in all MHC I proteins. The two proteins form a receptor with a four-domain structure, similar to one unit of the immunoglobulins (see below).

MHC II The MHC II proteins are less numerous, and are only found on a few cell types, such as B cells and **macrophages**, in the immune system. The MHC II receptor consists of two unlike polypeptide chains, an α and a β chain, loosely associated (Figure 8.17). The role of the MHC II receptors in the immune response is discussed below.

8.7.4.2 Receptors of the B and T Cells

The antigenic cell or substance first (1) binds to a particular macrophage carrying the appropriate antigen receptor [290] (Figure 8.18).

The macrophage (now called an antigen-presenting cell) absorbs the antigen by phagocytosis, and breaks it down into short peptides (2). The peptides (called **epitopes**) are then (3) placed on the surface of the macrophage in association with its own MHC II proteins.

The T Cells: The macrophage with its MHC II protein/epitope combination then is able to bind to and activate a **helper T cell**, which contains a type of receptor – the T-cell receptor – for the MHC II–epitope complex (Figure 8.17b).

The B Cells: These cells have naked antibody molecules on their surfaces; the cells may also take up antigen and express epitopes on their surfaces, but are not activated by this action alone. In order to become activated, they must bind both to the antigen and to the T helper cell (4) as presented by the macrophage.

Following binding, the cell enlarges and divides to form a clone of cells (called **plasma cells**) (5), each of which bears the memory of the initial exposure to antigen [291]. The plasma cells manufacture antibodies, and then secrete them into the blood (6). These antibodies are identical in amino acid sequence (and therefore in antigen-binding ability) to the antigen receptor that detected the antigen in the first place. The only difference between the antibody and the receptor is a sequence at the carboxy-terminus of the receptor that allows it to attach to the lymphocyte cell membrane.

The antibodies then bind and agglutinate more antigen, hastening the attack by blood-borne proteins (the complement system) that initiate phagocytosis of the antigenic cell [286]. The macrophage next secretes the peptide hormone **interleukin- 1** (IL-1 – see above), which diffuses to both the B cells and T cells to further activate them.

A given antigen presented on a macrophage will not bind to more than a very few B or T cells because its shape only happens to match a small number of antigen receptors.

The Cluster of Differentiation (CD) Molecules The T-cell receptor always seems to be associated with an additional set of related peptides, the cluster of differentiation (CD) molecules (Figure 8.19).

The CD molecules are a group of 80 or so proteins acting as signaling peptides of various immune system actions. CD4 and CD8 are T-cell surface proteins involved in activation of other cell types [286].

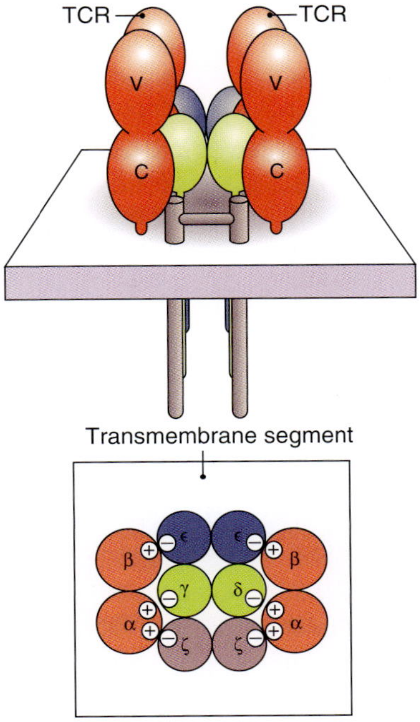

Figure 8.19 T-cell receptor (Ti) and its associated CD3 proteins. The receptor/CD3 complex is also shown in the lower frame. Figure 5.2 in Ref. [270]. Used with permission.

In the case of T cells, the CD3 proteins are a group of five to seven separate protein chains loosely associated with each other in the membrane (Figure 8.19). Each consists of a single membrane-spanning domain and extracellular domain for loose association with the Ti protein. Together, the Ti and CD3 complex recognize antigen and MHC II presented by macrophages. The intracellular domains of the CD3 are called immunoreceptor tyrosine-based activation motifs (ITAMs); they interact with intracellular soluble tyrosine kinases to transduce the binding signal to the nucleus, where cell division is stimulated [270]. CD2 is present on T cells; its ligand is another CD molecule, CD58, present on non-immune system cells. CD2–CD58 binding allows T cells to adhere to other body cell types [286].

These two types of CD possess intracellular serines that can be phosphorylated. The CD4 molecule is also, by an accident of molecular conformation, one of the receptors for the HIV envelope protein (env); the binding of env to the CD4 along with its coreceptor allows HIV to bind to and enter helper T cells [292].

9
Transduction IV: Nuclear Receptors

... Most of the animal kingdom is bathed in a sea of pollutants that act like estrogen ...

Janet Ralof [290]

9.1
Introduction

The classical era of receptor studies was the period up to the 1950s in which all receptors were thought to be at the cell surface, were inactive unless bound by agonist, had ligands that were easily classed as "active" agonists or "inactive" antagonists, did not engage in interactions with other proteins, and created instantaneous cellular effects when they were bound. All these classical receptor ideas gave way to subtleties over the next three decades; perhaps the most significant new idea was the discovery that a class of receptors resided in the cell cytoplasm or nucleus.

From the 1950s to the 1980s these nuclear receptors were added to the receptor list; in addition to their unusual location, they used a more complex set of agonists and antagonists because they interacted with multiple types of cellular proteins, and they operated slowly by activating transcription of genes for the synthesis of new proteins. Dealing with two types of receptors made things somewhat more complex, but at least for a while there appeared to be no functional overlap: cell-surface receptors bound hydrophilic agonists because those ligands existed in an aqueous environment; nuclear receptors bound hydrophobic agonists because the ligands could reach them only by diffusing through a hydrophobic cell membrane. Cell-surface receptors produced rapid responses and nuclear receptors developed their effects more slowly.

Researchers were thus surprised to find in the 1980s that the two apparent major groups of receptors did overlap in function. The hydrophobic agonists for nuclear receptors were discovered also to be agonists for cell-surface receptors of those very receptor types that had earlier been thought to be "wrong" because they were on the cell surface. In addition, the true nuclear receptors were discovered often to be activated without being bound by their hydrophobic agonists, but rather by either of two unlikely molecules: hydrophilic cellular proteins that bound to sites different from the active sites of the receptors, or hydrophobic fatty acids that bound to the active site itself and caused constitutive hormone action.

This chapter attempts to put these new discoveries into perspective. In the first section, we describe the traditional or classical actions of the nuclear receptors; they really do bind hydrophobic agonists (steroids, thyroid hormone, and fat-soluble vitamins A and D); binding does activate transcription, and these are called the **genomic actions** of agonists. But some actions of these ligands and their receptors were not predictable from early research in the field. In the second section of the chapter, we describe the **nongenomic action** of the steroid-like ligands, in which they act on G-protein-coupled receptors (GPCRs) and ligand-gated ion channels to activate processes other than transcription. In addition, we describe the ability of certain nuclear receptors to initiate genomic effects without binding traditional steroid agonists. Some of these, such as the hepatocyte nuclear factor (HNF4) receptors, have constitutively bound fatty acids and are constitutively active as transcription factors.

In sum, the receptors to be discussed here form a large superfamily of proteins that is structurally unrelated to the cell-surface receptors discussed in earlier chapters, and are best thought of as **tissue-specific transcription factors.** Indeed, these receptors make up the largest known group of eukaryotic transcription factors [293].

Receptor Biology, First Edition. Michael F. Roberts and Anne E. Kruchten.

9.2 Genomic Actions of Nuclear Receptors

9.2.1 Families of Nuclear Receptors

This superfamily is divided into six families, of which three are widespread in animals. This chapter covers these three important receptor families. Family 1: thyroid, retinoic acid (vitamin A), vitamin D, and insect ecdysis hormone receptors; Family 2: fatty acid and retinoic X receptors; and Family 3: steroid receptors for estrogens, progesterones, androgens, mineralocorticoids, and glucocorticoids.

9.2.2 Transcription Control

Gene activity leading to protein synthesis is a fundamental feature of all organisms; this activity must be controlled according to the nutritional or reproductive status of the organism. It is not surprising that complex intracellular and intercellular signaling mechanisms have evolved for the control of transcription and translation in multicellular organisms. A major function of the receptor tyrosine kinases (RTKs) is to control transcription of genes involved in metabolism and cell division (Chapter 8), and certain of the GPCRs transduce signals for control of gene transcription (Chapter 13). The nuclear receptors, however, are the only receptor superfamily devoted nearly entirely to ligand-based transcription control.

9.2.3 Constitutively Active Nuclear Receptors

Nuclear receptors are present only in animals (including the sponges), and though they have their greatest diversification in the vertebrates, researchers speculate that the multicellular heterotrophic lifestyle was initially allowed by the evolution of nuclear receptor signaling processes over 600 million years ago, coincidentally with the earliest metazoans [294]. Because the earliest animals did not have tissues or organs, nuclear receptors must have been useful for processes other than the sorts of endocrine functions seen only in animals with complex circulatory systems.

The oldest form of nuclear receptor appears to be a fatty acid receptor, HNF4. This protein binds fatty acids constitutively, rather than as facultative ligands; even so, it is a transcription factor, inducing synthesis of important liver and cytochrome P-450 enzymes [294]. It thus is a reasonable hypothetical early nuclear receptor, and the rest of the nuclear receptor superfamily, most of which have hydrophobic ligands, very likely evolved from an HNF4-like precursor.

9.2.4 Liganded Receptors

The receptors for steroids, thyroid hormone, vitamin A, and vitamin D reside in the eukaryotic cytoplasm and nucleus.

The ligands for the nuclear receptors are endocrine hormones or local modulators, but never transmitters released from nerves. They are produced in the gonads, adrenal cortex, or brain, and diffuse locally or travel in the blood to target cells. The genomic effect of these agonists is shown in Figure 9.1.

The classical (liganded) nuclear receptors fall into two families on the basis of amino acid sequence (Chapter 14): the steroid receptors for estrogens, progesterones, androgens, mineralocorticoids, and glucocorticoids; and the related but distinct receptors for thyroid hormone, vitamin A, and vitamin D. In the absence of ligand, both families of the receptor may be bound to DNA (at a specific sequence called the hormone **response element** – HRE) but in complex with corepressor proteins that prevent transcription. Agonist binding dissociates the repressor and allows binding of coactivator proteins that initiate transcription.

Once transcription is activated in the nucleus, the newly synthesized proteins go on to control growth, sexual behavior and gamete production (estrogens, testosterone), promote pregnancy (progesterone), produce adaptive responses to environmental and metabolic stresses (glucocorticoids), control water and electrolyte balance (mineralocorticoids), regulate Ca^{2+} and bone metabolism (vitamin D), regulate metamorphosis or metabolic rate (thyroid hormone), or direct aspects of embryonic development (vitamin A) [295].

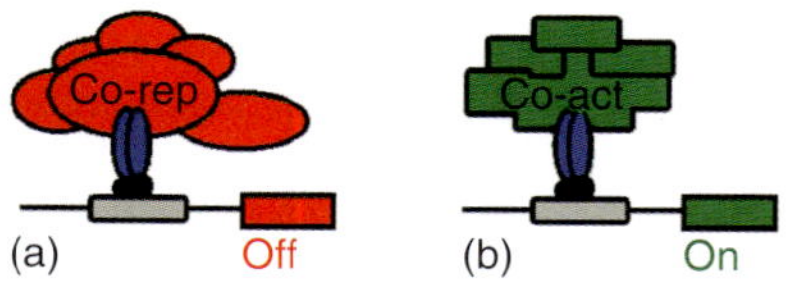

Figure 9.1 Overview of nuclear receptor action. The horizontal line is nuclear DNA; the gray rectangle is the DNA-binding region for the receptor (blue), called the hormone response element (HRE). (a) In the absence of ligand, receptors recruit multisubunit corepressor complexes (red) to promoter regions of genes. (b) When ligand binds, the receptor changes conformation to replace corepressors with coactivator complexes (green) that initiate transcription. Modified from Figure 1 in Ref. [294]. Used with permission.

9.2.5 History of Steroid Receptor Studies

Though the importance of steroids in physiology had been known for many decades, the mechanisms of steroid action were first understood only in the 1960s. Clever and Karlson [296] demonstrated that the insect hormone ecdysone caused activation of mRNA synthesis, leading to new proteins involved in metamorphosis. Jensen [297] suggested that steroids work by binding to an intracellular molecule. Karlson [298] borrowed from Jacob and Monod's model of bacterial gene control [299] and proposed that steroids operate in animals by activating gene transcription through inhibition of a repressor.

The actual mechanism was soon discovered to be somewhat different from what Karlson proposed, but his suggestion was prophetic, as steroids do activate genes, and they do it by controlling a protein transcription factor. Toft and Gorski [300] located steroid receptors by using density-gradient centrifugation to show that estrogen binds to an intracellular protein with a molecular weight of about 200 kDa; O'Malley and McGuire [301] showed that this binding event activates transcription. The use of radioactive steroid ligands during the 1950s had allowed localization of the receptors and studies of their functional properties, and Jensen *et al.* [302] proposed in 1968 that the binding of steroid to receptor changes conformation of the receptor so the complex can bind to DNA and thereby activate transcription. Jensen's concept followed Ptashne's [303] initial demonstration that proteins could bind DNA to affect transcription. Nuclear receptors combine a hormone-binding site and a DNA-binding site, so a single molecule acts as the recognition unit and the activating unit. They were cloned in the 1980s and were proposed to be direct activators of RNA polymerase (indeed, the receptors for glucocorticoids and estrogen were the first identified transcription factors for RNA polymerase [304]).

9.2.6 Receptor Structure

Steroids as hormones are produced only in eukaryotes [305], and are most diverse in animals, plants, and fungi. The steroids or steroid-like molecules seem to have originated as structural molecules rather than as signals, because bacteria, Archaea, and protists use them only to stabilize membranes. They are known to serve as signaling molecules only in plants [306] and animals.

In vertebrates, steroid hormones are produced primarily by the adrenal gland and the gonads, though the placenta and brain are also sources of steroids and sites of their action [304]. Thyroid hormone is produced by the thyroid gland in the neck of vertebrates, the homolog of the organ used in filter feeding in basal chordates; vitamin A is obtained in the diet, and vitamin D is both obtained in the diet and converted by sunlight acting on skin cells that contain the precursor steroid 7-dehydro-cholesterol.

The ancestral steroid receptor in animals seems to have been the estrogen receptor, and though this receptor is widespread, many invertebrate phyla have either developed different types of steroid signaling or have lost its gene entirely [295]. For example, the octopus (a mollusk) uses steroid signaling very much as do vertebrates, but arthropods only produce receptors for different steroid-like compounds that belong to the thyroid hormone receptor family [295, 307].

The nuclear receptors are all soluble proteins that exist in the absence of ligands either in the cytoplasm or in the nucleus; they have modular structures, with separate functional domains for ligand binding, dimerization, and DNA binding. The receptors for the nuclear-acting agonists all share a common structure [308] (Figure 9.2):

- *Domains A and B:* The N-terminal region is involved in regulation of transcription; its length and its sequence are variable across different receptor types; the activation function (AF) for transcription control (AF-1) binds to regulatory (**coactivator**) proteins (Figure 9.3b).
- *Domain C:* The central portion is the DNA-binding region, responsible for receptor dimerization as well as for specific binding to DNA enhancer sequences (**response elements**). This domain contains two cysteine-rich regions (the zinc fingers – see below) and is highly conserved across receptor types.
- *Domain D:* The variable hinge region, responsible for folding of the receptor and for its ability to travel to and operate in the nucleus. This region also adds the partial agonist activity of SERM-bound receptors (below).
- *Domain E:* The C-terminus is the hormone-binding region with the binding AF-2. Its sequence is also conserved across receptor types, though each receptor generally binds only one or a few naturally occurring hormones [308].
- *Domain F:* Recent investigators have suggested an additional domain, called F, exists in the receptor. This domain is involved in ligand binding in most steroid receptor types [304].

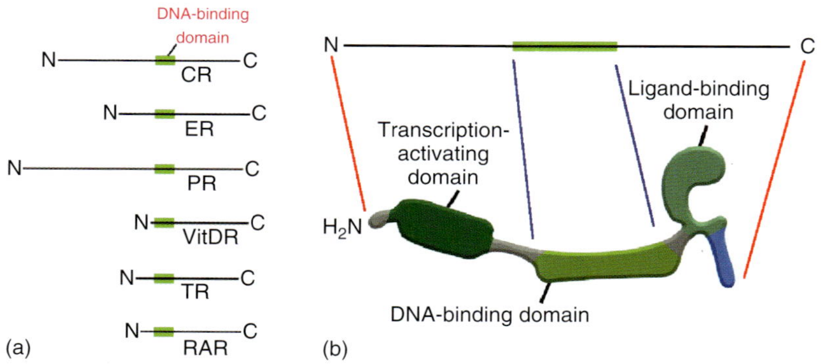

Figure 9.2 (a) Receptors for several steroid and thyroid hormones have similar organization: N-terminal region that controls transcription; DNA-binding region, C-terminal hormone-binding region. (b) Less diagrammatic representation of steroid receptor folding to expose DNA binding and ligand-binding domains. Figure 15-14 in Ref. [309]. Used with permission.

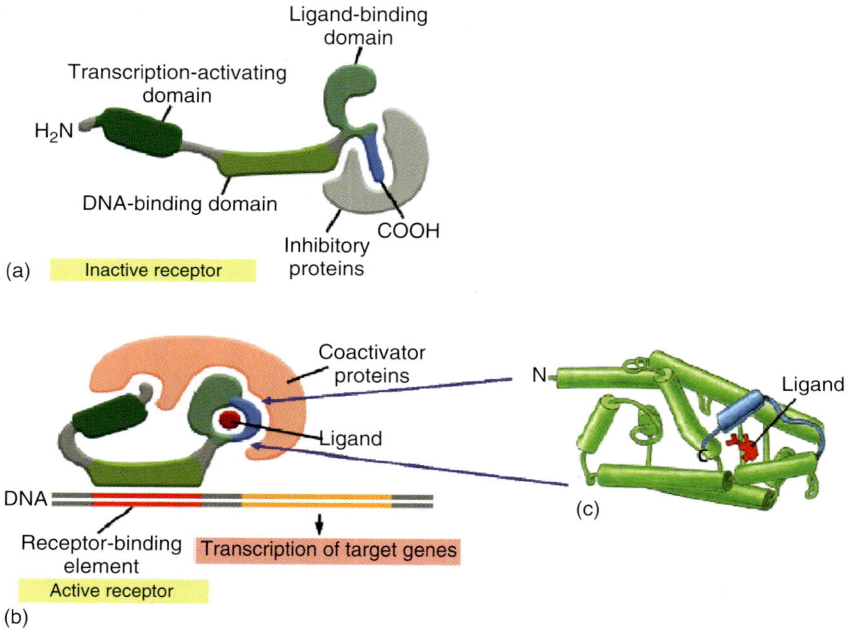

Figure 9.3 Steroid receptor in inactive (a) and agonist-bound (b,c) states. With no agonist bound, the receptor has its C-terminus covered by an inhibitory protein, for example, HSP90. This state prevents binding of receptor to DNA. When steroid binds, a coactivator also binds, exposing the DNA-binding site and allowing transcription. Panel (c) shows a more realistic view of the agonist bound by the ligand binding domain (LBD). Modified from Figure 15-14 of Ref. [309]. Used with permission.

9.2.7 The Ligand-Binding Module

The receptor consists of multiple helical domains [309]. In the inactive state, these domains are organized so as to bind to corepressor proteins such as **heat shock proteins** [310] (HSPs). In some cases, for example, the inactive glucocorticoid, mineralocorticoid, progesterone, and androgen receptors, the receptor-HSP complex is not bound to DNA [216], but is in the cytoplasm of the cell. In most other cases, the receptor and its repressors are bound to DNA (Figure 9.1). The E region of the receptor consists of 12 α helices arranged such that ligand binds in a pocket between 11 of them and is covered by the 12th. The covering helix binds to coactivators (below) involved in activating transcription [311] (Figures 9.3 and 9.4).

In the presence of the agonist, the receptor alters conformation to dissociate from the HSP and bind DNA and is phosphorylated. The first step is a dimerization of two receptors (Figure 9.5) and exposure of the DNA-binding site (Figures 9.3 and 9.5).

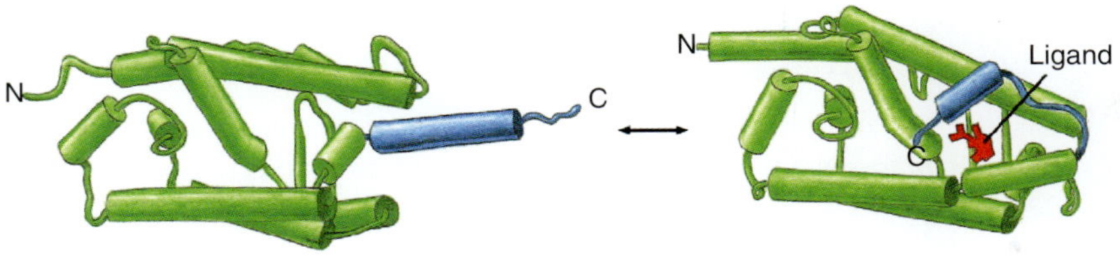

Figure 9.4 Steroid receptor in inactive (left) and agonist-bound (right) states. The coactivator and inhibitory proteins are not shown. The ligand-binding helix (blue) traps the ligand within the receptor, allowing coactivator binding (Figure 9.3b). Modified from Figure 15-14 in Ref. [309]. Used with permission.

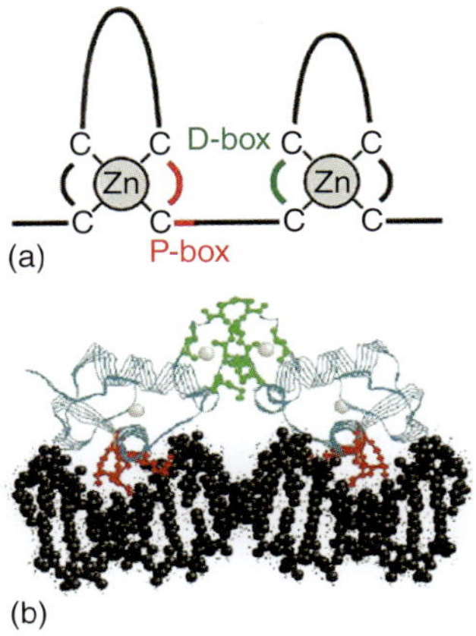

Figure 9.5 (a) Structure of two zinc finger loops (DNA-binding region in P box (red) and protein-binding region in D box (green)) of estrogen receptor. (b) Model of how a dimer of hormone–receptor binds DNA. Black: DNA double helix; blue: two receptors – red indicates P box Zn finger for DNA binding; and green indicates D box Zn finger controlling dimerization. Zn ions are the four gray spheres. Modified from Figure 5 of Ref. [304]. Used with permission.

Dimerization is generally accomplished by two identical receptors, but as most nuclear receptors have multiple isoforms, two members of the same agonist receptor group (e.g., estrogen receptor α and estrogen receptor β) may form **heterodimers** [304].

9.2.8 The DNA-Binding Module

The cysteines of domain C are involved in formation of the DNA-binding region or motif, called a **zinc finger**. This motif forms from four cysteines that make **coordination bonds** with zinc ions. The N-terminal zinc finger (**P**(roximal) box in Figure 9.5) binds to the major groove in the DNA molecule at the HRE, and the C-terminal zinc finger (**D**(istal) box) binds to the equivalent D box on the other receptor–agonist complex of the dimer. The P box is thus a DNA-binding motif and the D box is a protein-binding motif.

The P box binds at a characteristic sequence of the HRE. The various steroid receptors may have different HREs; the sequence of the androgen receptor (AR) response elements is

3′ACAAGANNNTCTTGT
5′TGTTCTNNNAGAACA

The N stands for any nucleotide. This pattern is thus a palindromic inverted repeat with three "any nucleotides" in the center [312].

The process of transcription control involves interaction of a receptor–ligand complex with several different types of nuclear proteins plus the promoter region of a DNA sequence. Not only do receptors facilitate the binding of standard cellular transcription factors, they also encourage the binding of coactivators. These proteins are bifunctional, combining transcription factor recruitment with an enzyme function, for example, modification of histone proteins to open the structure of DNA to allow access by transcription factors [311]. In order to act as a coactivator, a protein must have a conserved five-amino acid sequence (LXXLL – three leucines and two unspecified amino acids) that forms the binding site for the hormone–receptor complex [304].

9.2.9 Specific Nuclear Actions

9.2.9.1 Family 1 – Thyroid Hormone and Vitamins A and D Receptors

These receptors are homologous to the steroid receptors, and also act in the nucleus as transcription factors. The three main receptor types found in vertebrates are the

- thyroid receptor (TR);
- vitamin A receptor (VitAR, or retinoic acid receptor RAR);
- vitamin D receptor (VitDR).

In addition, a steroid-like hormone directs metamorphosis in insects via a receptor from this family.

Box 9.1 Vertebrate metamorphosis [314]

Juveniles of many phyla of animals have markedly different morphology, habits, and ecological status from adults. Metamorphosis is the conversion of the juvenile to the adult and involves alterations in gene expression in many parts of the body. This discussion focuses on vertebrates, and specifically on amphibians, but members of nearly all animal phyla go through this developmental process. Several mechanisms are common to most animal taxa: loss of certain larval body parts (e.g., the tail in tadpoles), gain of new body parts (e.g., the legs in adult frogs), and remodeling of persisting organs (e.g., shortening of the gut from tadpole to adult).

Metamorphosis is induced by binding of hormones to nuclear receptors, followed by DNA transcription and *de novo* protein synthesis. In vertebrates, the hormone thyroxine (TH) activates metamorphosis by binding to TR (see above); in insects, other nuclear receptors (ecdysone receptor) initiate metamorphosis via binding of a similar compound. Both TRα and TRβ are involved in metamorphosis (Figure 9.6). As usual, one TR dimerizes with RXR: in the absence of TH during larval life, TRα is present in the nucleus in larval life in association with RXR and with corepressors bound to the DNA in such a way as to inhibit transcription. As a larva approaches metamorphosis, corticotrophin releasing hormone (CRH) is released from the hypothalamus and causes the synthesis and release of thyroid-stimulating hormone (TSH) from the anterior pituitary*. This activates production of TH, which causes synthesis of new TRβ, and TH acts on both TR receptors to bind to TRα, to remove corepressors, and to initiate transcription. ACTH causes release of glucocorticoids from the adrenal cortex, and these also activate transcription via glucocorticoid receptors.

* The role of CRH in release of TSH appears to be unique to amphibians; the usual pattern of TSH control (in non-metamorphosing mammals) is via hypothalamic thyrotropin-releasing hormone (TRH).

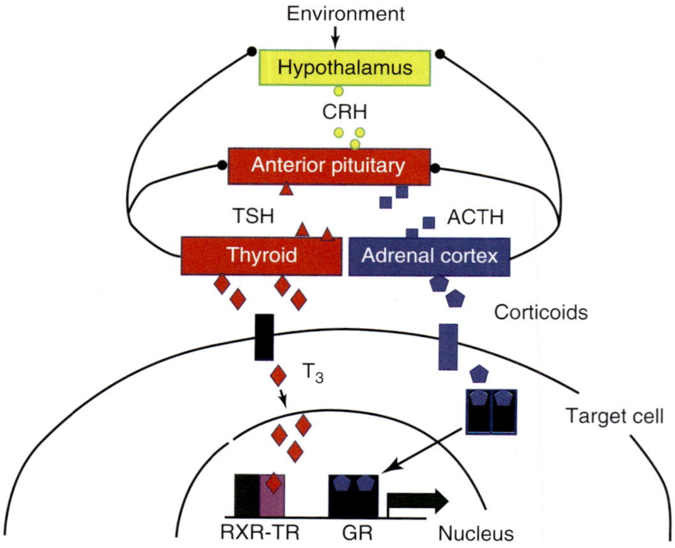

Figure 9.6 Environmental cues cause release of CRH from hypothalamus; anterior pituitary releases TSH and ACTH, which cause release of thyroid hormone (T3) and corticoids. These act on TR and GR to activate transcription of proteins for metamorphosis (actions of GR are not primary in metamorphosis, but are synergistic with those of TR). Modified from Figure 3 of Ref. [314]. Used with permission.

All these receptors are different from steroid receptors in that in the unliganded state they exist in the nucleus rather than in the cytoplasm and that in this state they form transcription **repression** complexes to prevent gene transcription. Binding of agonist is required for derepression and activation of transcription (Figure 9.1). Members of this family of receptors nearly always associate as heterodimers following ligand binding [307].

TR: Thyroid hormone actions are important in vertebrate development, metabolism, and metamorphosis. In mammals, hypothyroid conditions lead to lower metabolic rate and obesity, while hyperthyroidism leads to high metabolic rate and weight loss. The actions of the thyroid hormone on these and other functions derive from nuclear transcriptional effects, as well as the cell-surface effects [216], of thyroid hormone.

Two genes encode thyroid-binding nuclear receptors (TR). They are expressed in many tissues, including in the brain. In the absence of thyroid hormone, the receptors are located in the nucleus and repress transcription by binding DNA in such a way that prevents mRNA from being formed. If thyroid hormone is present, it binds to the TR and to the retinoid X receptor (RXR – below) and these two receptor–agonist complexes form a heterodimer that activates transcription [313].

Thyroid hormone increases metabolic rate by increasing transcription of genes for two different proteins: (i) a rate-limiting enzyme in the glycolytic pathway and (ii) a mitochondrial uncoupling protein. The uncoupling protein decreases the efficiency of ATP production, requiring greater glucose intake and breakdown to produce adequate amounts of ATP [313]. The effects of the two proteins are synergistic in increasing metabolic rate.

The response element for the TR, VitAR, and VitDR is a direct (not inverted) repeat with one to four Ns (any bases) in the center [304]. The difference between direct and inverted repeats may relate to the fact that steroid receptors always bind as homodimers, while TR-like receptors always bind as heterodimers. Inverted repeats on the DNA facilitate binding of homodimer receptor–ligand units because the receptors can bind regardless of the direction from which they approach the promoter.

Vitamin A Receptor: The RARs form heterodimers with the RXRs from family 2. Ligands are the long-chain hydrophobic molecules derived from vitamin A (various forms of retinoic acid (RA) and retinol). The RAR and RXR act in a manner similar to TR: in the absence of agonist, the receptor dimerizes with another RAR or with RXR and inhibits transcription of various genes. The RARs bind all-trans RA and 9-cis RA, while the RXRs bind 9-cis RA. The RXRs may operate on their own as RXR–RXR homodimers, but their main action is as heterodimers with TR, RAR, and vitamin D receptors (below).

Vitamin D Receptor: These receptors are targets of $1,25(OH)_2D$, the active form of vitamin D. The receptors are especially numerous in the intestine, bone, and kidney cells, where they regulate Ca^{2+} metabolism. The VitDR forms a heterodimer with the RXR (below) to regulate transcription. Transcription is inhibited by recruitment of protein corepressors that cause histone deacetylation and closing of the DNA helix. In the event of ligand binding, transcription is increased in certain genes when the VitDR–RXR dimer recruits coactivators that activate histone acetylation to open the DNA double helix [315] (Figure 9.7).

These receptors also have non-calcemic effects, in particular the ability to prevent cell proliferation (certain types of cancer) and also to reduce incidence of autoimmune diseases [315].

Insect Metamorphosis – Ecdysis: The hormones and receptors of insect metamorphosis are related to those of vertebrate metamorphosis. Ecdysone, the insect hormone, is similar in structure to steroids (Figure 4.22) but acts on family 1 receptors, also forming heterodimers with a receptor homologous to the RXR receptor (next section) [316].

9.2.9.2 Family 2 – Fatty Acid (HNF4) and Retinoic X Receptors (RXR)

These were the first two nuclear receptors to evolve (Chapter 14), and though the two do not form a monophyletic clade, are placed into a single family. The HNF4 protein is an unliganded transcription factor, but the RXR always operates as a heterodimer with each of the family 1 receptors (see above).

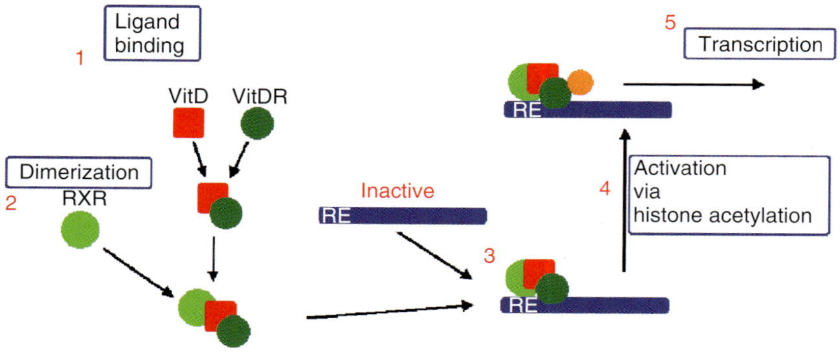

Figure 9.7 Pathway for vitamin D action. 1: Vitamin D binds to VitDR and 2: the VitDR and RXR dimerize. In the absence of binding, the vitamin D response element (RE) is unoccupied and the gene (blue bar) is inactive. 3: Ligand plus the two receptors binds to the RE, thus (4) recruiting other factors that allowing histone acetylation and opening of the DNA to promote (5) transcription.

HNF4: This intracellular protein does not bind exogenous molecules, but exists within the cell in complex with a molecule of linoleic acid (LA) in tight association [294]. Animals cannot synthesize LA or any of the polyunsaturated fatty acids, and they must obtain them from the diet. The ability of HNF4 to bind LA led to the suggestion that HNF4 functions as an "environmental sensor" (and attractor) for obtaining LA.

9.2.9.3 Family 3 – Steroid Receptors for Estrogens, Androgens, Progestogens, Mineralocorticoids, and Glucocorticoids

Androgens A single AR gene produces two receptors, AR-A and AR-B, by posttranslational processing (part of the C-terminus of the AR-B protein is cleaved to form the AR-A protein). The hormone-bound AR releases its HSP cofactors, forms a homodimer, passes to the nucleus, and binds to the HRE to activate transcription [312]. The receptor activates genes in the brain to synthesize proteins affecting behavior, **mood**, and cognition; it activates genes in muscles to affect cell growth and in bones to increase production of new bone cells [317].

Estrogens In vertebrates, the levels of estrogen are highest in the newborn and decline with age; thus, the estrogen response is quite different in infants compared with that in adults [114].

In brain cells of the basal forebrain, estradiol **induces** synthesis of the transmitter acetylcholine. In the hypothalamus, estradiol induces the synthesis of receptors for the hormone oxytocin, as well as oxytocin itself in nearby cells [318]. These actions initiate the neurotransmitter function of large groups of brain neurons, promoting proper brain developmental stages of both sexes. See Chapter 12 for a wider discussion of the roles of estrogen receptors in cell division.

Progesterones Nuclear progesterone receptors are located in the gonads and uterus of the reproductive tract, the brain, digestive tract, and the circulatory system [319]. They are mostly involved with reproductive cycles and maintenance of pregnancy in females. Their roles in non-reproductive tissues are not well understood.

Mammals have a single gene for the progesterone receptor, though different transcriptional promoters produce two isoforms of the protein itself. The two isoforms, PRA and PRB, have different activation domains, and activate different protein-synthetic mechanisms. The PRB receptor is a traditional nuclear receptor that acts as a transcription activator when bound by ligand; the mechanisms of transcriptional activation are the same for the B receptor isoform as for the estrogen receptor. The PRA receptor, in contrast, acts in some tissues as an inhibitor of transcription of the other steroid receptors. It even inhibits the action of the PRB receptor in causing uterine growth [320].

Progesterone acts on the uterus to promote its growth during the luteal phase of the menstrual cycle (after ovulation); it also prepares the uterus for embryo implantation, and during pregnancy decreases the contractile properties of uterine smooth muscle to reduce the chance of miscarriage.

Glucocorticoids The primary glucocorticoid steroid hormone is cortisol; it is released from the adrenal cortex in response to the peptide hormone adrenocortical tropic hormone (ACTH) from the anterior pituitary (Chapters 4 and 6). Cortisol is important in metabolic, developmental, and anti-inflammatory

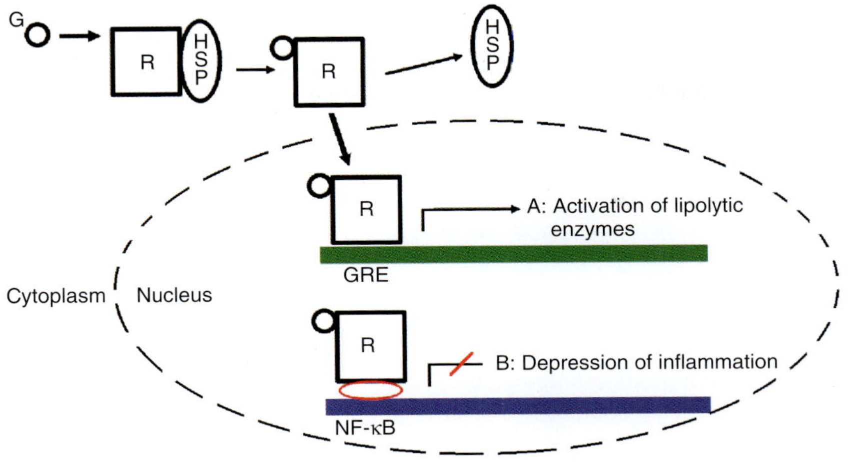

Figure 9.8 Effects of glucocorticoid action. Receptor binding of agonist (top left) causes translocation to nucleus, dimerization, and either activation or repression of transcription. Within the nucleus, two examples are shown: in **A**, transcription is activated by direct binding of GR to response element. In **B**, GR binds to NF-κβ transcription factors and inhibits transcription of immune system cytokines.

pathways, and it acts in either of two ways: by increasing transcription of anti-inflammatory proteins or by decreasing transcription of proinflammatory proteins. The single glucocorticoid receptor (GR) gene synthesizes two different isoforms of the receptor, GRα and GRβ, by alternate splicing [321].

Activation of Gene Transcription: The GRα isoform activates nuclear glucocorticoid signaling in a manner equivalent to that described for the sex steroid receptors. The metabolic effects are primarily those of **gluconeogenesis**, the release of glucose into the blood in response to stress. This is accomplished, for example, by synthesis of enzymes that stimulate lipolysis [322] (Figure 9.8, example A).

Other newly synthesized proteins suppress the immune response to wounds and invading cells, in part by causing **apoptosis** of immune-system T cells involved in fighting these invaders (Figure 9.8, example B).

Repression of Gene Transcription: GRβ is a modified receptor protein that does not bind steroids. It is therefore "unliganded," and is able to bind DNA in the absence of cortisol. When it binds to the promoter of a gene normally activated by GRα, it inhibits binding of the glucocorticoid–GRα complex, thus preventing normal GR action. In this manner, the GRβ receptor is able to prevent many of the metabolic and stress-related effects of GRα [323].

Negative Feedback Effects on Steroid Synthesis: The steroid hormones also repress synthesis of those hypothalamic and pituitary hormones responsible for their own synthesis. A good example of this type of repression is seen in the pro-opiomelanocortin gene, which produces the pituitary hormone ACTH as well as other peptide hormones (Chapter 4). The production of ACTH is itself under negative feedback control by cortisol, which inhibits transcription of the mRNA for ACTH. Cortisol diffuses from the blood to pituitary cells, where the cortisol–receptor complex binds to the promoter of the ACTH gene. This binding prevents the normal binding of pro-opiomelanocortin transcription factors, and hence prevents gene transcription for ACTH [308].

Mineralocorticoids The primary mineralocorticoid is aldosterone, responsible for salt (mineral) homeostasis, which in turn affects water balance and blood pressure. The primary effect of aldosterone is to act on a traditional nuclear receptor to activate transcription of genes encoding sodium channels (epithelial Na+ channel (ENaC) – see Chapter 6) in the distal tubule of the kidney [324].

9.3 Actions of Receptor Antagonists

Antagonists of steroid receptors bind to the receptors but do not activate transcription [293, 308]. Antagonist binding affects the conformation of the DNA-binding site such that the 12th helix does not cover the ligand (unlike the case shown by agonist binding in Figure 9.4). Thus the receptor–antagonist complex cannot achieve the correct conformation to activate transcription [304].

Receptor antagonism is not simple, however, because different cell types produce different isoforms of each receptor, and because different cell types also produce different activation cofactors. Thus, antagonists often act as **selective receptor modulators** (SRMs) rather than pure antagonists. Selective modulators exist for estrogen and progesterone (selective estrogen receptor modulators (SERMs) and selective progesterone receptor modulators (SPRMs) respectively). After binding, they recruit cofactors that cause their overall actions to be excitatory or inhibitory [325].

The SERMs function as estrogen agonists in the bone, brain, and cardiovascular system, and as estrogen *ant*agonists in the breast, uterus, or ovary. A clinically important SERM is tamoxifen, which binds to the estrogen receptor in breast tissue, preventing it from binding to DNA and activating protein synthesis. Tamoxifen was thus initially classed as an antagonist of the estrogen receptor, as it is useful in preventing the division and spread of breast cancer cells. However, tamoxifen activates proliferation in uterine cells (the cancer-related actions of steroid receptors are covered more fully in Chapter 11).

The SPRMs are important modulators of pregnancy, as interfering with progesterone action causes disruption of the nutritive and quiet uterine lining. The main SPRM so far developed is mifepristone (formerly called RU-486), a powerful abortifacient because of its antagonism of the actions of progesterone (Figure 9.9).

Mifepristone (**X** in Figure 9.9) competes with progesterone for the PR; it binds to the receptor, causes dissociation of the HSPs, and allows dimerization. It may even allow the receptor to bind to DNA, but it

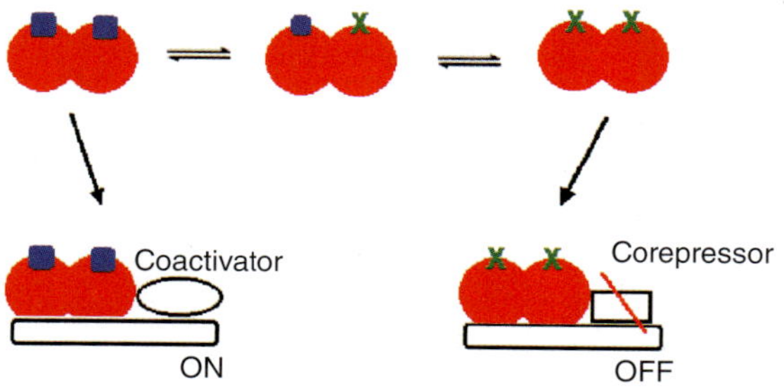

Figure 9.9 Actions of PR agonists (blue squares) and SPRMs (X). Left: The agonist PR binds, causes dimerization, recruitment of coactivators (oval), and transcription. Right: The SPRM Mifepristone causes dimerization and recruitment of corepressors (rectangle) that prevent transcription.

also recruits corepressors that prevent transcription of new proteins. Thus, it antagonizes progesterone-activated aspects of the female reproductive cycle, in particular the preparation of the uterus for implantation of a newly fertilized embryo.

9.4 Non-Traditional Actions of Steroid-Like Hormones and Their Receptors

The **nongenomic** effects of the steroids, thyroid hormone, and vitamins A and D do not result from protein synthesis. Instead of diffusing through cell membranes, these agonists may bind to cell-surface receptors such as ligand-gated ion channels or GPCRs, thereby modifying channel or receptor behavior or activating intracellular second messenger systems [326].

The presence of cell-surface steroid receptors was first suggested by data showing that progesterone could affect sperm behavior in ways not explained by what was known about its genomic actions [327]. In these and other experiments, the evidence for a non-genomic effect of steroids is shown by the following types of evidence [328]:

- The effects occur within seconds or minutes after application of the steroid.
- The effects are not inhibited by nuclear receptor antagonists, nor are they mimicked by nuclear receptor agonists.
- The effects are often dependent on extracellular Ca^{2+}.
- Radioligand binding assays have identified likely receptor sites on cell surfaces.
- Transduction of steroid effects operates through known receptor-activated systems, such as ion channels, phospholipase C (PLC) activation of inositol trisphosphate and diacyl glycerate.

These observations have led to the three-part concept of steroid hormone action: (i) classical genomic transcriptional effects; (ii) transcriptional effects from intracellular kinases not normally part of the genomic effect; these may be mediated by steroid binding to either intracellular or cell-membrane receptors of the classical steroid receptor type; and (iii) non-transcriptional effects from binding of steroids to cell-surface receptors that are nonclassical steroid receptors. The two latter (nongenomic) effects [329] are shown in Figure 9.10 for the influence of testosterone on testicular Sertoli cells.

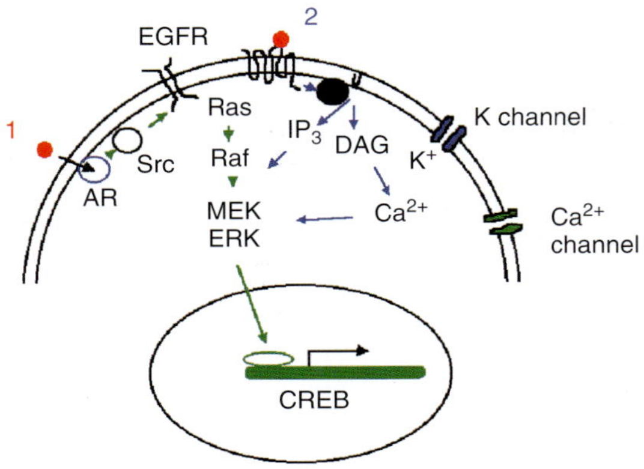

Figure 9.10 Nongenomic effects of androgens in spermatogenesis, a process controlled by sertoli cells. 1: Androgen binds to traditional androgen receptor (AR) attached to membrane. This activates Src, which activates intracellular portion of an epidermal growth factor (EGF) receptor and the cascade Ras-Raf-MeK-ERK for transcription activation of CREB (cAMP response-element binding) protein. 2: Androgen binds to G-protein-coupled receptor; PLC signaling produces IP_3 and DAG to open K^+ and Ca^{2+} channels. Ca^{2+} initiates intracellular signaling. IP_3: inositol trisphosphate and DAG: diacyl glycerate.

Androgens activate at least two membrane-based receptor systems (1 and 2 in Figure 9.10), and these receptors activate membrane channels and cellular kinases indirectly. Similar processes are seen in estrogen activation of membrane receptors.

The membrane AR in Figure 9.10 has not yet been characterized; perhaps it is a traditional AR but with a side-group that allows it to be anchored to the membrane. The GPCR for androgens is also uncharacterized, though the structure and functions of the GPCRs activated by estrogen and progesterone (below) have been studied.

In the case of membrane-mediated estrogen effects, the classical estrogen receptor itself is attached to cell membranes of many cell types. Binding of estrogen at the cell surface mediates immediate behavioral effects in brain neurons and changes in fluid balance in the gastrointestinal (GI) tract cells [330].

9.4.1 Cell-Membrane Progesterone Receptors

Non-nuclear receptors for progesterone have been defined on cell membranes of vertebrate reproductive tract cells. Progesterone acts on a GPCR to influence maturation of oocytes [323]. This receptor has been labeled GPR3; sequence homology shows it belongs to family A of the GPCR superfamily, and is homologous to the melanocortin receptor (Chapter 7).

Some non-nuclear actions of steroids have also been found to occur at the gamma-amino butyric acid, type A ($GABA_A$) receptor (Chapter 6), which in its normal action inhibits neural activity by opening Cl^- channels. Steroids enhance neural inhibition by also binding to the $GABA_A$ receptor and facilitating Cl^- channel opening when GABA is released. The action of steroids thus resembles that of barbiturates. Two important effects are lessening of anxiety and reduction of epileptic symptoms [318]. These agonists belong to a group of steroids that are metabolites of gonadal or adrenal steroids rather than the original androgens or estrogens themselves [331].

9.4.2 Cell-Membrane Mineralocorticoid and Glucocorticoid Receptors

Both aldosterone and cortisol have been shown to have nearly immediate (1 min) effects that are not reduced in the presence of classical mineralocorticoid receptor antagonists. Though **corticosteroids** are known to bind and inhibit membrane transporters for catecholamines in vascular smooth muscle cells [332], other receptors belonging to GPCR or RTK superfamilies are being sought for most membrane-mediated effects [320].

9.4.3 Cell-Membrane Thyroid Hormone and Vitamin A/D Receptors

Metabolites of thyroid hormone bind at the cell surface to GPCR of family A (rhodopsin-like, closest to the adrenoceptors) called trace-amine-associated receptor (TAAR1) or trace-amine-associated receptor 1. Binding causes increased cyclic adenosine monophosphate (cAMP) and modulates neuronal activity in brain cells carrying this receptor [333]. Thyroid hormone also acts on cell-surface receptors to modulate neuronal excitability via Na^+ channels, Na^+/H^+ transporters, and Ca^{2+} pumps [216].

9.4.4 Ligand-Independent Activation of Transcription

Recent experiments have shown that steroid receptors can be acted on in the nucleus by cellular kinases and can activate protein synthesis in the absence of agonist binding. This is still a genomic effect of the receptor, though not a traditional one.

Two additional features of coactivator interactions have been noted: first, coactivators direct transcription in a manner that promotes adaptive responses to given environmental stresses. Membrane receptors for other ligands (e.g., those that activate intracellular kinase pathways – Chapter 8) initiate phosphorylation of activator proteins for particular receptors, and this phosphorylation causes the coactivators to be directed to particular genes [334] (Figure 9.10). Second, coactivators regulate the amount of transcription that occurs: the coactivators are rapidly dissociated from the agonist–receptor complex by the ubiquitin pathway soon after transcription is completed [334]. This, rather than hormone dissociation from receptor, is a major process in the regulation of steroid hormone action as the hormone–receptor–coactivator–DNA complex may be too complicated for easy dissociation of subunits on their own.

In addition, phosphorylation of the unliganded receptor may activate it to become an effective transcription factor [335]. The first ligand-independent action of a steroid receptor was suggested in 1990 [335] when chick oviduct cells were activated to synthesize new proteins by kinases in the absence of progesterone. The mechanism of activation seems to be phosphorylation of the unliganded PR, causing it to become a transcription activator. This observation was followed by others [336] describing estrogen-related receptors (ERRα and β) that did not bind estrogen, and in fact were activated by binding intracellular proteins.

Part IV
Applications

Introduction to the Applications of Receptor Knowledge

Previous chapters have outlined the mechanisms by which the four receptor groups transduce cellular signals. Those chapters have focused mostly on fundamental physiological processes, keeping each receptor type separate from the others. The final four chapters show several important biological features that derive from the principles of receptor communication and the interactions among different receptor types:

- Receptor activation is transduced by complex intracellular signaling processes – Chapter 10.
- Differentiation during embryonic development is influenced by gene regulation and by the transmitter/receptor interactions between groups of cells – Chapter 11.
- Diseases often result from quirks of or mistakes in processes of cellular communication – Chapter 12.
- The workings of the mind emerge out of localized neurotransmitter–receptor interactions – Chapter 13.
- The evolutionary history of receptors is shown by the families of related protein sequences within each group – Chapter 14.

10
Signaling Complexity

Cell signaling is all of biology.

John Torday [337]

10.1
Introduction

The central theme of this book has been the three-step process by which an external signal (ligand) activates a cell via interaction with a receptor, followed by transduction of the binding event to the interior and the cellular response (Figure 10.1).

Chapters 6–9 described the structures and mechanisms of the various receptor types and their individual transduction mechanisms. This chapter describes briefly some of the cascade-like processes that occur after the signal transduction event: these processes consist of the various cellular signaling systems that result from the initial receptor–ligand interaction and involve cellular actions of dozens of intermediary molecules, including lipids, proteins, and nucleic acids. Each cascade is not restricted to a single receptor type, and complex biochemical events may result when two receptors are activated in such a way that more than one cascade is activated (see below). Some cascades are fast while others are slow. With every new experiment, researchers learn more about the complexity of these cascades and their purpose in cellular regulation.

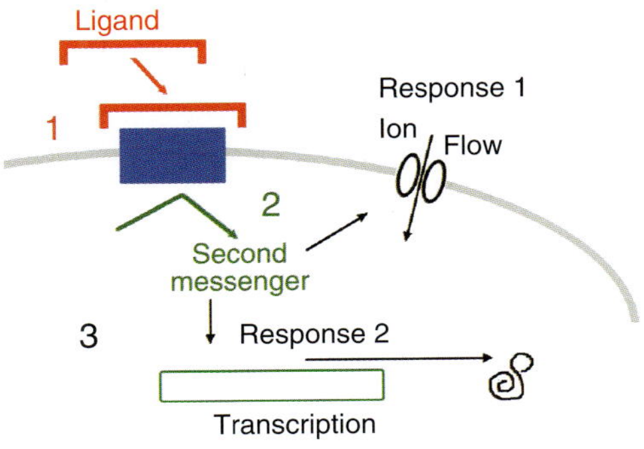

Figure 10.1 Signal transduction linked to signaling cascades. A ligand or other molecule binds to a receptor (1) and transduces its message through an intracellular signaling process (2) to ultimately result in some type of cellular effect, in this case either ion flow into the cell or transcription of new proteins.

Each signaling process now extant in eukaryotes came from a particular evolutionary event; origins of the major systems are shown in Figure 10.2.

10.2
Experimental Determination of Signaling Cascades

10.2.1
Glycolysis

To begin, it is important to understand how signaling cascades are experimentally determined. We introduce the idea of cascades with the example of glycolysis, one of the earliest cascades to be experimentally described. The six-carbon glucose molecule is oxidized through a series of enzyme-mediated reactions into two three-carbon molecules: glyceraldehyde-3-phosphate (G3P) and dihydroxyacetone phosphate (DHAP). Further reactions result in the complete oxidation of glucose into two molecules of pyruvate plus electrons captured in NADH (Figure 10.3).

The glycolytic pathway was determined in large part by the work of Meyerhof and colleagues during the 1930s and 1940s. Their work built on that of Pasteur, whose experiments on the metabolism of prokaryotes led to initial understanding of the fermentative process. In 1897, the brothers Hans and Eduard Buchner expanded on this work by recognizing that the addition of sugar to a yeast extract resulted in the release of carbon dioxide. This was the first time it was observed that fermentation could exist *outside of* an intact cell.

At the same time, muscle physiologists were discovering that lactic acid fermentation of sugars in the muscles shared many of the same properties

Receptor Biology, First Edition. Michael F. Roberts and Anne E. Kruchten.

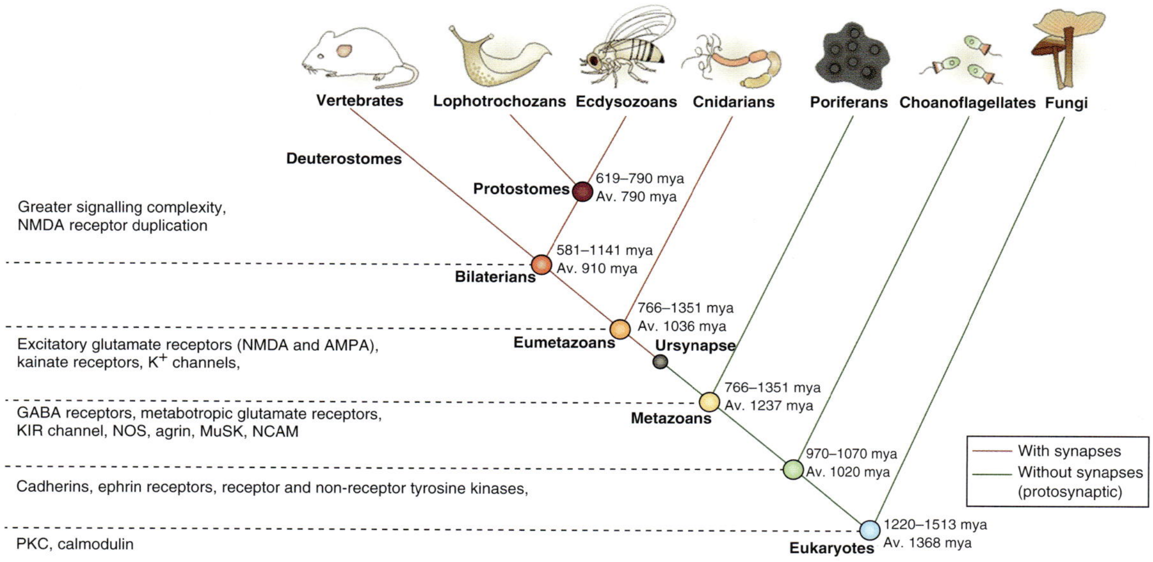

Figure 10.2 Eukaryotic history of signaling systems, showing receptors and processes involved in cell–cell communication. Dates represent approximate times at which molecules or processes evolved. Figure 1 in Ref. [338]. Used with permission.

as alcoholic fermentation in yeast. This was accomplished by understanding the organic chemistry of the enzyme–substrate reaction. By the early 1930s, Meyerhof proposed the importance of inorganic phosphate (which is now understood to be supplied by ATP) in the progression of glycolytic reactions. Thus, Meyerhof and colleagues elucidated the glycolytic pathway over the course of 50 years by identifying and understanding the substrates, products, intermediates, enzymes, and cofactors necessary for the many steps. Eventually, these steps were integrated into one coherent pathway. Yet, lest a student think that pathways are static, dogmatic knowledge, knowledge of the glycolytic pathway is continuing to be developed further today. For example, glucokinase, which was once thought to be minimally regulated, has been found to be a "glucose sensor" in pancreatic beta cells, and dysregulation of the enzyme can contribute to "glucokinase disease" in states such as **diabetes** [339].

10.2.2
MAPK: a Phosphorylation Cascade

The above discussion of glycolysis indicates the emerging importance of phosphorylation in cellular biochemistry, as initially described by Krebs and Fischer (see also Section 7.3.1.1). The next discovery was that of phosphorylation cascades, such as the mitogen-activated protein kinase (MAPK pathway), as discovered by Krebs, Ahn, and Fischer. In this system, a series of kinases phosphorylate each other in sequence. This system of enzymes is also linked to receptor tyrosine kinases (RTKs) (see below and Chapter 8).

For example, the MAP kinase (MAPK, which is also known as ERK) phosphorylates its substrate, the S6 kinase ribosomal protein, but only after it has been activated by an upstream kinase, the Map kinase kinase (MAPKK, which is also known as MEK). MAPKK, in turn, is also phosphorylated by an upstream kinase, MAP kinase kinase kinase (MAPKKK – Figure 10.4).

The process begins as the MAPKKK protein is first activated by a signal transduction event originating from a number of different receptors and ligands, often including the (epidermal growth factor) EGF receptor and EGF ligand pair.

The MAPK pathway was elucidated by first determining the final step of the pathway and then moving backward to the step closest in space and time to the membrane signaling event. The first step was to determine that MAPK phosphorylated S6 kinase as its target, and this was accomplished using **phosphatases** that removed phosphates from S6, thereby inhibiting its function until rephosphorylated by MAPK. Using similar techniques, MAPK could be inactivated using phosphatases until it was reactivated by MAPKK. Through this work, the entire MAPK pathway was elucidated.

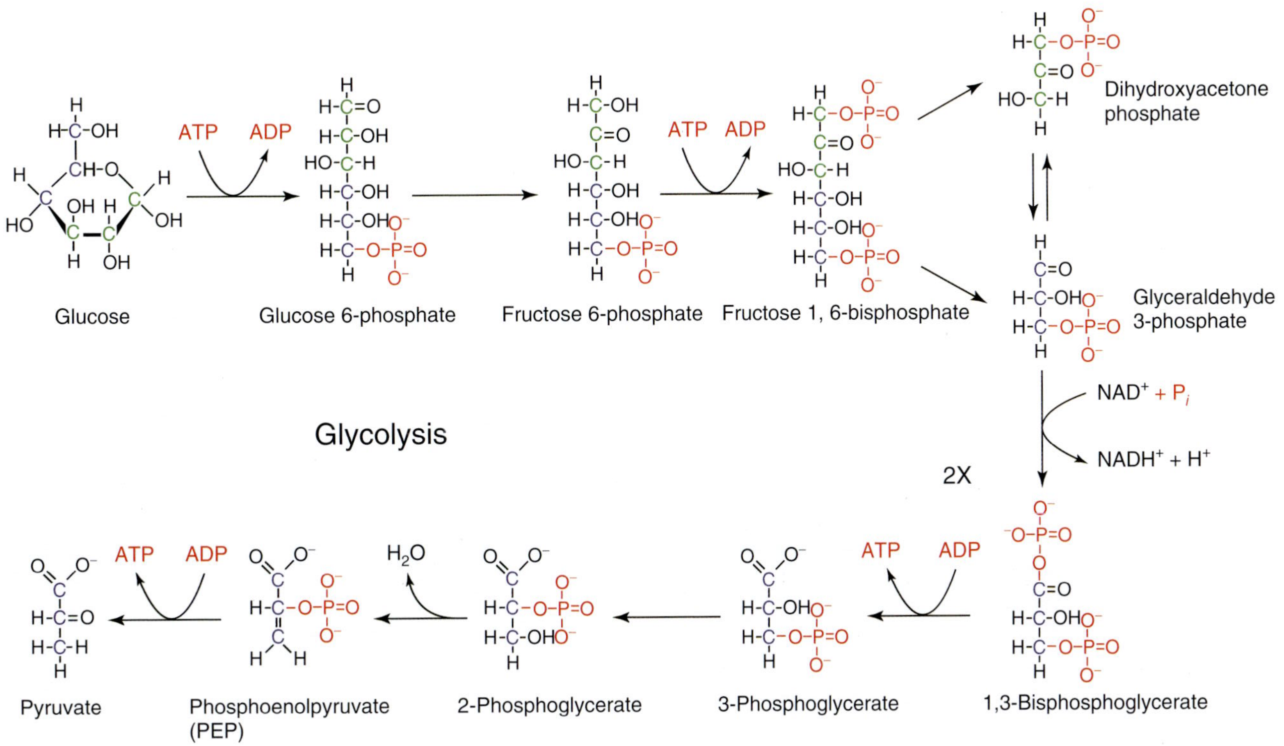

Figure 10.3 The glycolytic pathway. The steps of glycolysis were elucidated by examining the substrate and product sugars, the enzymes (or "ferments") catalyzing each step, and the necessity of cofactors such as inorganic phosphate (ATP) in the process. http://chemwiki.ucdavis.edu/Biological_Chemistry/Metabolism/Glycolysis.

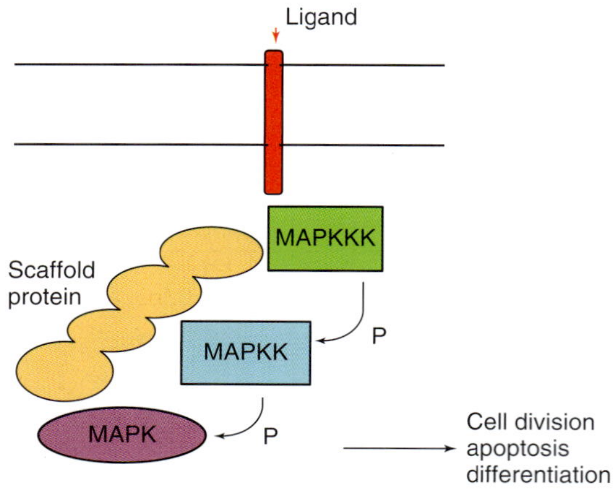

Figure 10.4 Signal transduction via three sequential phosphorylation steps following ligand binding (e.g., to an RTK). The scaffold protein may bind the kinases to promote the phosphorylation cascade.

10.3 Transduction across the Membrane

Cell-surface receptors are the primary mechanism for cross-membrane signaling, and the next section expands on processes described in earlier chapters (e.g., ion channels, G-protein-coupled receptors (GPCRs), and receptor kinases). In addition, we describe non-receptor proteins such as cell aggregation molecules and integrins that are also important for cell–cell signaling.

10.3.1 Ion Channels

The primary material dealing with ion channel signaling is found in Chapter 6; here, however, we provide additional information on the general signaling aspects of one type of ion channel, the voltage-sensing Ca^{2+} channel. Calcium ions have long been known for their importance in neuronal and muscle excitability, but additional influences include modulation of secretion and enzyme function. These latter functions mean that the ion does more than simply depolarize the cell – it alters the function of intracellular metabolic systems [176] by binding to and activating various Ca^{2+} binding proteins (Figure 10.5). Thus, Ca^{2+} is often termed a "second messenger." Increased cellular Ca^{2+} occurs via several types of channels, some gated by voltage changes and some not [340].

10.3.2 G-Protein-Coupled Receptors

As described in Chapter 7, GPCRs are activated by a ligand binding to the receptor's external domain. This action, in turn, results in the activation of an internal G protein via GDP–GTP (guanosine diphosphate/guanosine triphosphate) exchange and dissociation of the G protein subunits. Release and subsequent actions of the subunits pass the signal on to one of a number of signaling cascades

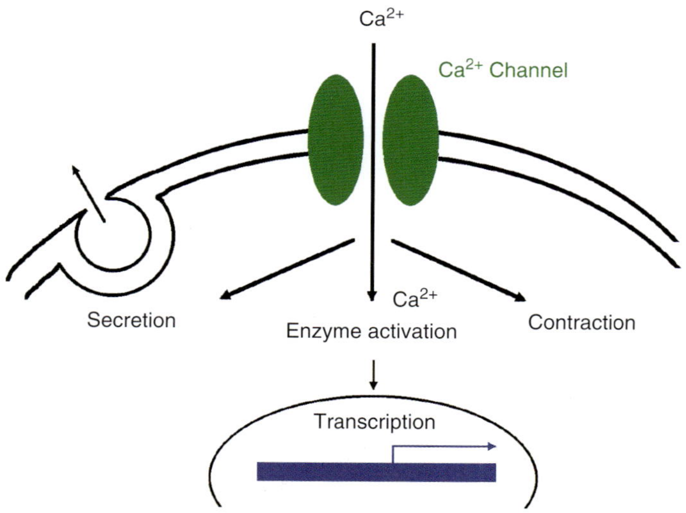

Figure 10.5 The multiple actions of Ca^{2+} channels. Here, the voltage-gated channel is shown as a means of increasing intracellular Ca^{2+}. The rise in Ca^{2+} can activate the diverse processes of muscle contraction, secretion, protein phosphorylation, and gene transcription.

through enzymatic activation and protein–protein interactions.

Box 10.1 Criteria for Invoking G Proteins

G proteins are involved in many important cell signaling pathways, but not in all. Thus, it is important to develop criteria by which one evaluates evidence for G protein involvement before accepting that any particular cellular process involves G proteins [225].

1) GTP must interact with the receptor-ligand-effector system.
2) The response must be elicited by the non-hydrolyzable analogs of GTP (GTPγS or Gpp[NH]p), or by fluoride (F^-) and aluminum (Al^{3+}) ions.
3) Receptor binding of agonist must be reduced in the presence of GTP.
4) Cholera and/or pertussis toxins must have characteristic effects on cellular or purified components of the system.
5) Purification of the receptor, G protein, and adenylyl cyclase components of the transduction system must produce appropriate responses in reconstituted membrane systems.

10.3.2.1 Other G-Protein-Like Transducers – Ras

The Ras proteins (discovered in **Ra**t **S**arcoma cells) are guanine-nucleotide-binding proteins (homologous to but smaller than the alpha subunit of G proteins). They are normally involved in regulating cell growth, probably through transduction of receptor signals of the mitogen-activated pathway. Ras proteins activate kinases that, in turn, activate other cellular enzymes. Ras action is shown in Figure 10.6.

Inactive Ras is bound by GDP; another cellular protein called Sos is activated by binding of the agonist to an RTK. This activation allows exchange of GTP for the GDP, putting Ras into the activated state.

In certain mutated forms, the *Ras* genes become **oncogenes** that direct improper cell growth and rates of division, thus causing cancer. The mutated form of the gene encodes a Ras protein that continuously activates kinases involved in cell division (see below and Chapter 12 for discussion of the cancer consequences of Ras oncogenes).

10.3.2.2 Other G-Protein-Like Transducers – Ran

These proteins (**Ras**-related **N**uclear proteins) assist the movement of other molecules into and out of the nucleus. They operate by transitioning between GDP-bound and GTP-bound states. The GTP-bound state is the "nuclear" state in which they bind proteins to be taken to the cytoplasm. After entering the cytoplasm, GAP (GTPase activating protein) (Chapter 7) hydrolyzes the GTP, which allows dissociation with the exported protein. A protein to be imported binds to the GDP state and the complex migrates to the nucleus, where GTP replaces the GDP and the imported protein dissociates [309] (Figure 10.7).

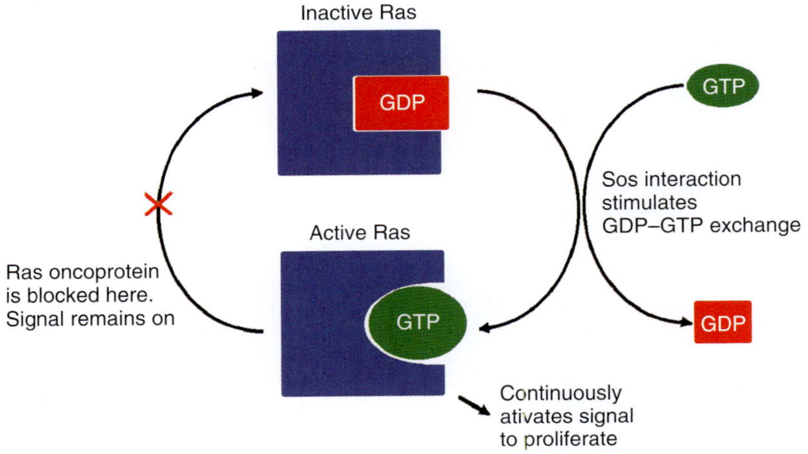

Figure 10.6 Ras activation of serine/threonine kinases. Ras is a G protein but is not associated with a receptor. In the inactive state it binds GDP. Sos causes exchange of GTP for GDP, allowing Ras to activate kinase activity.

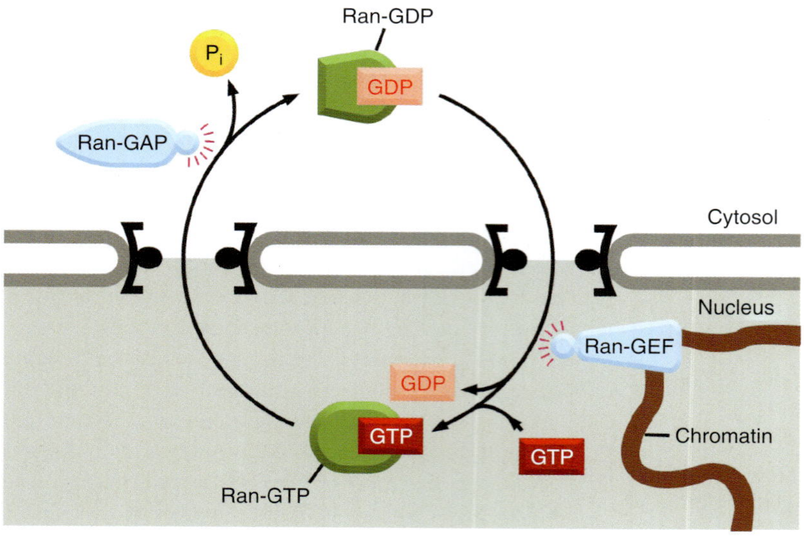

Figure 10.7 The Ran import–export cycle. (i) Ran-GTP-importin is in the cytoplasm, after emerging from the nucleus. (ii) GAP hydrolyzes GTP to GDP and the importin dissociates; "cargo" to be imported can now bind to importin. Ran-GDP-import complex moves to the nucleus through nuclear pores. (iii) In the nucleus, GEF (guanine nucleotide exchange factor) exchanges GTP for GDP, causing dissociation of import-cargo in the nucleus. (iv) The Ran-GTP state returns to the cytoplasm. Figures 12–14 of Ref. [309]. Used with permission.

10.3.3 Cell Aggregation and Development

Cell aggregation is the basis for multicellular life; all organisms that are not single celled maintain a body consisting of cells held together by special cell-adhesion molecules. In addition to these specific cell-adhesion receptors, other receptors are involved in signaling between cells to promote movement, aggregation, and the morphogenetic processes that lead to embryonic development. Thus, the processes of cell communication that are central to the maintenance of multicellular organisms are also those that promote development during the embryonic state [341].

10.3.3.1 Coaggregation in Bacteria

Cell aggregation can also occur in prokaryotes, though not for production of permanent multicellular structures. Bacteria *coaggregate* under certain circumstances, for example, on the surfaces of teeth; this process occurs most easily between species rather than within a species. The receptors for bacterial coaggregation act in ways similar to those in eukaryotes to accomplish adhesion [342] (Figure 10.8).

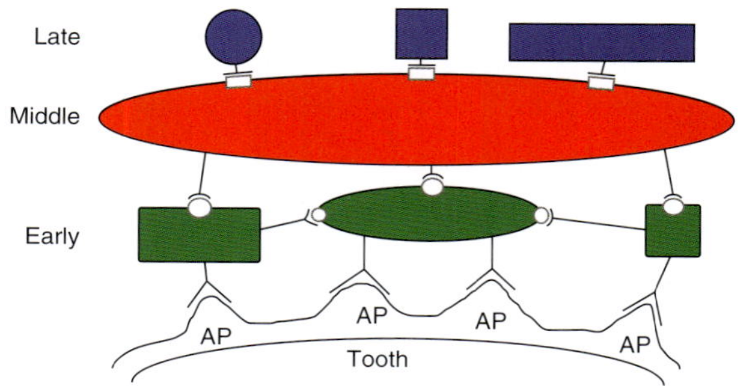

Figure 10.8 Possible temporal sequence of bacterial colonization of teeth in dental plaque. Bacteria coaggregate in multispecies groups, unlike in eukaryotes where best tissue adhesion is between similar or identical cells. AP: acquired pellicle, or body material containing proteins adhering to tooth surface. Early bacteria have receptors for the pellicle; middle bacteria have receptors for early bacteria; and late bacteria have receptors for Middle bacteria. Based on data in Ref. [66].

10.3.3.2 Aggregation in Eukaryotes

General Ideas Cell–cell contact within a eukaryote underlies the formation of tissues. Different cell types in the developing embryo express different types of **cell-adhesion molecules** (CAMs) and cell–cell binding is encouraged between cells bearing identical surface proteins. This preferential binding of identical molecules is called **homophilic binding**, and it differs from the receptor–ligand mechanisms covered so far in this book; most hormone-like ligands are *complementary* to their receptor rather than identical.

Sponges In the early 1900s, Wilson [343] passed living sponges through silk cloth, disrupting the

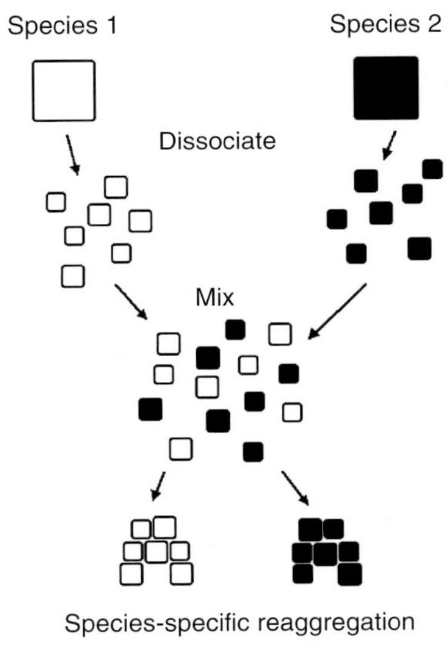

Figure 10.9 Reaggregation of sponge cells after dissociation. Cells of two species separate from each other to form species-specific reaggreagates. Later experiments showed aggregation factors to be glycoproteins.

cellular organization without damaging the cells themselves. After the cells settled to the bottom of the container, they migrated back together to re-form the intact sponge. If the disaggregated cells of sponges from two different species were mixed, the cells sorted themselves out and re-formed the two original sponges (Figure 10.9).

Moscona [344] and Humphreys [345] studied the mechanisms for this species-specific aggregation. They found that sponges could also be disaggregated by placing them into Ca^{2+}- and Mg^{2+}-free sea water. The individual cells could reaggregate if the medium contained the original fluid from the sponges after their previous disaggregation. These investigators proposed that the disaggregating cells released molecules that were responsible for their adhesion, and later determined that the molecules were **glycoproteins**.

10.3.3.3 The Molecules of Cell Adhesion

Two large groups of glycoproteins are involved in cell contact. Both have a single transmembrane domain, an intracellular domain with sites for phosphorylation, and an extracellular domain with adhesive sites for cell contact. They are termed cell-adhesion molecules, or CAMs. One group is the **cadherins**, or calcium-dependent CAMs (Figure 10.10); the other is the calcium-independent CAMs, often also called the immunoglobulin superfamily CAMs because their molecular structure resembles that of the immune system molecules (discussed in Chapter 8).

The Calcium-Dependent CAMs (Cadherins) These proteins are divided into several smaller groups, depending on the mammalian embryonic cell type in which they were first found. Each has the same fundamental structure: a single transmembrane domain, a short intracellular domain making contact with other proteins involved in cell movement, and an extracellular domain consisting of a chain of repeated units (the EC2 units in Figure 10.10). The adhesive recognition site is a short portion of the EC2 molecule (the tripeptide histidine-alanine-valine – HAV) that binds to another HAV on an identical cadherin from another cell. All cadherins, regardless of the tissue from which they come, use the HAV recognition sequence; the tissue specificity originates in the amino acids that surround the HAV site [346].

The Calcium-Independent CAMs These molecules have the same general structure as the cadherins. They also closely resemble the immunoglobulin proteins (Chapter 8) in having several globular extracellular domains that are formed by disulfide linkages, and it is likely that the immune system proteins were originally derived from the CAMs at some time in the past [347].

Role in Development Tissues change the pattern of CAM expression during development [346]. These changes reflect alterations in gene expression that occur with differentiation, as well as the patterns of migration and the associations that migrating cells make with other cell types. For example, migrating embryonic mesenchyme cells do not express CAMs, but once they reach the site where they will remain, they begin to express their CAMs. Muscle cells express N-CAM over most of the surface while nerves are growing out toward them, but once the nerves arrive, the CAMs are expressed only at the neuromuscular junction.

10.4 Complexity in Cross Talk – Roles of PIP3, Akt, and PDK1

While the examples of glycolysis and the MAPK pathways above seem very straightforward, in reality, signaling within a cell involves a number of cross-talking pathways. The carbons in a molecule of glucose do not always move directly down the glycolytic pathway in a linear fashion but rather can

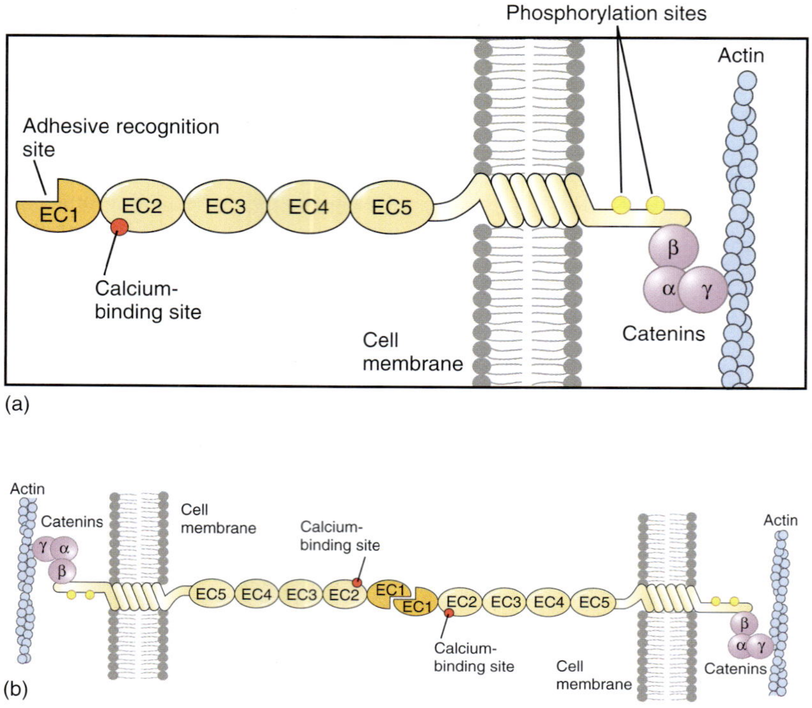

Figure 10.10 Model of cadherin-based cell adhesion. (a) One cell with a cadherin extending away from the cell membrane. (b) Cadherin from one cell binds an identical cadherin on an adjacent cell. Figure 3.5A in Ref. [346]. Used with permission.

be siphoned off to reactions occurring to produce many different cellular components such as amino acids and nucleic acids. In addition, these same molecules can be funneled toward the glycolytic pathway to be oxidized and used for energy needs. Thus, the pathway maps that are often presented in textbooks are merely a starting point to understand the complexity of signaling within a cell.

Figure 10.11 shows an example of this complexity: multiple receptors sit on the membrane of a cell. At the center of the cell sits Akt, also known as protein kinase B (PKB), a serine-threonine kinase that has multiple regulatory functions in protein synthesis, antiapoptotic survival, proliferation, glucose metabolism, and many others. By following the arrows from the membrane inward, it is possible to trace the action of each of the receptors to a positive regulation of Akt via PIP3. This is an excellent example of cross talk between signaling pathways that is described in more detail below.

10.4.1 Signaling Cascades Using PIP3

PIP3 is an abbreviation for the membrane phospholipid **P**hosphatidyl**I**nositol (3,4,5)-**trisP**hosphate produced via the phosphorylation of phosphatidylinositol (4,5)-bisphosphate (PIP_2) by the enzyme PI3 kinase (phosphatidylinositol-3 kinase). Specifically, PI3-kinase, or PI3K), adds a phosphate group to the third carbon position of the inositol ring of PIP_2, resulting in a tri-phosphorylated lipid PIP_3. PIP_3 should not be confused with IP_3 (inositol triphosphate), which is composed of only the inositol head group (and no fatty acid tails) and is an important second messenger in the cell created from the *hydrolysis* of PIP_2. The phosphoinositides are important lipid signaling molecules and can be found localized in the membrane and utilized in a number of signaling pathways (Figure 10.12).

PIP3 is an extremely important lipid signaling molecule, and therefore its formation needs to be tightly regulated. Because of its location as a membrane lipid, PIP3 is activated by receptors and their associated complexes after ligand activation outside the cell.

Once PIP3 is formed, it associates with PDK1, a serine-threonine kinase that is constitutively active at low levels. When complexed with PIP3, PDK1 and PIP3 recruit Akt to the membrane, resulting in the formation of a highly active PIP3/PDK1/Akt signaling complex. This signaling results in the cascade of

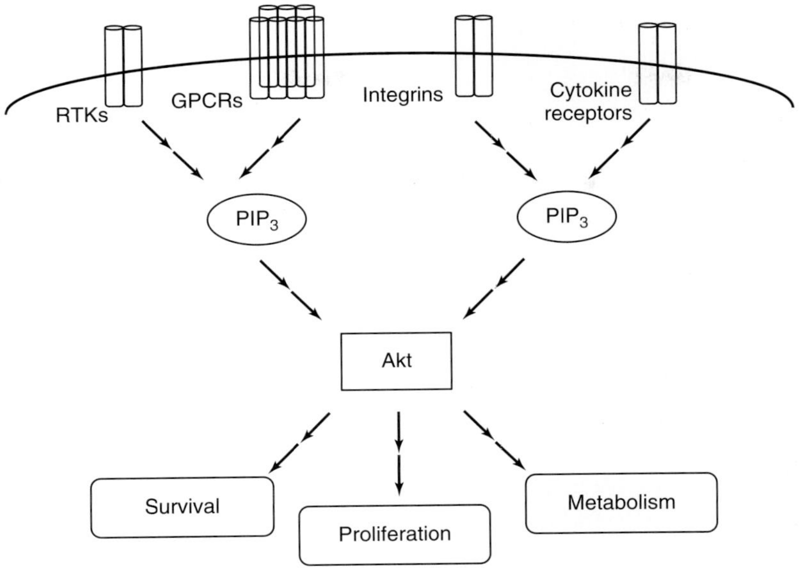

Figure 10.11 Akt is a downstream target of RTKs, GPCR, integrins, cytokine receptors, and other pathways (upper half of figure). AKt continues the propagation of extracellular signals toward a host of downstream targets, playing major roles in cellular synthesis, survival, proliferation, metabolism, neural function, and other functions (lower half of figure).

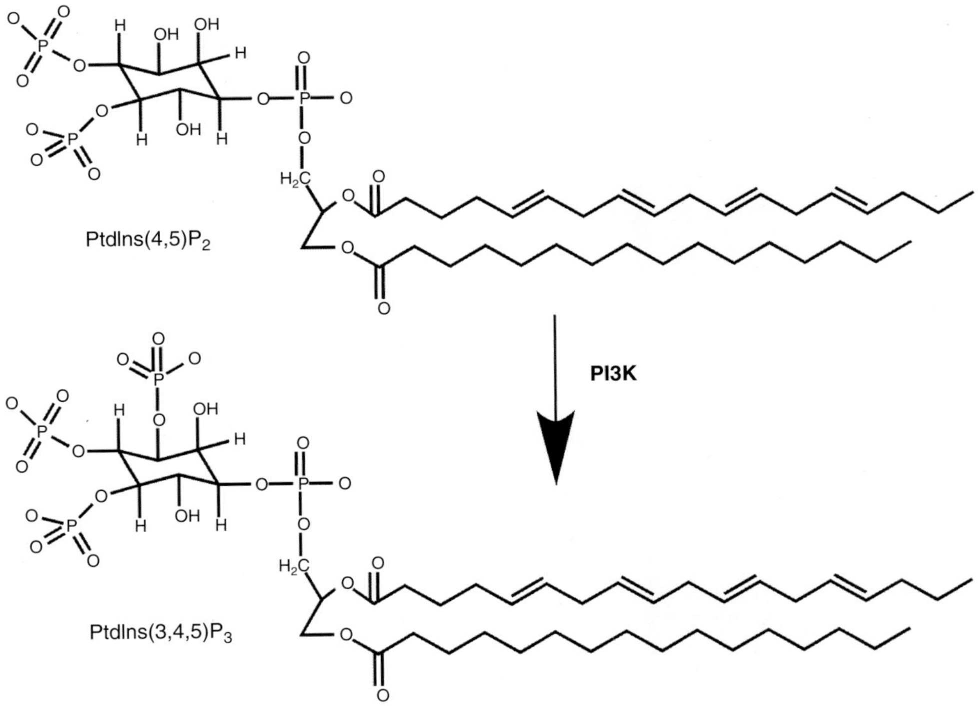

Figure 10.12 Conversion of PIP_2 to PIP_3 via phosphorylation by PI3K.

signals seen in the lower half of Figure 10.11 in which activated Akt phosphorylates a wide array of downstream targets, effectively spreading the signal from the receptors at the surface of the cell to myriad functions in the cell.

10.4.2 Integrins

Integrins are transmembrane proteins that function as dimers in a bidirectional manner. They can signal from the "inside out" and cause the cell to anchor to the extracellular matrix (ECM), and they can also signal from the outside in, often by associating with and activating RTKs. In the outside-in direction, activated integrins recruit a complex of FAK, PI3K, ILK, paxillin, and other proteins that ultimately result in the production of PIP3 via PI3K and activate Akt as described above. The importance of integrin signaling in cancer and other diseases is becoming increasingly apparent, and because of their importance, integrins are targets for several types of cancer therapies [348].

10.4.3 Receptor Tyrosine Kinases

RTKs can be activated by a number of growth factors, as outlined in Chapter 8, resulting in dimerization and autophosphorylation of the receptor tails. This phosphorylation results in a conformational change providing a docking site for PI3K to bind. Localized conversion of PIP2 into PIP3 can occur, resulting in activation of Akt. Alternatively, autophosphorylation of the RTK tails can also result in the recruitment of an adaptor protein, such as Gab1. Gab1 associates with PI3K, leading to an activation of Akt similar to that in the direct RTK-PI3K association. The inclusion of an adaptor protein simply adds another layer of potential regulation to a highly important signaling pathway (Figure 10.13).

10.4.4 Cytokine Receptors and the JAK/STAT Proteins

The JAK (Janus kinase) family consists of intracellular non-receptor kinases that are bound to cytokine receptors that have no kinase domains. They activate STAT (signal transducer and activator of transcription) pathways for transcription control after binding of cytokines to their receptors. JAK thus are the activating factors for the (otherwise inactive) cytokine receptors. When bound to cytokine receptors, JAKs can also recruit additional factors that result in the activation of PI3K and subsequently production of PIP3.

10.4.5 Combined Cellular Signaling – GPCR and RTK Actions

Organs have different rates of metabolic activity depending on their physiological state. The enzyme phosphofructokinase (PFK – Figure 10.3) converts fructose 6-phosphate to fructose 1,6-diphosphate, but the reaction is normally slow. It can be sped up in times of organismal activity by hormonal influences from epinephrine (adrenaline) and insulin. These agonists operate on different receptors, but their effects synergize, as both activate enzymes able to promote the action of PFK [350] (Figure 10.14).

10.5 Role in Cancer

A key component of signaling is the ability to transduce an external signal across a membrane and into a cell in a regulated manner. Disregulation can occur via a malfunction of the signaling cascade itself, resulting from an overabundance of molecules within the cascade, and absence of molecules in the cascade, an error in the activity of a molecule, or a number of other potential problems. As an alternative, the disorder can arise from a problem with the transduction event itself, often arising from **constitutive** activation of the receptor.

10.5.1 Constitutive versus Inducible Activation

Constitutive activation refers to different phenomena in different fields. In the field of ecology, for example, constitutive defenses can be contrasted with inducible defenses. Some plants have the constitutive (always present) defense of thorns or bark while also having the ability to induce the synthesis and secretion of a chemical defense in response to a certain stimulus. While the thorns and bark present a continuous barrier to potential dangers such as herbivores and microbes, an inducible chemical defense that has a foul taste or smell can prevent future attacks from these predators. Inducibility allows a plant to reduce the cost of making these chemicals at high concentration all the time; rather, stimuli induce the production of these chemicals.

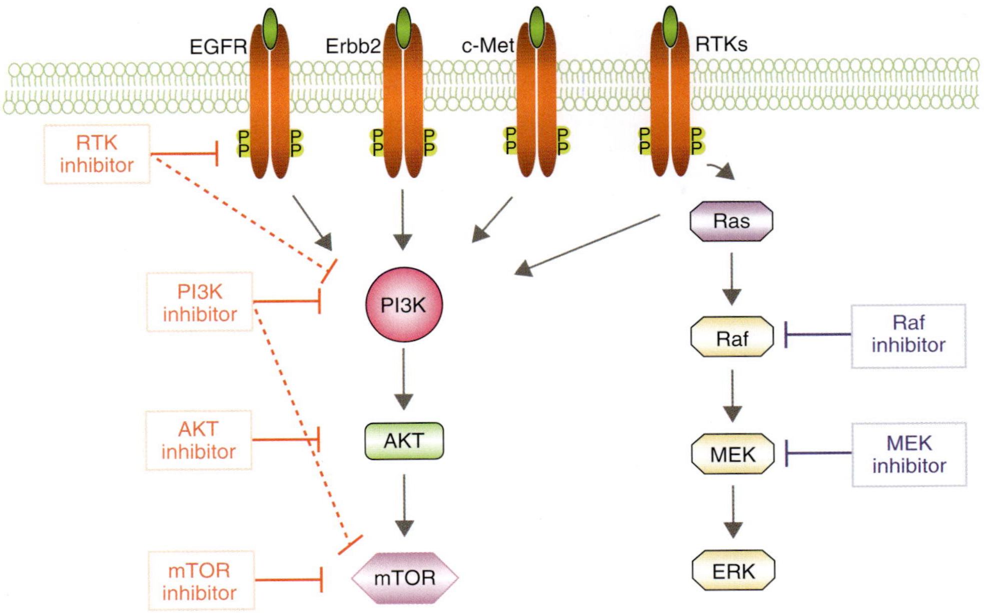

Figure 10.13 Receptor tyrosine kinases and PI3K activation. A number of different RTKs and their ligands, including EGFR, HER2, Met, and other RTKs, activate PI3K, resulting in the downstream activation of Akt via PIP3 (not shown) and ultimately mTOR, a master regulator protein involved in various mechanisms of cell growth. Note the alternative names of the MAPK pathway (RAF, MEK, and ERK) on the right of the figure. Modified from Figure 3 in Ref. [349]. Used with permission.

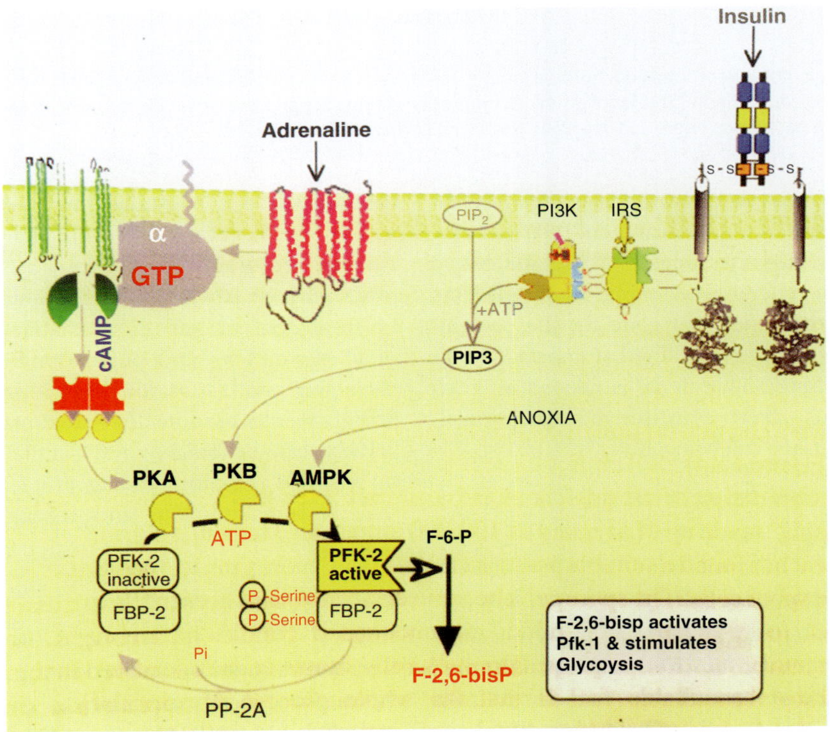

Figure 10.14 Actions of GPCR and RTK on the same processes in cell signaling. In the heart, glycolysis (and therefore energy availability) is enhanced by adrenaline activation of Protein Kinase A (PKA), as well as by insulin activation of Protein Kinase B (PKB). Figure 1.13 in Ref. [350]. Used with permission.

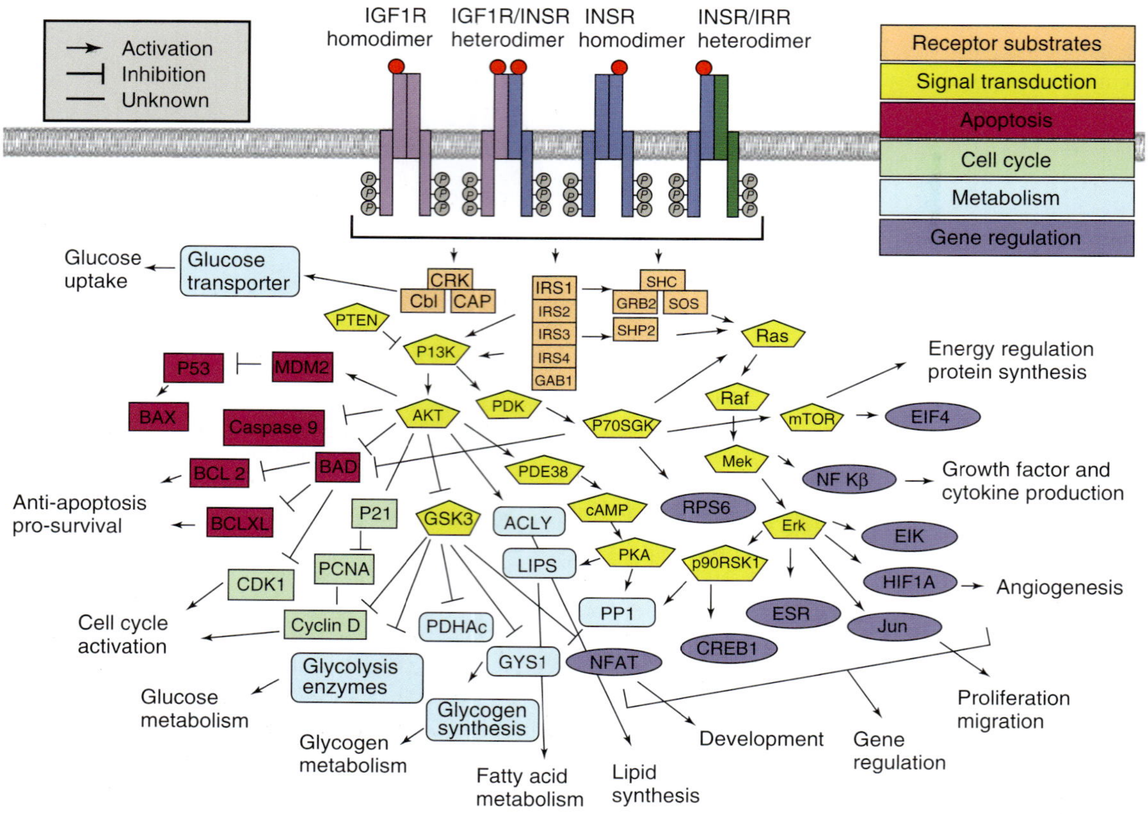

Figure 10.15 The complexity of signaling. In this example, researchers have compiled and integrated known signaling pathways in cancer into one complex figure. This highlights the complexity of signaling and the cross talk occurring in both convergent and divergent signaling pathways. Figure 2 in Ref. [351]. Used with permission.

At the cellular level, constitutive expression has much the same meaning. Researchers refer to some genes as "housekeeping genes." These genes are required for the general functioning of the cells, and many such as the genes that code for the proteins actin and GAPDH are maintained at constant levels at all times. No particular signal is needed for their expression; thus, their constitutive expression is simply the result of a cell being alive.

10.5.2 Cancer Pathways

Many signaling cascades are inducible by the initiation of receptor binding and activation. While some receptors can have constitutive low-level activation, activation of the receptor by a ligand leads to a significant increase of signal transduction across the membrane and into the cell (or from the cytoplasm to the nucleus via a nuclear receptor). Some signaling cascades have constitutive low-level activity, such as the MAPK pathway, because the enzymes of the cascade are scaffolded in close proximity to each other (Figure 10.4). While individual enzymes can be in an "on" or "off" conformation, the likelihood of all enzymes within a population being on or off is extremely small. Thus, low-level constitutive cascade activity results in some downstream activity, and this activity can be quickly and significantly amplified by a single receptor binding event at the receptor that is transduced to and through the cascade (Figure 10.15).

10.6 Signaling Mediated by Gas Molecules

In addition to the ligands discussed in the preceding chapters, small gas molecules also serve as signals; the primary signaling gases with relevance to animals

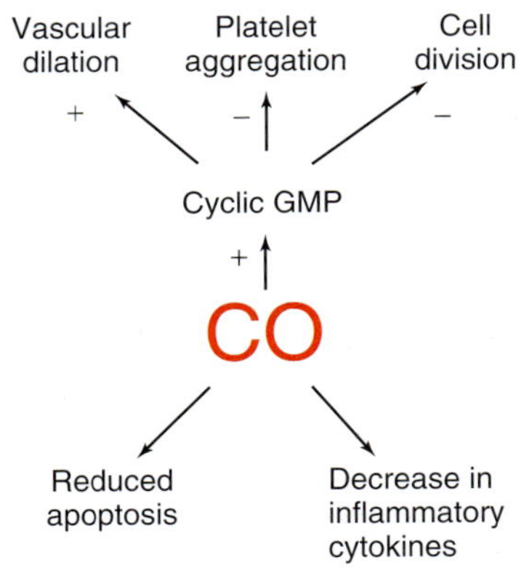

Figure 10.16 Signaling pathways regulated by carbon monoxide (CO). CO activates production of cGMP, which promotes vascular dilation, reduces inflammation, and inhibits cell division. CO also inhibits cell death via apoptosis and decreases the production of inflammatory proteins.

are carbon monoxide (CO), nitric oxide (NO), and hydrogen sulfide (H_2S). Though these gases may come from environmental sources, they also are produced in the body in a manner that suggests their function is adaptive. Research is in the early stages, but so far indicates that they influence enzymes, signaling cascades, vascular reactivity, and inflammatory responses.

10.6.1 Carbon Monoxide

CO is produced in two ways that are relevant to animals, by the incomplete combustion of fuels in the environment and by the enzymatic degradation of hemoglobin in the body [352]. In high concentrations, CO interferes with O_2 binding to hemoglobin in the blood, and is therefore potentially lethal; however, in lower concentrations it regulates inflammation, apoptosis, cell division, and vascular constriction.

The gas binds to and inhibits enzymes in several cellular locations, resulting in diverse physiological effects [353] (Figure 10.16).

10.6.2 Nitric Oxide

NO is produced in the environment by the action of lightning or in internal combustion engines, though its primary availability in the body comes from enzymatic modification of the amino acid arginine by the enzyme nitric oxide synthase (NOS). It is produced in neural cells and vascular endothelial cells and diffuses easily to other cells such as smooth muscle. There, it causes vasodilation by the activation of guanylyl cyclase (GC) and production of cyclic

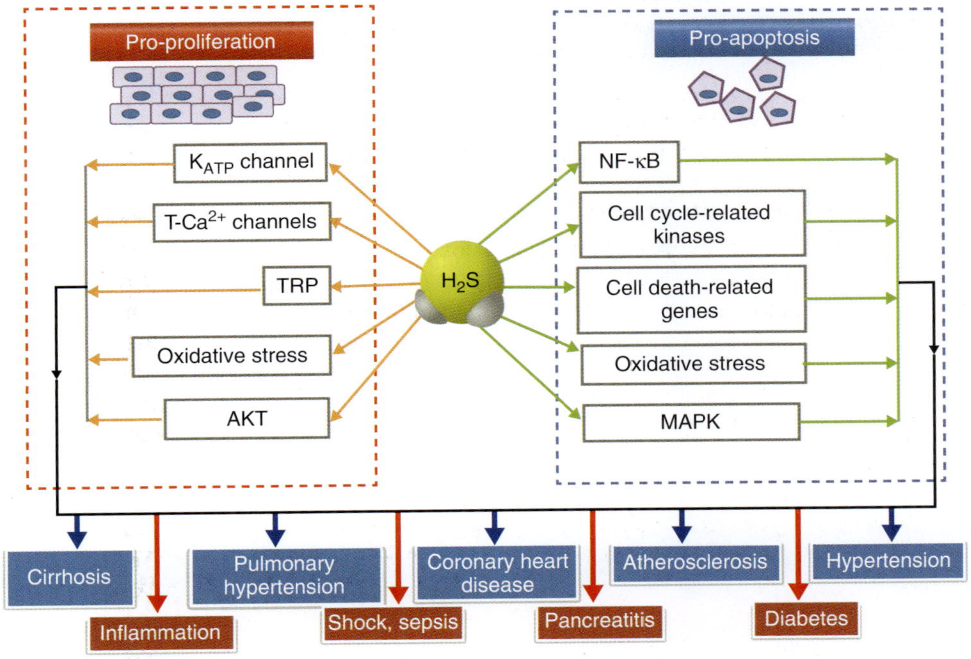

Figure 10.17 Potential involvement and application of the targets of H_2S, including observed dual effects. Figure 2 in Ref. [355]. Used with permission.

GMP, which inhibits catecholamine release from sympathetic neurons and enhances vasodilator actions of $beta_3$ adrenoceptors [354].

10.6.3 Hydrogen Sulfide

H_2S is also produced in the environment – by anaerobic breakdown of organic matter – and biologically by enzymatic action on the amino acid cysteine [355]. In high concentrations it is poisonous as an inhibitor of mitochondrial enzymes, but in low concentrations it activates several cellular pathways involved in vasodilation, cellular proliferation, and apoptosis (Figure 10.17).

H_2S is produced in vascular endothelial cells, and vasodilation is caused by its binding to and opening of potassium channels, leading to cellular hyperpolarization; it also modulates the activity of the MAPK and AKT signaling cascade enzymes discussed elsewhere in this chapter.

11
Cellular Interactions in Development

> The developing embryo (is) a community of cells whose integrity and activities depend on a well-developed system of intercellular communication.
>
> *Bruce Carlson [356]*

> All I really needed to know I learned during **gastrulation**.
>
> *Scott Gilbert [353]*

11.1
Introduction

Reproduction occurs in single-celled and multicellular organisms, but because single-celled organisms do not change morphology after division, their reproduction is not accompanied by a process of embryonic development. True development, in contrast, occurs as a part of the reproduction of diploid multicellular organisms: each of two multicellular adults releases single-celled haploid gametes, and these combine to form single-celled diploid zygotes. Development of the new adults consists of two major processes: first, division of these zygotes into two cells, followed by continued rounds of division that increase the numbers of cells; and second, the **differentiation** of those cells into multiple types. Multicellular organisms thus are not simply larger versions of single-celled organisms: the many cell types of which they consist are all descendants of the first body cell, but development creates a characteristic morphology and function for each cell type within the multicellular adult.

The many types of cells in a multicellular organism occur because of cell-specific gene expression; this pattern of expression leads to different protein populations in each cell type, and thus to the different cell phenotypes of the multicellular organism. The pattern of gene expression is determined by extracellular signal molecules or hormones; these are released by one embryonic cell and bind to surface or nuclear receptors of a nearby cell. Binding of signal molecules and the biochemical transduction steps that follow alter the direction of differentiation of the recipient cell.

The focus of this chapter is on the ways that cell signaling leads to developmental processes that produce adults within each generation. Most of the time, development leads to a population of adults with only minor differences in structure, physiology, and behavior among them. However, the genes that govern development are subject to mutation, and changes may, in particular individuals, alter major developmental pathways such that new structures or processes result. If these novelties are neutral or advantageous, new body types may arise. Thus, this chapter also discusses how alterations in signaling pathways have led to the multiplicity of body types seen in multicellular organisms. The evolutionary history of eukaryotic organisms is, to a large extent, a history of such developmental modifications.

The first two sections discuss the roles played by receptor mechanisms in the origins of multicellularity and symmetry in different eukaryote taxa; the next two illustrate how those same signaling systems operate with others to promote individual developmental stages. The sections are as follows:

- The origins of multicellularity
- The origin of symmetry and axes – the multicellular **body plan**
- Fertilization and early development
- Differentiation of **triplobastic** embryos – organogenesis to apoptosis.

In all, these four levels relate the development of an individual organism to the evolutionary history of the group to which it belongs.

Receptor Biology, First Edition. Michael F. Roberts and Anne E. Kruchten.

11.2 The Origins of Multicellularity

11.2.1 Multicellular Lineages in Prokaryotes

All organisms, whether single celled or multicellular, large or small, must engage in the same life-sustaining functions such as reproduction, food intake, metabolism, and excretion. Most organisms on earth are single celled [357] and each single-celled organism operates as a complete and successful organism. Single-celled organisms are thus not at a fundamental disadvantage as compared to multicellular organisms. Continuing this thought, if any particular single-celled organism alters its reproductive pattern to form a lineage of multicellular descendants, that change is not an "advanced" state compared to the fundamental cellular function that already existed. Rather, the change is either a selective response to an environmental challenge or an accident that happens to be successful within the existing environment.

Several times in evolutionary history, organisms have produced multicellular associations within their original single-celled lineages. These alliances may sometimes be temporary, environmentally related aggregations, as in the biofilms of prokaryotes (Chapter 10), or they may be permanent and obligatory, as in animals. Among eukaryotes, various forms of multicellular associations have evolved in the algae and plants, slime molds, fungi, and animals, though not at all or only rarely in other groups. Thus, the question is not when, or why "*the multicellular state*" arose, as no single state exists that can be defined as multicellular. Further, in each lineage in which it arose, multicellularity may have been in response to a different selective challenge.

11.2.2 Multicellular Lineages in Eukaryotes

The eukaryotes today consist of five major taxa (Figure 11.1).

In two of these, the Archaeplastida (red/green algae and plants) and the Unikonta (amoebas, fungi, and animals), most members are multicellular and use complex developmental pathways to achieve the adult state. The remaining three major eukaryotic clades illustrate a primarily unicellular lifestyle. The Excavata (including the photosynthetic *Euglena*), the Chromalveolata (including the malaria-causing *Plasmodium*), and the Rhizaria (including the planktonic

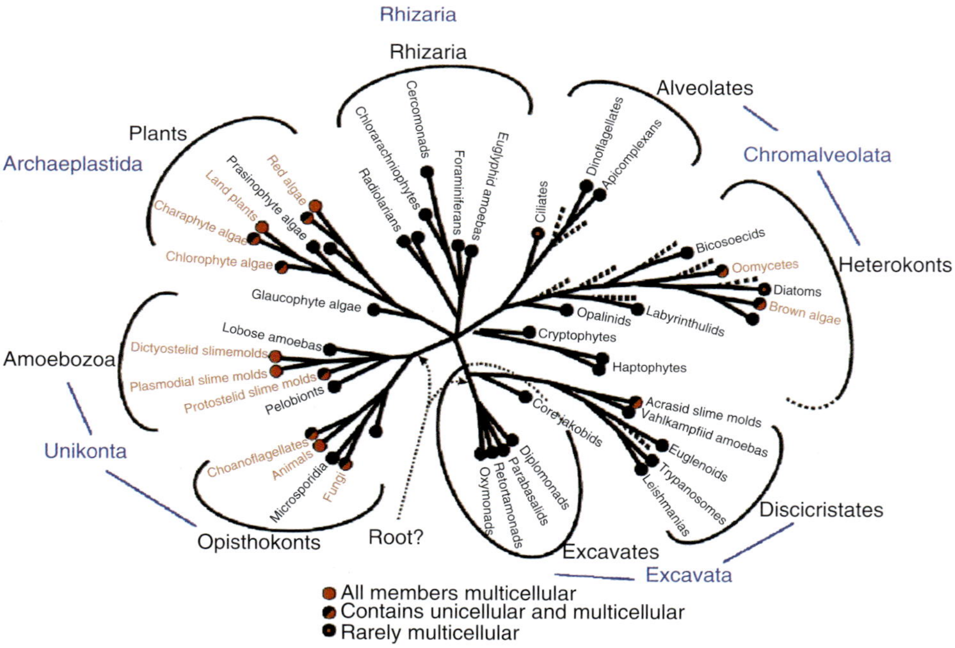

Figure 11.1 Tree of extant domains of life, showing that multicellular forms exist within most taxa. The clustering of multicellular taxa in related groups – sometimes called kingdoms – suggests that a limited number of genes and proteins are responsible for higher order cellular interactions. Modified from Figure 2 of Ref. [358]. Used with permission.

Foraminifera) all belong to major eukaryotic taxa that are unicellular.

All major eukaryote taxa have flagella as part of their locomotive function; all but the Unikonta possess multiple flagella, and these are often arranged in the center of the cell or at the cell's anterior. This flagellar arrangement is important for understanding relationships among the various eukaryotic taxa.

11.2.2.1 Chromalveolates – Generally Unicellular but with One Multicellular Clade

This group contains the multicellular brown algae, also known as kelp. It is instructive to mention the brown algae here because they develop from a zygote in a manner similar to that of plants, and some of the same hormones, discovered initially as plant hormones, are involved in defining algal developmental stages [359]. Thus, they exhibit some of the features considered to be important predictors of the cellular interactions required by multicellular organisms [360]. Although the kelp are called brown "algae," they are not part of the large taxon including the true plants and red or green algae.

11.2.2.2 Archaeplastida – Algae and Plants

Many members of the Archaeplastida share with the animals the appearance of being "multicellular." The plants with which most people are familiar are complex multicellular organisms that have many cell types, **tissue** types, and **organ** types, as well as developmental pathways that use complex receptor-hormone-signaling systems to achieve the adult state. The evolution of multicellularity in plants can be followed by looking at its hypothesized stages as seen in one group of modern green algae, the order Volvocales.

The Volvocales comprise organisms from single-celled *Chlamydomonas* to 50 000-celled colonial *Volvox*. It is not certain how the colonial state evolved in these algae, but because they are aquatic, they most likely evolved multicellular complexes by the failure of mitotic products to separate [357]. Thus, it is useful to compare the single-celled *Chlamydomonas* with the colonial *Volvox* to identify the genetic basis for colonial associations that might give clues to the process of multicellularity.

Volvox and *Chlamydomonas* have roughly the same number of genes (14 500), so the multicellular colonial state is brought about not by the invention of new genes but by modifications of **lineage-specific proteins** that are involved in altering two aspects of morphology and cell division. For example, *Volvox* has more cell cycle proteins and proteins for the construction of the extracellular matrix (ECM) than does *Chlamydomonas*; both of these protein sets are important for maintenance of its colonial state. Indeed, the ECM contains a pheromone-based signaling system responsible for maintaining the integrity of the multicellular state and directing the transition from asexual to sexual reproduction [361].

11.2.2.3 Amoebozoans, Fungi, Choanoflagellates, and Animals

The remaining eukaryotes are unusual in having a single flagellum, and are thus called the Unikonts. Within this group, animals and fungi both have a rear flagellum and are called Opisthokonts. The choanoflagellates evolved from early opisthokonts, and one choanoflagellate clade evolved into the animals [360] (metazoans – Figure 11.2).

The evidence includes (i) the near identity of the single-celled free-living choanoflagellates with a sponge cell type called a choanocyte; (ii) the existence in choanoflagellates of several proteins homologous with those in animals, including the receptor tyrosine kinase (RTK) receptor group, the ECM proteins, and a large number of transcription factors [362]. In addition, the animals have evolved proteins for bulk transport, large numbers of transcription factors for gene expression, and regulatory RNA species [360], all of which control transcriptional and translational processes in development [363]. The abundance of transcription factors and regulatory RNAs in genomes are proportional to the degree of multicellularity achieved.

The choanoflagellate level of organization is unicellular or colonial, while the first metazoan was a multicellular organism with embryonic development and a **gastrulation** process that produced the differentiated adult state [362].

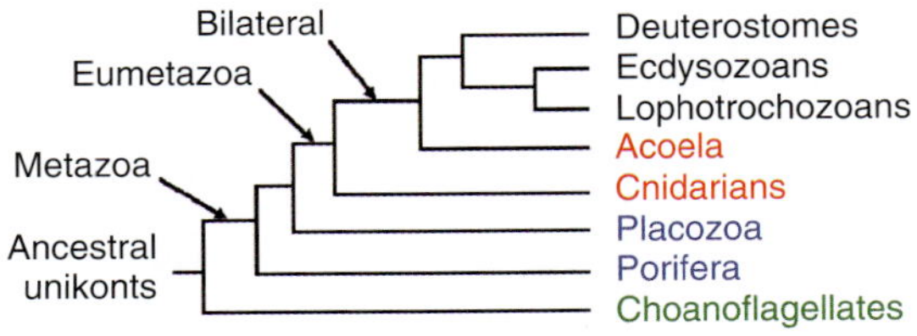

Figure 11.2 Generalized phylogeny of the metazoa and their closest relatives, the choanoflagellates. Colors refer to whether each taxon contains proteins, genes, and RNAs that contribute to the complex multicellular habit. Green: adhesion receptors, cell signaling, and bHLH transcription factors; blue: the former plus micro RNAs that regulate transcription; red: the former plus gap junctions; and black: all the above.

Since the early 1900s, sponges have been regarded as a model for the earliest animals – they have either radial or no symmetry; few cell types; few, if any, tissues; and no organs, no true gut, circulation, or nervous systems. This apparent lack of features present in other animals has led to their occupying for many decades an uncertain evolutionary position at the base of the animal phylogeny. Even so, Wilson's [343] early studies showed they have species-specific cell–cell recognition and adhesion molecules that indicate the existence of proteins used by metazoans to hold cell types together to form tissues and organs. The adhesion proteins of sponges are homologous to the adhesion molecules of mammals; in addition, the extracellular proteins collagen and fibronectin are also homologous to mammalian proteins. Taken together, these similarities suggest that sponges, even with a non-symmetrical structure, could be similar to the ancestors of today's radially or bilaterally symmetrical animals.

The next section outlines the receptor-based mechanisms present in animals that led to the evolution of several important aspects of the animal body plan.

11.3 The Origin of Symmetry and Axes

11.3.1 The Multicellular Body Plan

The concept of a body plan is central to thinking about development and evolution. Each of the various eukaryotic kingdoms (e.g., Archaeplastida, Unikonta, and the three unicellular protist kingdoms) has a unique body plan defined by the actions of signaling systems interacting with genetic transcription pathways. The correspondence between individual development and population-level evolution is an important theoretical program of study in modern biology, and a major part of the research in this area involves the signaling molecules and gene regulatory molecules of each organism's developmental program.

The next section of this chapter focuses on the metazoans and on the central role of signaling in animal evolution: the same signaling systems and the same proteins determine the axes of (asymmetric) sponges, (radially or bilaterally symmetric) hydras, and (bilaterally symmetric) worms. As this section develops, the various body plans are arranged by phylum-specific organization of a single signaling system.

11.3.2 The Porifera – Asymmetric with a Single Cell Layer

Sponges have traditionally been thought to consist of a few cell types with no tissue organization. They do lack many cell types that are characteristic of other animals (e.g., true neurons and distinct endoderm, mesoderm, and endoderm) but for two reasons the classical view is inadequate. First, sponge *adults* have genes that are responsible for several developmental pathways characteristic of **eumetazoan** animals (Figure 11.2). The most important of these are the receptors and signaling peptides of the Wnt and TGFβ (transforming growth factor) pathways [364, 365] (Figure 11.3; Chapters 7 and 8), as the interaction between these signaling proteins is responsible for development of symmetry and anterior–posterior patterning in animals.

In addition, sponges express the Notch–Delta pathway (Chapter 8) that is used by eumetazoans to construct the nervous system. Though they do not have neurons as such, sponges do have voltage-gated Ca^{2+} channels and some sponges are able to generate action potentials in certain cells [366]. These action potentials are used to control feeding, and though they are slow, they are adequate to affect water movement through the sponge. Further, sponges use the same ion channel receptors (Chapter 6) and transmitters for signaling as do other animals: glutamate activates receptors to cause Ca^{2+}-induced contraction of water flow channels, and gamma-amino butyric acid (GABA) activates receptors to inhibit the effects of glutamate receptors. Actions of both receptor types are opposed by specific glutamate and GABA receptor antagonists [367]. Evidence is

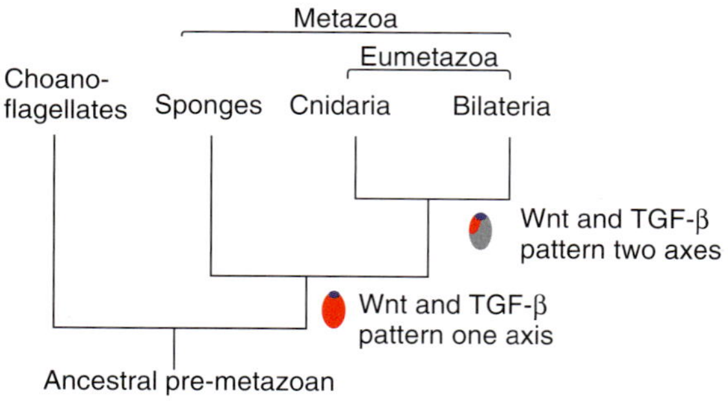

Figure 11.3 Phylogeny of animals and their chonoflagellate ancestors, showing origin of Wnt (blue) and TGFβ (red) signaling systems. Wnt and TGF-β originated after the split of sponges from choanoflagellates. In sponges, Wnt is expressed posteriorly and TGF-β is expressed anteriorly. In eumetazoa, Wnt is still posterior, but TGFβ establishes a new axis perpendicular to Wnt. LCA: last common ancestor.

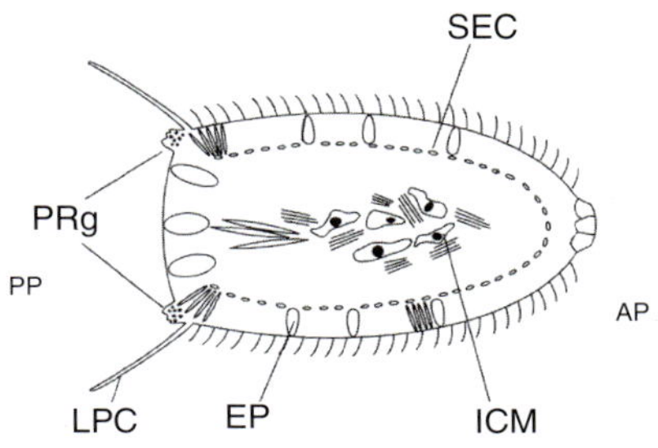

Figure 11.4 Larva of the demosponge *Reniera*, showing clear anterior/posterior polarity. AP: anterior pole; PP: posterior pole; PRg: pigmented ring (photosensitive); EP: epithelial layer; SEC: subepithelial cell layer; and ICM: inner cell mass, containing spicules and collagen. The surface is covered with short cilia and a ring of longer cilia is associated with the pigmented ring at the posterior pole. Modified from Figure 11 of Ref. [365]. Used with permission.

thus strong for typical animal signaling systems within sponges.

Second, sponge *larvae* show a number of cell types and tissue-like layers lost in the adults: the larvae of demosponges have bilateral symmetry with a distinct anterior–posterior axis [364] and contain several different cell types organized into three layers – an outer epithelium, a middle subepithelial layer, and an inner cell mass (ICM) (Figure 11.4).

The epithelial layer contains three cell types devoted to locating a settling spot and coordinating metamorphosis to the adult. Thus, larvae resemble **triploblastic** animals [365], suggesting that adults appear to have lost cell types and perhaps even tissues originally possessed by larvae.

The origin of the anterior–posterior axis depends on the relative extents of Wnt and TGFβ signaling: the transcription factor β-catenin is activated by Wnt binding its receptor; this activation synthesizes proteins that cause the formation of posterior sponge structures. If Wnt signaling is inhibited, β-catenin is not activated and anterior proteins are produced instead [364]. In contrast, TGFβ signaling occurs in a direction opposite to Wnt signaling, with greatest expression of TGFβ in the anterior. This polarity of signaling sets up anterior/posterior patterning in the larva. Settling of the larva is associated with loss of expression of the genes for these signaling molecules and receptors; thus, both anterior and posterior polarity are lost in the adult (Figure 11.5).

Thus, adult sponges have a secondary, not primary, **asymmetry**. The fact that sponges have symmetry, polarity, and multiple cell layers as larvae means that their absence in adults is not a true taxonomic

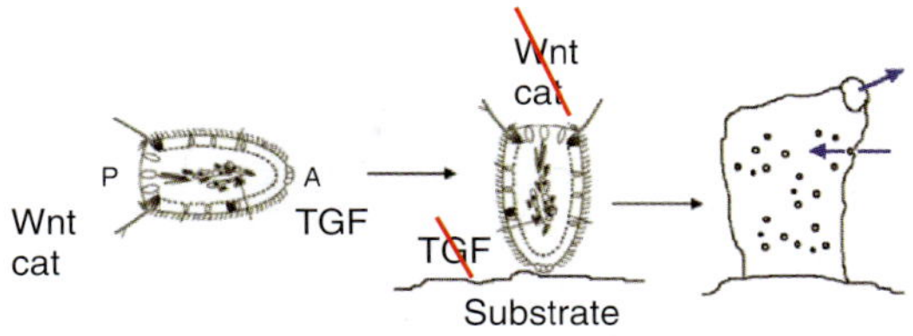

Figure 11.5 Loss of anterior–posterior patterning and bilateral symmetry in sponge at the time of larval settlement. In larva (left), TGFβ regulates anterior structures and Wnt regulates posterior structures through β-catenin (cat). These genes are turned off at the end of the larval stage, leading to the asymmetrical sessile adult (right). Blue arrows show direction of water flow, in through **ostia** and out through the **osculum**. Larva from Ref. [365]. Used with permission.

character; the boundary between the traditional taxa Porifera and Eumetazoa is therefore somewhat shakier than it has been thought to be [368]. The next step in the evolution of Eumetazoa was likely started in an ancient sea by a signaling error in a sponge larva – instead of losing Wnt and TGFβ signaling and settling down to a sessile existence, it continued as a free-swimming active organism, and Wnt and TGFβ signaling directed the formation of new cell types.

11.3.3
Cnidaria – Radial Symmetry, Two Cell Layers, Tissues

The Cnidaria are considered the stem group of the Eumetazoa; their radial symmetry, **diploblastic** cell structure, and true tissues indicate their position intermediate between Porifera and Bilateria (Figure 11.3). These characteristics develop under influence of the same two signaling systems as exist in Porifera, but in a perpendicular fashion: Wnt signaling establishes anterior/posterior axes while TGFβ signaling establishes a perpendicular axis that appears to be the predecessor of the left/right axis [369]. The expression of Wnt is highest in the polyp and in particular on the opening to the gastrovascular cavity, which is the combined mouth and anus in Cnidaria. In bilateral animals, Wnt is expressed in the **blastopore**, which becomes either the mouth (in **protostomes**) or the anus (in **deuterostomes**).

It is now time to consider the third axis (dorsal/ventral), which originated along with the third cell layer (mesoderm) as bone morphogenetic protein (BMP) signaling (next section) and converted the Cnidaria to the Bilateria. The signals that define the opening to the gastrovascular cavity in diploblastic Cnidarians are homologs of those that define the mesoderm in triploblastic eumetazoans. The Ctenophora are a group of animals that are superficially similar to the Cnidaria, but with an important difference, the presence of a layer of mesoderm that forms in the mouth area during development. This small region of mesodermal cells suggests that the signaling process that would later make the full layer of mesoderm had already begun by the time of the earliest radially symmetric animals.

11.3.4 Mesoderm

Bilateral symmetry and mesoderm are defining features of Bilateria [370]. Researchers have discovered several secreted proteins and their receptors that act to pattern dorsal tissues and differentiate mesoderm from ectoderm. For example, the protein discovered in vertebrates and called bone morphogenetic protein is secreted from many cell types in the embryo. It binds its receptor, a membrane serine-threonine kinase (STK) (Chapter 8), which activates transcription factors to synthesize proteins that form ventral mesodermal structures in vertebrates (e.g., bone rather than neural structures, ventral heart mesoderm rather than dorsal kidney mesoderm).

Where BMP is secreted, the cells in the vicinity become "ventralized." Because BMP is secreted throughout the embryo, it may seem surprising that parts of the embryo can become "dorsalized." The mechanism for dorsalization turns out to be a simple ligand-level antagonism: in ventral areas of the embryo, BMP is secreted alone; in dorsal regions, peptides such as chordin (BMP-binding proteins) are secreted; these bind to and alter the shape of BMP so it can no longer bind its receptor; the cells thus receive no ventralizing BMP signal, and become dorsal [371]. Chordin is thus a signaling molecule that dorsalizes the embryo.

Though Cnidaria have two cell layers and radial symmetry, BMP signaling may set the stage for the invention of a new bilateral body plan. BMP is found in Cnidaria and Bilateria but not in sponges; it seems to have altered its expression pattern to set up the third body axis perpendicular to the axes specified by Wnt and TGFβ; this third axis later became the dorsal/ventral axis and created the bilaterally symmetrical body [372].

The remaining sections of the chapter show how the signaling systems already present in the earliest animals (mono- or diploblastic and organless) are also important parts of the pathways of **morphogenesis** in animals with three tissue layers and organs.

11.4 Fertilization and Organization of the Multicellular Body Plan

11.4.1 Sperm–Egg Recognition

Fertilization is most easily studied in organisms in which the egg and sperm meet outside the body. Echinoderms and amphibians have thus been the mainstay of fertilization and developmental studies for over a century. Mammalian fertilization occurs within the body, but since the advent of *in vitro* fertilization techniques and intrauterine photography, mice and humans have also been added to the group of species whose development can be easily followed.

11.4.1.1 Sea Urchin Fertilization

Sperm Activation and Chemotaxis Eggs of several marine species have been shown to release peptides that activate or attract sperm. One urchin species releases speract, a 10-amino acid peptide that binds to a receptor on sperm tail membranes and increases sperm swimming activity. Another species releases resact, a similar peptide that acts on sperm cell membrane receptors to attract sperm of that species [373].

The Egg Surface The urchin egg is surrounded by two external protective surfaces, a thick jelly coat and the membrane-like **vitelline layer** (Figure 11.6).

Both of these layers are glycoprotein compounds that protect the egg. Both also contain receptor proteins that act as specific binding sites for sperm [374]. Specificity of receptors on the jelly layer allows only sperm from the same species of urchin to approach the egg surface.

Sperm Surface Proteins The initial binding of sperm to jelly coat receptors is achieved by the receptor for egg jelly (REJ) [375], a glycoprotein in part resembling epidermal growth factor (EGF) (Chapter 8). In addition to the EGF region, the receptor contains two other modules as well, a lectin (protein that binds carbohydrates) and a unit that activates ion channels. The binding of the REJ to the egg jelly opens a sperm ion channel that initiates the sperm **acrosomal reaction**, in which the membrane of the sperm tip (the **acrosome**) dissolves and releases enzymes that allow it to burrow through the jelly toward the egg surface (Figure 11.7).

Another specific binding reaction occurs as the sperm reaches the vitelline layer. In the sea urchin, the specific sperm protein for binding the egg plasma

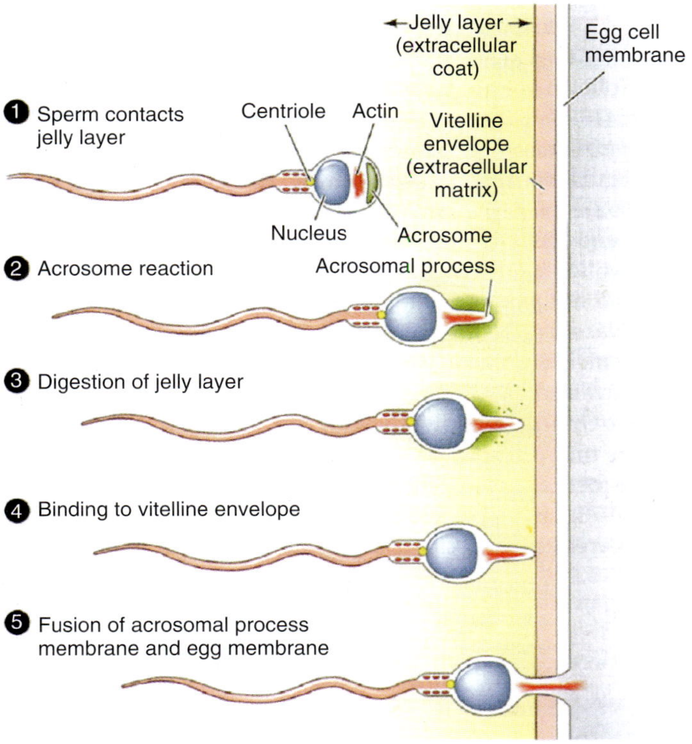

Figure 11.6 Sea urchin egg during fertilization. Cell membrane at right, covered by jelly layer (yellow). Sperm penetrates the jelly layer, then binds to egg plasma membrane receptor. Modified from Figure 4.8 of Ref. [346]. Used with permission.

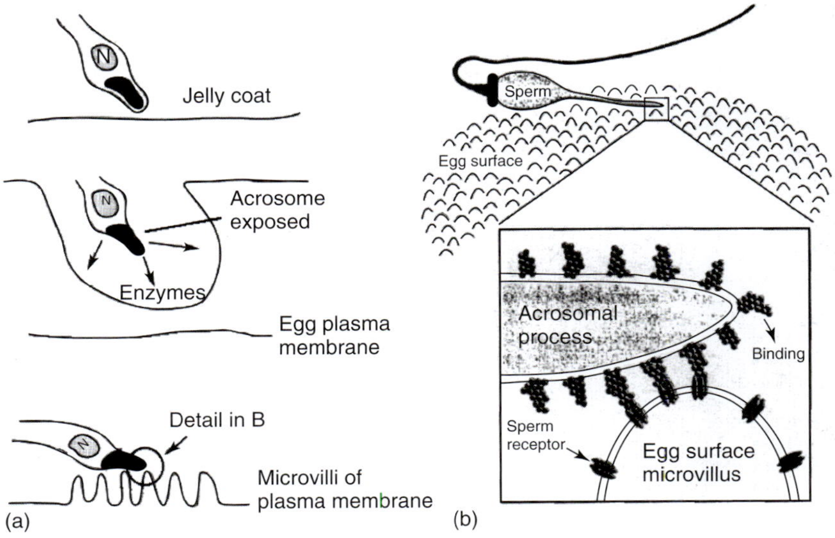

Figure 11.7 (a) Sperm binds to REJ of jelly coat. This ruptures sperm cell membrane and releases acrosomal (a) enzymes. Released enzymes dissolve jelly. N = sperm nucleus. (b) (Detail of part of a): Sperm protein (bindin) anchors sperm to receptors on egg cell membrane. Modified from Figure 3 of Ref. [376]. Used with permission.

membrane is called **bindin**, and is the major protein component of the acrosomal vesicle. Bindin is a peptide of about 250 amino acids, of which 18 form a core binding motif that is perfectly conserved across sea urchin taxa [377]. It seems to have two functions in fertilization: to promote sperm-egg attachment and to allow the fusion of the gamete cell membranes.

Though a soluble protein within the acrosome, bindin associates into multimolecular aggregates and attaches both to the sperm and to the sperm receptor on the egg surface, holding them together until the acrosomal membrane of the sperm can fuse with the egg cell plasma membrane (Figure 11.7b). The interaction of bindin and its egg cell receptor is species specific. In low concentrations, bindin from one species will agglutinate eggs of the same species (homologous), whereas much higher bindin concentrations are required to agglutinate eggs from another species (heterologous) of urchin.

Egg Surface Receptor for Sperm The receptor for bindin in the sea urchin egg is a glycoprotein and in its specific binding domain shows relationships to other cell surface proteins of the growth receptor class, specifically the major histocompatibility antigens [378] major histocompatibility (MHC class I – see Chapter 8).

Early Events After binding of sperm to the egg surface, the first events to occur in the egg are membrane depolarization, fusion of the sperm and egg cytoplasms, rise in cytoplasmic Ca^{2+}, increase in egg cell pH, and extrusion of cortical granules that reside in the oocyte just under the cell membrane. Release of these granules into the space between the egg cell cytoplasm and the vitelline membrane causes a **fertilization membrane** to form around the newly fertilized egg. The initial depolarization and raising of the fertilization membrane are considered to be the blocks to polyspermy, whereby the egg prevents more than one sperm from entering [346].

Late Events Upon binding of sperm, the receptor is rapidly phosphorylated by a membrane-bound tyrosine kinase [379]. In addition, inositol (1,4,5) trisphosphate (IP_3) levels increase in the fertilized egg. These events may be partly responsible for the increase in cytoplasmic Ca^{2+} concentration that activates the zygote to increase its metabolism. These two reactions (phosphorylation and IP_3 production) imply the existence of G-protein-coupled receptors (Chapter 7) and receptor kinases (Chapter 8) in the egg cell membrane. The pathways for zygote activation thus may operate as shown in Figure 11.8.

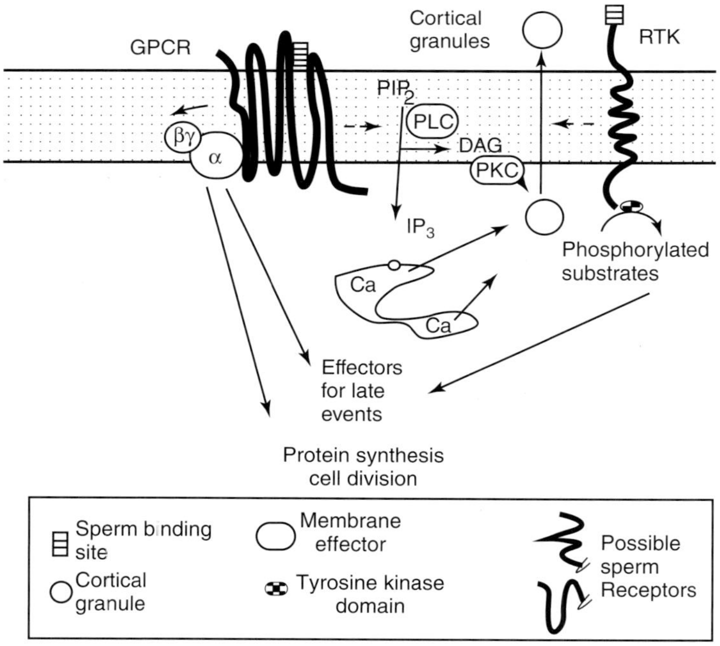

Figure 11.8 Activation of sea urchin egg by sperm binding. Evidence suggests participation of G-protein-coupled receptor mechanisms and receptor kinase mechanisms. The result is an increase in Ca^{2+} within the cell, leading to release of cortical granules and an increase in cell metabolism.

Receptor Types Involved: G-Protein-Coupled Receptor Action Sperm binding causes dissociation of the $\beta-\gamma$ subunit from the α subunit of a membrane-bound G protein. The α subunit can activate adenylyl cyclase and phospholipase C (PLC). Activation of PLC causes the splitting of a membrane phospholipid (phosphatidylinositol 4,5-bisphosphate – PIP_2) into two intracellular second messengers – diacyl glycerol (DAG) and IP_3. DAG activates Ca^{2+} channels and IP_3 acts as an intracellular messenger to cause Ca^{2+} release from the mitochondria. The two systems thus increase intracellular Ca^{2+} concentration. The increase of Ca^{2+} is responsible for the extrusion of cortical granules that raise the fertilization membrane so that additional sperm cannot bind.

RTK action: Sperm binding to its receptor produces phosphorylation of tyrosine residues on the intracellular domain of the receptor; certain phosphorylated intermediates may activate membrane-bound PLC, causing release of cortical granules.

Additional phosphorylated intermediates may alter gene transcription or cell division. The interactions between sperm and its possible receptor are not sufficiently well known to give a complete explanation of zygote metabolism and cell division in the first stages of development. It is also possible that the binding event activates other membrane effector proteins, or that the sperm introduces a substance that activates the egg, or that a combination of processes initiates development [373].

11.4.1.2 Mammalian Fertilization

Egg Surface Proteins The mammalian oocyte is surrounded by a glycoprotein matrix called the **zona pellucida**, equivalent to the vitelline layer of urchins. The binding of sperm to the zona involves the interaction of sperm proteins with a protein component (**ZP_3**) of the zona. ZP_3 is a globular protein of about 400 amino acids containing several disulfide linkages and several sites for carbohydrate attachment. It also has a "hinge" region, similar to that in the RTKs [380], suggesting that it is a member of that receptor group (Chapter 8). Binding of sperm to the zona results in the acrosomal reaction – release of sperm enzymes that dissolve the zona and promote binding of sperm to oocyte plasma membrane receptors (Figure 11.9).

Sperm Surface Proteins and the Acrosomal Reaction One of the sperm surface molecules is a 95 kDa transmembrane protein that has an extracellular ZP_3-binding site and an intracellular tyrosine kinase domain. The events that follow binding of sperm to ZP_3 are increased tyrosine kinase activity and aggregation of the receptors. The acrosome reaction (and fertilization) will not occur if tyrosine kinase activity is inhibited [346].

Early Events in Metabolic Activation of the Mammalian Egg The first detectable event is the increase in Ca^{2+} ion concentration within the egg. This ionic event is similar to that seen in urchins (Figure 11.8), and its molecular basis may be similar, with an increase in Ca^{2+} concentration triggered by G-protein-coupled

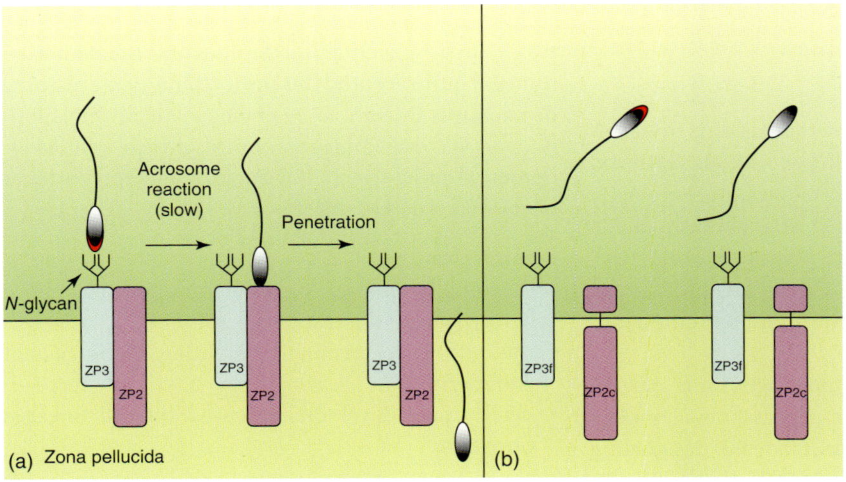

Figure 11.9 Sperm penetration of mouse egg. (a) Pre-acrosomal reacted sperm have red heads. ZP3 and ZP2 are closely associated; this keeps ZP3 in state to bind sperm and cause acrosome reaction (1–2). Sperm can now bind ZP2 and enter egg (3). (b) After sperm initial penetration, an enzyme causes separation of ZP2 and ZP3 and also alters both so they cannot bind sperm. This action prevents polyspermy (4). Figure 2 of Ref. [381]. Used with permission.

receptors and/or RTKs. The detailed events following initial sperm binding to ZP_3 are uncertain, but a likely hypothesis proposes the following (Figure 11.9): In the unfertilized egg zona, ZP_2 and ZP_3 egg proteins are closely associated. Sperm binds to ZP_3, which causes the acrosome reaction and preparation of sperm for the final binding step. Acrosome-reacted sperm binds to ZP_2; the sperm can then penetrate the zona and reach the egg. At this point is released an enzyme that alters ZP_2 (to ZP_{2c}) and ZP_3 (to ZP_{3f}) so neither can associate, nor can they bind sperm. This alteration thus prevents additional fertilization [356].

Late Events Both G-protein-coupled receptors and receptor kinases have been implicated in mammalian egg activation [382], although the specific metabolic and cell division mechanisms that each receptor might initiate are unknown. Following fertilization, the zygote produces new proteins and activates genes needed for increased metabolism and cell division. Cell division and differentiation of the many cell types of the adult follow the steps of fertilization.

The most recent innovation in the evolution of development was the origin of true organs, found only in triploblasts. The remainder of the chapter focuses on organogenesis, the signaling and genetic control mechanisms of development in these animals.

11.5 Differentiation of Triploblastic Embryos – Organogenesis

11.5.1 Introduction

Most Bilateria have similar overall patterns of development: each individual begins as a single fertilized cell; a period of cell **cleavage** follows, during which cells proliferate and form a ball of apparently undifferentiated cells (the blastula). Within blastula cells, genes in some cells produce signals for initiation of the next stage (the gastrula) where differentiation into multiple cell types begins.

During gastrulation, cells move from one part of the embryo to another; in the process, new contacts are made and **induction** occurs as a result of signaling between cells. Since the early 1900s, developmental biologists had known that differentiation of certain tissues required the presence of other tissues nearby. Induction was suspected to result from a chemical interaction of the two tissues, but the mechanism of this interaction was unknown. Spemann and Mangold [383] first demonstrated that contact between "organizing" cells and undifferentiated cells could cause differentiation; Crick [384] and McMahon [385] in the early 1970s predicted the roles of diffusible substances as inducers, and in 1978 TGFβ was among the first inductive ligand–receptor interactions to be demonstrated [386].

Induction leads to two major processes in development – differentiation and morphogenesis. Differentiation is the specification of a particular cell to alter its phenotype from stem cell or some otherwise non-specialized character to having a particular function. That alteration occurs as a result of ligand–receptor binding and activation of intracellular biochemical pathways. Morphogenesis is the result, in the embryo as a whole, of all the individual cellular differentiation events [387].

11.5.2 The Origin of Triploblastic Animals

Triploblastic and bilaterally symmetrical animals arose from a Cnidaria-like ancestor that had apparent radial symmetry. Each new phylum of animals evolved by redefining the actions of preexisting signaling and genetic processes. For example, Wnt signaling with its receptor frizzled (Chapter 7) is responsible for construction of the cnidarian gut opening, which in these animals is both the mouth and anus (an incomplete gut). Alterations in Wnt signaling were involved in the evolution of Bilateria (with complete guts) from Cnidaria [388].

Today, several bilateral phyla exist, and they fall into two main groups, the protostomes and the deuterostomes. Until recently, the protostome/deuterostome division was thought to be phylogenetically meaningful, and was based on anatomical criteria and developmental processes [346]. For example, protostome phyla were said to have

- spiral cleavage,
- the mouth (stome) forming from the blastopore before (proto) the rest of the gut,
- the mesoderm forming by splitting of the cells of endoderm,
- determinate development based on cell lineages (**mosaic**).

In contrast, deuterostome phyla were said to have

- radial cleavage,
- the mouth forming second (deutero) at the opposite end of the gut from the blastopore,

- the mesoderm forming from outpocketing of the endoderm,
- indeterminate development based on cell–cell communication (**regulative**).

Recent evidence indicates that these characters are not consistent enough to be used as a basis for phylogeny, so the terms protostomia and deuterostomia (Figure 11.10) are now used as convenient **grade**-level designations instead of phylogenetic clades.

Even so, it is useful to distinguish between the deuterostomes and the protostomes in the mechanisms underlying developmental patterns (Figure 11.10). In deuterostomes, regulative cell signaling is the primary factor in differentiation; regulation refers to the fact that the fate of a given cell depends on alterations in transcription in response to signaling from nearby cells. In contrast, protostome-like differentiation progresses via a more stereotyped mosaic pattern in which, as each cell divides, it passes different cytoplasmic factors to its daughter cells. These two ways of causing differentiation suggest that protostomes use signaling to a lesser degree than do deuterostomes. Though that may be the case in general, this chapter presents enough examples of developmental signaling in protostomes to show that receptor–ligand interactions are of general importance in development.

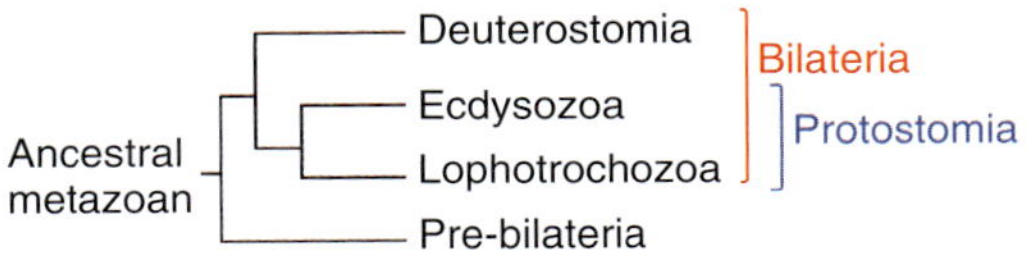

Figure 11.10 Phylogenetic chart of metazoans, based on DNA sequence data. The first split at left separates the animals with bilateral symmetry as adults from the radial or asymmetric animals. Subsequent evolution produced at least 15 phyla belonging to three large groups, the Deuterostomia, Ecdysozoa, and Lophotrochozoa. The latter two taxa are sometimes combined to form a supergroup called the Protostomia.

Indeed, the sections below emphasize that a small number of signaling molecules and their receptors are involved in protostome and deuterostome development. Several of the signal-receptor pairs are given in Table 11.1, and will form the basis for our discussion of developmental mechanisms in the remainder of the chapter.

11.5.3 Development in Protostomes

11.5.3.1 Segmentation and Organ Formation in *Drosophila*

Development in *Drosophila* involves conversion of the original amorphous unicellular zygote into a multicellular segmented adult (Figure 11.11).

The first several rounds of nuclear division occur without production of cell membranes; this allows the spread of developmentally important molecules within the syncitial blastoderm of the embryo. These molecules are the mRNAs for a number of transcription factors, and are secreted into the oocyte before fertilization by ovarian cells of the mother. They control the expression of the first set of genes to be active in the later zygote [346], including those that set up major body axes such as front and hind ends and dorsal and ventral sides (Figure 11.11). Once cell membranes are formed, substances can no longer diffuse long distances through the embryo, and purely local cell–cell interactions dominate. The normal segmentation of the later fly embryo arises only if the initial interactions proceed properly. The next stage involves the inductive interactions of cell–cell signaling.

Table 11.1 Important developmental signaling peptides and their receptors.

Signal families	Receptor	Action	Source
Wnt	Frizzled	Anterior–posterior	[364]
TGF-β	Ser-thr kinase	Dorsal–ventral	[364]
Nodal		Left–right	[364, 365]
BMP		Ventral mesoderm	[389]
Vg1, activin		Dorsal mesoderm	[346]
decapentaplegic			[372]
FGF	FGFR	Ventral mesoderm	[390]
Hedgehog	Patched	Limbs	[391]
Chordin, noggin	BMP	Nervous system	[391]
Delta	Notch	Nervous system	[391]
Agrin	Musk	Muscle	[391]
BOSS	Sevenless	Muscle	[392]

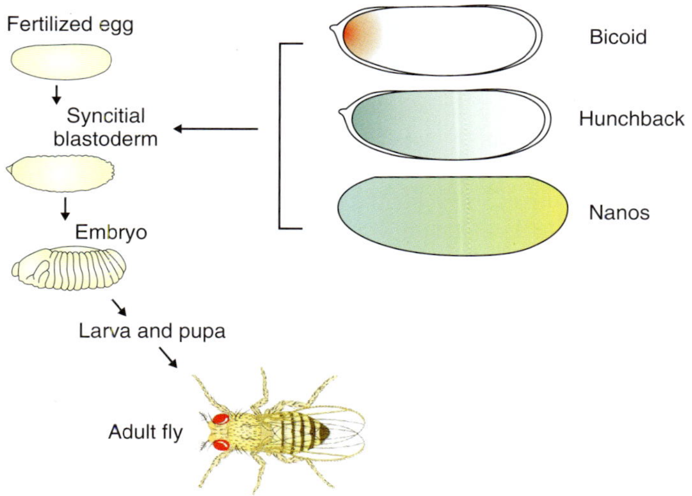

Figure 11.11 Early Drosophila development results from gradients of **morphogens** within the newly fertilized egg at the syncitial blastoderm stage. The morphogens define major axes of the body and provide for later segmentation patterns. The morphogens shown are transcription factors for anterior structures (bicoid and hunchback) and for posterior structures (nanos). The intensity of the color in each embryo is proportional to the concentration of each morphogen. Modified from Figures 2.1, 2.8, and 2.12 in Ref. [391]. Used with permission.

Induction. Genes of the zygote produce two types of molecules involved in control of differentiation: the developmental signals (agonists) and the receptors for those signals. One such pair of signal/receptor molecules are the products of the two genes *Hedgehog* and *Patched* (Box 11.1).

Box 11.1 Terminology of Pharmacological and Developmental Receptors

Pharmacological Receptors

During most of the classical era of pharmacology, receptors (e.g., adrenoceptors, the insulin receptor, the ion channels, and many others) were discovered by experiments in which a drug was added to a tissue and a response measured. From the nature of the response, investigators inferred that a receptor mechanism was responsible. The receptors themselves were also isolated, their genes cloned, and mutant genes artificially made for the study of the effects of altered receptor structure on physiological function. These receptors were named by the type of drug or hormone known to bind to them – adrenergic receptor (or adrenoceptor), insulin receptor, and so on. Their genes are simply called "gene for the ___ receptor."

Developmental Receptors

Receptors important in developmental processes were discovered in a different way and have a different nomenclature. They were first noticed through the accident of developmental mutation: if a fly embryo developed improperly, a specific mutation was inferred, the gene responsible was mapped to a particular chromosome, and it was given a genetic name, for example, *hedgehog*, based on the investigator's fanciful notion of the shape of the mutated embryo. After genetic and molecular study, the gene responsible for the mutation was found, cloned, and its protein product identified. Even if the protein is a hormone or a signaling molecule, as the hedgehog protein is, it is still called by its genetic name. In similar fashion, the *patched* mutant was shown to result from a different faulty protein, this time a mutated form of a receptor. Developmental biologists do not rename these proteins according to the conventions of pharmacology, however, and the rest of us must use the names as we are given them by the investigators.

The existence of a mutation indicates that a normal gene must be present, and it must produce a functional protein (though that functional protein was not known before the discovery of its mutated version). The functional gene is given the same name as the originally discovered mutated gene, for example, *Hedgehog*. Thus, Hedgehog is not the mutant protein, it is the functional one. Gene names are written in *italics*. Their protein products are written in normal text: the *Hedgehog* gene produces the Hedgehog protein, a signal molecule that binds to its receptor the Patched protein, product of the *Patched* gene. Despite their names, when *Hedgehog* and *Patched* are not mutated, the animal has normal development.

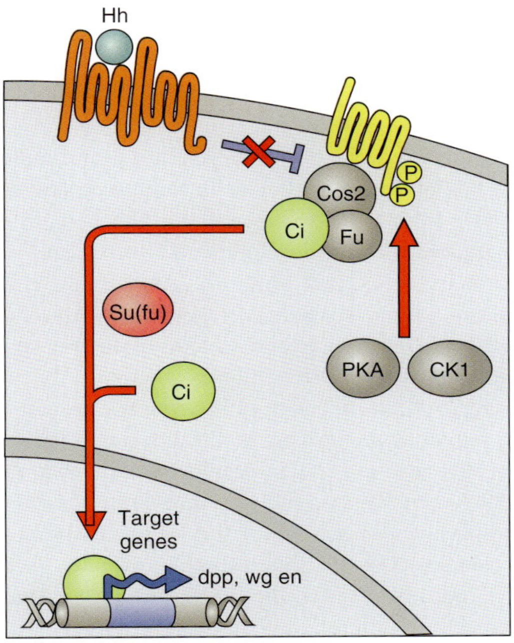

Figure 11.12 Patched, Hedgehog, and Smoothened actions. In the unstimulated state, the membrane protein Smoothened (yellow), maintains the Ci (cubitus interruptus) protein, a transcription factor, in bound state. This state prevents Ci from allowing DNA transcription. The Hedgehog protein (Hh) is synthesized and released by a nearby cell. It diffuses to the target cell, and binds to the Patched protein (orange). Binding of Hedgehog to Patched inhibits Smoothened from keeping Ci inactive; thus Hedgehog binding causes transcription of genes under control of Ci. Modified from Figure 2.37 in Ref. [391]. Used with permission.

Hedgehog and *Patched* are **segment polarity genes** needed for the proper organization of each segment of the Drosophila embryo. *Hedgehog* produces a signaling peptide that is secreted by embryonic cells of the **imaginal discs** that will later form legs and wings. Binding of Hedgehog to its receptor, Patched, on nearby cells promotes fly development by activating transcription of other genes specifically involved in building wings and legs. Gene induction occurs following the activation of a transcription factor when Hedgehog binds Patched [391] (Figure 11.12). A homolog of the *hedgehog* gene, called *sonic hedgehog*, also occurs in vertebrates, and has functions similar in development.

11.5.3.2 Cellular Interactions in Later *Drosophila* Development

Mesoderm Induction by Fibroblast Growth Factor (FGF) Signaling FGF signaling is responsible in both invertebrates and vertebrates for proper migration of cells within multicellular embryos and the formation of complex mesodermal organs. Flies have just two FGF receptors, termed Heartless (Htl) and Breathless (Btl) for reasons analogous to the naming of Hedgehog (Box 11.1). They, along with their ligands, direct nearly all mesodermal signaling in the early *Drosophila* embryo: Htl is activated by two ligands, Pyramus (Pyr) and Thisbe (Ths), and controls migration of mesodermal cells that will form heart and visceral muscles; Btl is activated by a single ligand, Branchless (Bnl), and controls formation of the highly branched respiratory system of tracheas [393] (Figure 11.13).

Heart Formation Htl is expressed in central mesoderm cells, while Pyr and Ths are expressed on more external neurectoderm cells. After gastrulation, mesodermal cells spread out over the inner surfaces of neurectoderm cells, where Htl-Pyr and Htl-Ths interactions occur. Activation of Htl causes migration of mesoderm cells toward dorsal ectoderm and the ultimate formation of dorsal mesodermal structures such as the heart [393].

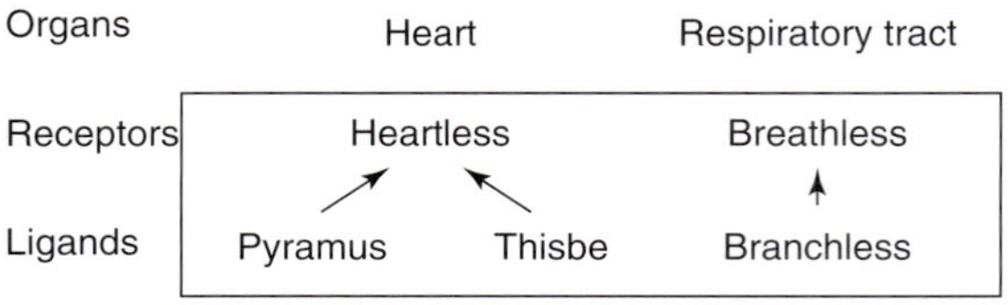

Figure 11.13 Mesodermal induction in *Drosophila* under influence of FGF receptors Heartless (Htl) and Breathless (Btl). These receptors control much of early mesodermal induction, with Htl setting up the central mesoderm and Btl organizing the respiratory system. Pyramus and Thisbe are the two peptide ligands for Htl and Branchless is the peptide ligand for Btl.

Respiratory Tract Mesenchymal cells within the embryo release Bnl, which attracts tracheal cells expressing Btl on their surfaces. The tracheal cells respond by altering epithelial cell structure into tubular shapes and by sending cytoplasmic protrusions out to form the highly branched respiratory tree [393].

The Eye The compound eye of the fly has about 800 light-sensitive units called ommatidia, each containing eight light-sensitive photoreceptors with different spectral sensitivities and sending a neuron to the brain. Six of the photoreceptors (R1–R6) are located in a ring around the two inner ones (R7 and R8). The eye develops because of inductive influences from cells within the disc (Figure 11.14).

Many genes are responsible for complete differentiation of all eight photoreceptors; so far, the known proteins fall into three classes of molecules: receptors, extracellular agonists for these receptors, and intracellular transcription factors.

In overview, the process of eye induction seems to proceed as follows: within the eye disc, the inducing cell (probably the future R8 photoreceptor) produces two types of signaling molecules: (i) paracrine proteins called spitz that diffuse to the nearby R2 and R5 cells (Figure 11.14), and bind to EGF receptors (Chapter 8). The R2 and R5 cells then become signaling units and induce R1, R3, R4, and R6 by releasing additional spitz signals. (ii) In the final step, R8 releases a cell-surface protein called Boss (for Bride of sevenless) that binds to its receptor (sevenless) on the undifferentiated R7 cell. The binding event activates intracellular transcription factors, which turn on selected genes for differentiation into the seventh photoreceptor cell (R7 – responsible for ultraviolet light reception) [392].

11.5.4 Development in Deuterostomes

11.5.4.1 Early Frog Development

Axis formation in the frog embryo is a good example of the role of signaling systems in the conversion of symmetry from radial to bilateral – in this case, in a single organism during development. The frog egg has two primary cytological zones: the animal pole with cytoplasm and nucleus and the vegetal pole with yolk. Near the yolk are intracellular proteins of the Wnt signaling pathway (e.g., Dsh – Figure 11.15).

At fertilization, the entry of the sperm causes the outer region of cytoplasm of the oocyte to rotate about 30°, carrying the molecules of Dsh with it. The movement creates the **gray crescent**, a region of cytoplasm near the site of the future blastopore that contains Dsh as well as other proteins (see below). Because Dsh is the activator of β-catenin (a transcription factor), the result of rotation is to bring the Wnt signaling pathway proteins into the gray crescent. The cells in the vicinity of the gray crescent will become the blastopore, a structure with two distinct levels of importance: it is the embryonic location of gastrulation and the cells surrounding it

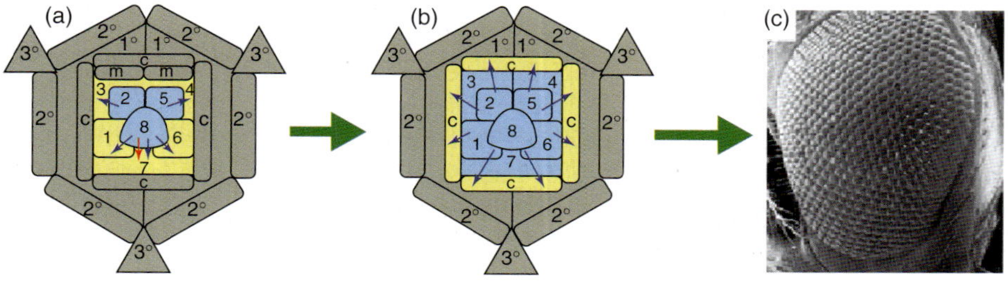

Figure 11.14 Eye development in *Drosophila*. (a) The eight receptor cells of the ommatidium; the central cell (R8) releases a protein signal (spitz) that binds to a TGFR to induce "eye" identity of the R2 and R5 cells; these then release spitz, which activates the R1, R3, R4, and R6 cells. The R7 cell expresses a receptor, "sevenless," an epidermal growth factor receptor (EGFR). (b) R8 then produces the BOSS signaling protein, and BOSS binds sevenless to induce R7 to become the final ommatidial cell. (c) View of the mature compound eye with several hundred ommatidia. Modified from Figures 9.40 and 9.42 in Ref. [391]. Used with permission.

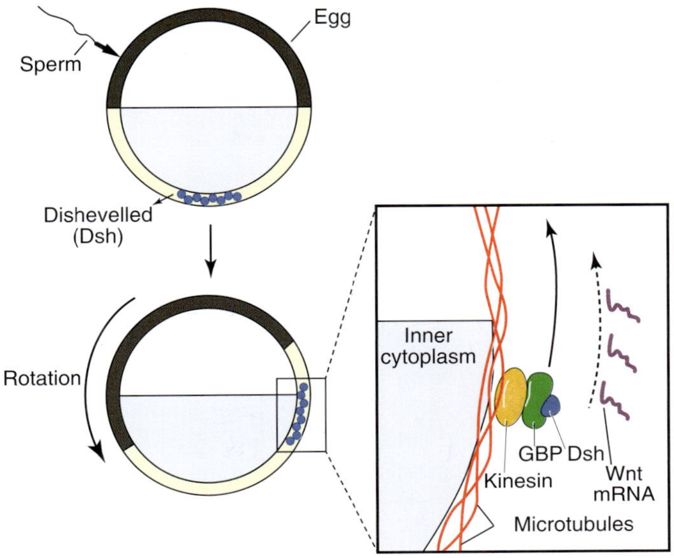

Figure 11.15 Relation of sperm entry point to future body axes. VegT and Vg1 are mRNA molecules for future proteins of the embryo that will influence location of dorsal and ventral regions. Modified from Figure 8.22 in Ref. [346]. Used with permission.

form the **organizer**, or the inductive control center for early cellular migrations [383].

After several rounds of cell division, the original zygote cytoplasm is divided up into a few hundred cells. Cells all over the embryo release BMP, the ventralizing signal. The organizing functions of the blastopore are shown in the next steps: organizer cells release noggin and chordin into the extracellular space dorsal to the blastopore. Noggin and chordin are BMP-binding proteins [389], binding of either to BMP causes BMP to alter its conformation and lose its receptor-binding ability. The ultimate effect of this initial stage of organizer signaling (via noggin and chordin) is thus the dorsalization of the region above the blastopore. In the process, the noggin and chordin signals allow the expression of neural signals such as FGF; this process sets up the dorsal pole as the region of the future nervous system [346].

Because the blastopore becomes the anus (the posterior), the localization of dorsalizing factors at the blastopore creates a new symmetry for the frog: the zygote now has an anterior/posterior axis and a dorsal/ventral axis. It is on its way to becoming a bilaterally symmetrical animal.

All the substances mentioned in the previous section are ligands of the RTKs or STKs. In the RTK group, ligand binding activates the receptor to autophosphorylate, producing a binding site for certain intracellular proteins. This binding event activates Ras, which begins the cascade eventually resulting in replication of new DNA for cell division and induction of new genes for differentiation. In the STK group, receptor binding activates the intracellular kinase domain to phosphorylate a different group of proteins (the Smads – Chapter 8), which migrate to the nucleus and bind to other proteins to form transcription complexes that initiate gene activity. One important role of Smad2 is in the transcription of the signaling proteins noggin and chordin, which, as earlier indicated, bind to and inactivate BMP, the ventralizing signal molecule. To start this dorsalizing process, noggin and chordin are synthesized by binding of the signaling protein Vg1 to its receptor, a member of the TGFβ family of receptor kinases (Chapter 8).

In deuterostomes such as the frog (Figure 11.15), the division of cytoplasm in the first cleavage is symmetrical so that the first two blastomeres are identical and either one can produce a complete organism if the two are separated. In other words, the first two cells have all the materials needed for complete developmental instructions to the genes. Indeed, in some deuterostomes, the first few dozen cells are of uniform composition and indeterminate fate [346]. It is only their relative locations that will determine what each one will become. Thus, a given cell may become skin or nerve depending on what signals it receives from nearby cells.

The next sections provide a few of the many examples of regulative interactions in vertebrates. These will suffice to give an idea of the importance of molecular signaling in developmental pathways.

11.5.4.2 **Nerve Growth**

Neurons begin life within the neural tube, the embryonic structure that forms the brain and spinal cord. Each future neural cell divides for a time, and then at a moment called its "birthday" [346], it stops dividing and either migrates within the central nervous system (CNS) or sends an axon out to innervate muscles or glands. The cell body remains in the CNS and its axons forge the connections with distant tissues. Within the CNS, axons grow to make connections with other brain regions. Thus, the nervous system develops from a few undifferentiated cells of the simple neural tube into a complex three-dimensional structure. Ligand–receptor interactions are the mechanisms responsible for the development of the arrangement of neurons in the brain and its connections to the rest of the body. This process implies that cells follow cues located on tissues through which they migrate. The process of neuronal outgrowth is remarkably similar in vertebrates and invertebrates, and occurs in three phases [394]: **pathway selection**, **target selection**, and **address selection**.

Pathway Selection In the early 1900s, Cajal and Harrison observed that the growing tips of neurons, called **growth cones**, followed definite pathways to find their targets. In the 1960s, Sperry proposed the **chemoaffinity hypothesis** to explain the pattern of outgrowth of nerves. He proposed that chemical codes exist on the surfaces of neurons, and that these codes serve as identifiers to guide neurons in growth and development [395]. This hypothesis has been substantiated and expanded in recent years as researchers have studied the interactions between neurons and two elements of its environment: the ECM and other cells.

Interactions with the Extracellular Matrix Growing axons require a substrate to guide extensions out from the cell body. This substrate can be any solid surface, such as a protein layer between cells, and though it may influence the type of growth pattern of the axon, the substrate does not necessarily give the axon specific clues about the target cell. In addition to non-specific extracellular proteins such as collagen, several types of proteins attached to cell surfaces allow the cells to recognize and move along collagen substrates.

Two main types of cell-surface molecules allow interaction with proteins of the ECM. The **cadherins** and the **integrins** are members of the immunoglobulin superfamily (Chapter 10). They are proteins with extracellular calcium-binding domains and intracellular connections to the cytoskeleton. They serve as adhesion molecules and promote axonal growth [394]. It is not yet clear whether either of these cell-surface molecules acts in a specific manner to promote directional growth of axons along the ECM of collagen, or whether they simply promote growth in general [394].

Interactions with Other Cells Highly specific cellular interactions may operate via ligand–receptor interactions:

- Growth cones of vertebrate neurons have receptors for proteins of the **basal lamina** (a layer of collagen, laminin, and proteoglycans secreted by epithelial cells). The interaction between growth cone receptors and basal lamina proteins guides the cone. Laminin is a good specific guide because it is expressed in only a few brain areas, and it is expressed only when nerve fibers are growing on it.
- In frogs, retinal cells send axons to brain interpretive regions via the optic nerve. Even when they are far from their destinations, neurons travel in the correct direction because neural-cell-adhesion molecules (N-CAMs) (Chapter 10 and below) on their surfaces identify glial cells on which to grow. A gradient of chondroitin sulfate proteoglycan may provide additional directional cues.
  The basis for such specificity lies in the cell-surface molecules expressed by each type of neural cell. Among the most important members of the group called the cell-adhesion molecules (CAMs) is N-CAM , a cell membrane protein found on many types of cells, not just neural cells. It contains several immunoglobulin (Ig) domains and can form **homophilic bonds** (it binds to another molecule of N-CAM on a different cell, thus holding two cells together). Part of the specificity resides in the chains of **sialic acid** that are added enzymatically to the N-CAMs as the cells grow out from the brain. This means that the binding abilities of N-CAMs change as development progresses [394]. Another part of the specificity resides in structural differences caused by alternative splicing of precursor mRNAs for N-CAM, thus producing different polypeptide chains with different binding abilities.
- Ng-CAM (**neuroglia**-cell-adhesion molecule): This CAM is largely restricted to neural and glial cells; it performs homophilic binding, but will also bind to other cell-surface proteins. These molecules are found in both invertebrates and vertebrates. In insects, N-CAM (called fasciclin II when it was

first discovered) plays a similar role in guidance of early axons [394].

Target Selection Specific cues to growth of neurons can also come from diffusible molecules. When ganglia of the mouse are cultured near (but not touching) several tissues, some of which are targets and some of which are not, the ganglia send axons only to the appropriate ones [396]. When a neuron reaches its target, it first stops growing, then makes specific functional contact. The region of contact is called the **synapse**. Contact is often very close, but a gap is always present between the neuron and its target cell, across which the transmitter diffuses when the neuron later becomes mature and active. At the time the axon reaches its target, neither the target nor the axon is obviously specialized.

Soon after contact, however, vesicles of transmitter arrive at the axonal **presynaptic** membrane by transport along the axon itself. In addition, the postsynaptic membrane on the target cell thickens as it produces receptors for the transmitter. Before growth cones express surface proteins, they either do not migrate or they move in a non-specific manner. When the surface proteins become expressed on the surface of the cone, they bind to nearby extracellular markers to guide the growth cone. This initial contact may be made by several neurons with separate axonal growth cones. Each adult target cell has only one neuron innervating it, however, so all but one are lost during development. This process of neuronal cell death is mediated by competition between the arriving neurons for a set of substances released from the target tissues. These substances are called neurotrophic factors. The best known is nerve growth factor (NGF); it is a glycoprotein, and is a ligand of a member of the RTK family. At one time, NGF was thought to guide neurons to their targets, but it is now seen as being responsible for final target selection instead. The evidence for this role is that target cells do not release NGF until axons reach them, and axons do not express NGF receptors until they reach their target [346].

Address Selection When the neuron has reached the target tissue, it must make functional connections with a specific cell in the tissue, generally a particular muscle or gland cell. In skeletal muscle, this specificity is achieved by the signal molecules agrin and neuregulin, plus their receptors [391] (Figure 11.16).

Before a neuron arrives at a skeletal muscle cell, the muscle contains ACh receptors evenly distributed on its surface. As the neuron approaches, it releases three signal molecules, ACh, agrin, and neuregulin. ACh is important in allowing the neuron to compete with other neurons for a muscle cell; agrin binds its receptor (MuSK – muscle-specific kinase) on the muscle cell and causes aggregation of the widely spread ACh receptors so they come to occupy only the space where the neuron makes its connection. Finally, neuregulin binds its receptor, which activates transcription of new ACh receptor molecules [391].

11.6 Programmed Cell Death (Apoptosis)

The end of programmed cell differentiation is programmed cell death. Eukaryotic cells have mechanisms for self-destruction (apoptosis) that operate at various times in life. Apoptosis is programmed rather than random or accidental, indicating it is part of the adaptational machinery of the organism; it is generated from within the body itself; it is present in all eukaryotes so far investigated; it is localized rather than wholesale; and it is important both in adults and in the developmental sculpturing of the animal body plan. The ultimate process leads to changes in cell shape, destruction of mitochondria and nucleus, and phagocytosis of cells by the immune system.

Bacteria also have intracellular means of self-destruction, though these are not chemically related to those of the eukaryotes. However, even in the eukaryotes, apoptosis is generated by mitochondrial (i.e., bacterial) proteins; it is thus likely that multicellular organisms have modified preexisting bacterial proteins for the apoptotic process [397].

Self-destruction seems at first to be a maladaptive practice, especially for a single-celled protist, but considering that parasitism and infection are frequent occurrences in eukaryotes, it is reasonable to hypothesize that mechanisms to eliminate a few infected cells may promote survival of the organism as a whole. The mitochondrial origin of the destructive enzymes suggests that cell death might have originated as a means whereby an early endosymbiotic bacterium could rid itself of a host that might be inhospitable, and then use the host materials for its own metabolism and for association with a new host [397].

The proteins of apoptosis are enzymes of the cysteine **asp**artate prote**ase** group (caspases) that are also involved in immune responses to inflammation. They are activated by cytokine signaling through the JAK (Janus kinase)/STAT (signal transducer and activator of transcription) pathway (Chapter 8), by

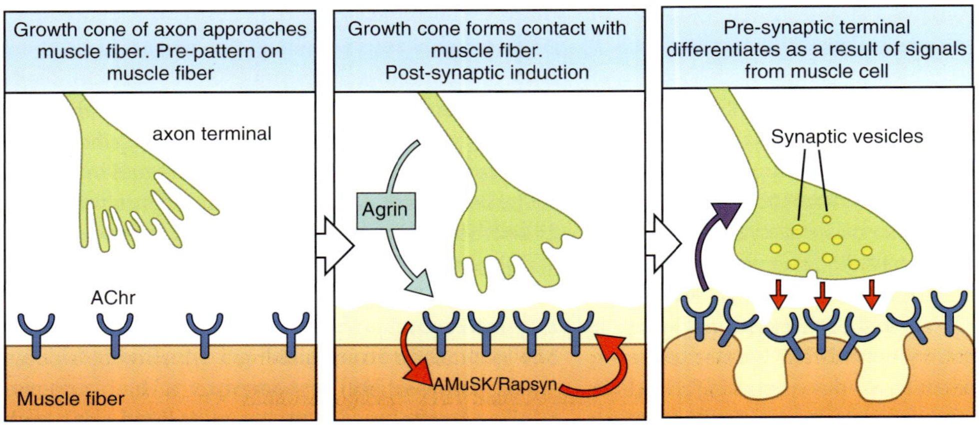

Figure 11.16 Address selection in nerve–skeletal muscle connection. As the neuron approaches a muscle cell, it releases the protein agrin; binding of agrin with its receptor (MuSk) causes aggregation of receptors for the neurotransmitter ACh. The neuron also releases neuregulin. Binding of neuregulin to its receptor activates transcription of new ACh receptors. Figure 10.32 in Ref. [391]. Used with permission.

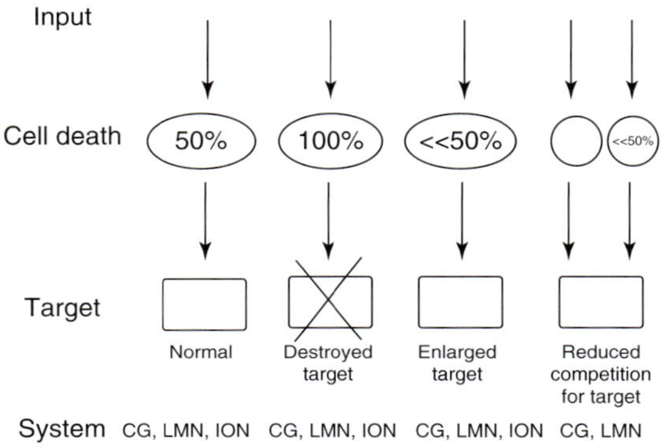

Figure 11.17 Regulation of cell death during neural development. CG: ciliary ganglion; LMN: lumbar motorneurons; ION: isthmo-optic nucleus. Based on data in Figure 2 of Ref. [398]. Used with permission.

carbon monoxide signaling (Chapter 10), and catalyze several types of reactions: the conversion of inactive cytokines to active signaling forms, regulation of transcription, and regulation of cell division.

11.6.1 Apoptosis During Development

In the embryo, a large proportion of cells degenerate in place once they have begun to form organs. The clearest examples are the regression of tails in frog tadpoles and the formation of separate digits in humans through loss of cellular webbing between the bones of the hands and feet.

Evidence for the role of intercellular communication in cell death is shown in Figure 11.17.

In several brain areas, about 50% of incoming neurons die after making initial contact. If the target is destroyed, all the neurons die. If the target is artificially enlarged to permit more neurons to find permanent connections, fewer neurons die. If neurons from one source are eliminated, fewer neurons from another source will die [398].

Apoptosis is required for proper development, at least of neurons, as inhibition of apoptotic pathways causes major deformations of the nervous system. These deformations occur in part because of continued signaling of cells that would normally have been eliminated.

11.6.2 Apoptosis During Adult Life

Cell death is controlled by communication at two levels: intercellular signaling and intracellular signaling. Intercellular signaling proteins belong to the tumor necrosis factor (TNF) peptide family and their receptors to the RTK family (Chapter 8). TNF receptors (TNFR1) are trimeric and bind three molecules of the ligand TNF (Figure 11.18).

Intracellular signaling is accomplished by large complexes of proteins activated by receptor binding. The intracellular systems involve "suicide enzymes" that operate unless they are inhibited by outside signals [396]. Both types of signaling are shown in Figure 11.18.

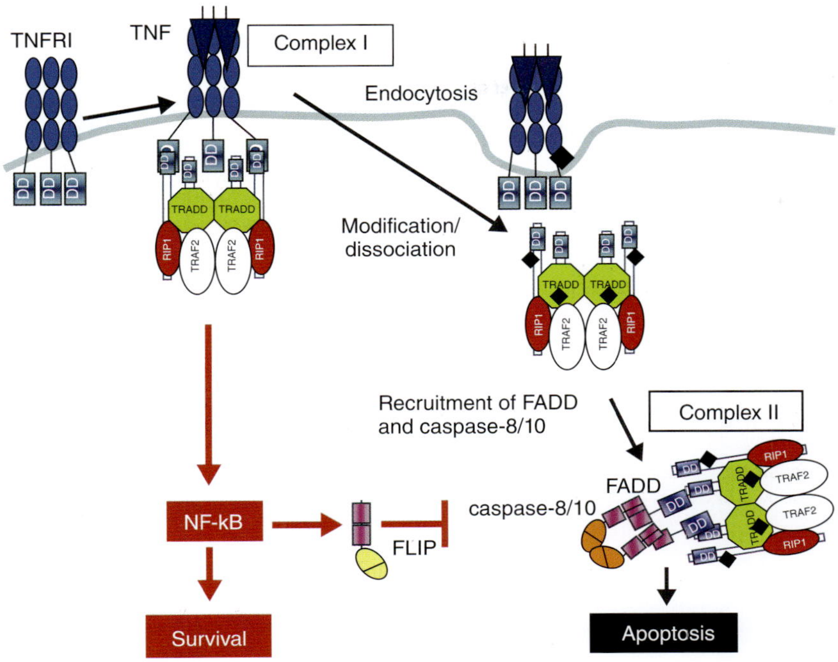

Figure 11.18 After binding of TNF to TNFR1, recruitment of TRADD, RIP1, and TRAF2 occurs (complex I). The complex then dissociates from TNFR1 and recruits caspase-8/10, forming complex II and resulting in apoptosis. Figure 8 in Ref. [399]. Used with permission.

The activator of apoptosis in mammals is caspase 8. It is located in the cytoplasm in inactive form until acted on by a multimolecular complex that is itself activated by binding of the apoptotic ligand TNF. In the unstimulated cell, FLIP interferes with caspase 8 activation, thus promoting cell survival.

It thus appears that for a cell to remain in place, it must actively produce apoptosis-inhibiting signals. Once a cell stops producing these signals, it loses the ability to keep apoptotic enzymes inactive. It is now clear that cell–cell signaling is involved not only in normal development but also in simply staying alive.

12
Receptor Mechanisms in Disease Processes

> Growth for the sake of growth is the ideology of the cancer cell.
>
> *Edward Abbey [400]*

> Dying is a very dull, dreary affair, … and my advice to you is to have nothing whatever to do with it.
>
> *W. Somerset Maugham [401]*

12.1
Genetic Basis for Receptor Function

Several kinds of diseases result from mistakes in cellular communication. Mutations in the genes encoding proteins such as hormones, receptors, and effector systems are the most common sources of these disorders. Even small changes in the molecules of communication can produce serious alterations in function, sometimes resulting in loss of a normal process, other times resulting in a new, pathological process. In addition to structural changes, alterations of receptor function can be produced by environmental chemicals or by pathogenic organisms. This type of cellular alteration may even result from bacterial infection or water pollution. In addition, minor brain damage, stress, or drugs may alter neural activity in brain areas responsible for behavior, thus altering perception and bringing on changes in mood.

Examples of receptor-related diseases given in this chapter involve malfunctions of all four receptor types, and scientists gain two types of value from these studies. Often the uncovering of pathologies improves understanding of the fundamental processes underlying communication. In addition, by understanding how defects in the molecules of communication cause disease, researchers gain a sounder basis for medical treatments.

12.1.1
Genotype and Phenotype

The biological basis for a phenotype, whether it is a normal characteristic or a disease, is generally described in terms of a "gene for" that characteristic. This approach is unrealistic in implying that a one-to-one correspondence always exists between gene and trait, and it oversimplifies the actual pathway by which gene expression leads to phenotypic expression. Thus, researchers do not find single genes that cause cancer or schizophrenia. However, because genes specify proteins, and proteins are the functional molecules of the cell, it is important to understand the genetic basis for any characteristic, normal or abnormal. The initial sections of this chapter outline some of the complexities of genotype–phenotype expression, giving the basis for later discussions of disease mechanisms.

12.1.2
Classical Dominance Mechanisms

Mendel in the 1860s defined genes as being **dominant** or **recessive** based on their outward expression. During the twentieth century, geneticists expanded Mendel's purely descriptive concept by recognizing that the terms refer not to the genes themselves, but to the proteins they encode: dominant genes encode functional proteins and recessive genes encode non-functional proteins. These concepts have been modified with the addition of more complicated patterns of expression; for example, incomplete dominance (below) describes a condition of protein function that is less intense than complete dominance.

These classical concepts are shown in Figure 12.1 in relation to receptor function:

Complete Dominance (Figure 12.1a): This pattern is characteristic of receptors that have a high

Receptor Biology, First Edition. Michael F. Roberts and Anne E. Kruchten.

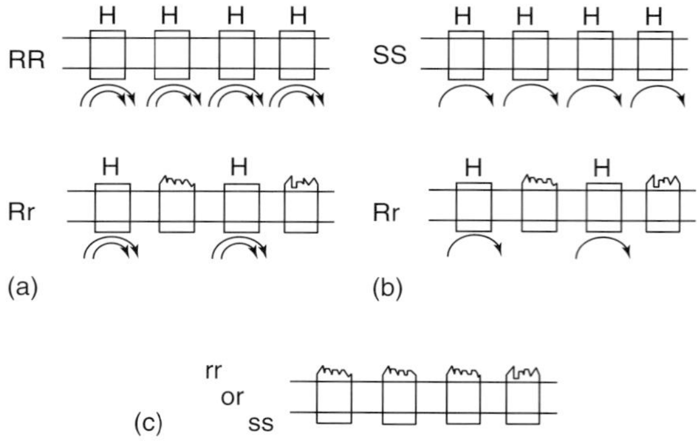

Figure 12.1 (a) Complete dominance (high-efficacy agonist): cell membranes (horizontal lines) with receptors (rectangles) on the surface. Transduction steps are shown by arrows. Mutated receptors have jagged surfaces. Assume that four arrows are sufficient to give a full cellular response. (a) RR, all receptors normal and Rr, half receptors normal but response within normal range. (b) Incomplete dominance (low-efficacy agonist): SS, all receptors normal and Ss, half normal and response half normal. (c) Recessive: all receptors rr or ss; no receptors function.

efficacy interaction with their ligands, with strong coupling to intracellular processes. These receptors activate cells even if only a small fraction of receptors are bound by the agonist (Chapter 5). In the figure, gene R produces a receptor; in the homozygous dominant condition *RR*, all receptors are normal and functioning. The heterozygote *Rr* produces half as many receptor molecules as the homozygote, but that number is adequate to provide proper ligand activation of cells. In this case, we say that the phenotype of the *heterozygote* in complete dominance is the same as the *homozygous dominant.*

Incomplete Dominance (Figure 12.1b): The gene S encodes a slightly different receptor, with lower efficacy than the receptor encoded by R. For receptor–agonist systems with lower efficacy, the homozygous dominant *SS* will be normal because all receptors can be activated, but the phenotype of the heterozygote *Ss* will not be the same as in the homozygous dominant. In Ss heterozygotes, half as many functional receptors exist on the cell surface as in SS homozygotes, but this number is inadequate to give a normal response to ligand binding because of the lower efficacy of S itself. This mechanism is also called a loss of function mutation, or a haploinsufficiency [402].

Recessive (Figure 12.1c): This condition is independent of the efficacy of the receptor–agonist interaction. All copies of the gene are abnormal, and all the receptors produced are nonfunctional. Receptors may not be synthesized at all, or they may not properly bind the ligand, or they may not activate the cell after binding.

12.1.3 Other Levels of Gene Expression

A new set of descriptive terms for gene expression emerged out of molecular genetic studies of development (Figure 12.2). Each term describes a mutation that alters gene expression:

Gain-of-Function: This type of mutant represents the overexpression or hyperactivity of a protein. An example is the *Notch* mutation in *Drosophila.* The Notch protein is a cell-surface receptor involved in differentiation of neurons. It normally acts only in response to an agonist, but a gain-of-function mutation allows it to function constitutively even if it does not bind the agonist. Thus, it activates its cytoplasmic tyrosine kinase domain even in the absence of the ligand [403] (Figure 12.2a).

Dominant-Negative: This mutation reflects the interaction between two proteins. An example concerns modifications in expression of certain growth factor receptors described in Chapter 7. The growth factor TGF (transforming growth factor) β binds to the TGF β II receptor; binding causes dimerization of types I and II receptors. The type II receptor is an active kinase that, after dimerization, causes the type I receptor to become an active kinase. If the type II has a mutated kinase domain, it cannot cause the type I to become active, yet it can bind ligand and thereby compete with functional type II receptors. It thus acts as an inhibitor of the intracellular action of TGF β, and its effect is dominant to the case where no mutation has occurred [404] (Figure 12.2b).

12.1.4 Pre-receptor Mutations

Hormones and transmitters are of two kinds: some (e.g., epinephrine (EP), acetylcholine, steroids) are not proteins, but rather are small molecules that are the products of enzyme-catalyzed biochemical pathways. Others (e.g., insulin, glucagon,

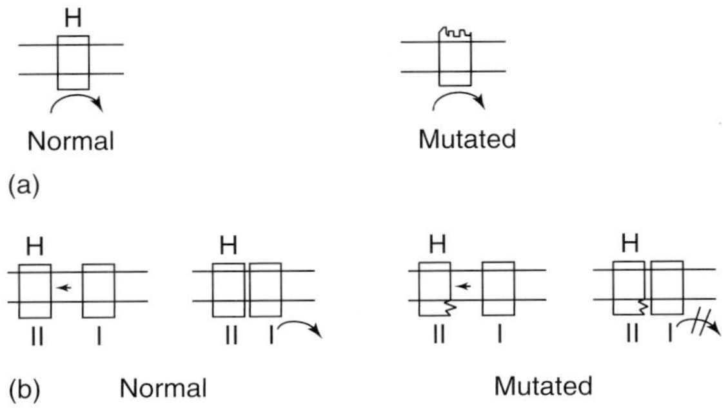

Figure 12.2 Genetic mutations that alter protein production and function. (a) Gain-of-function mutant, in which receptor functions without binding hormone (H). (b) Dominant-negative mutation, in which receptor binds H but does not activate intracellular functions.

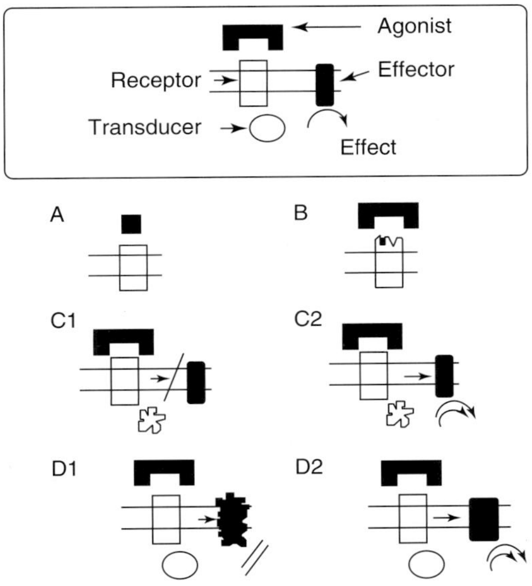

Figure 12.3 Receptor mutations. Box: normal function of agonist, receptor, transducer, and effector. (A) Mutation in protein agonist. (B) Mutation in receptor. (C) Ineffective (C1) or hyperactive (C2) transducer. (D) Ineffective (D1) or hyperactive (D2) effector. The hyperactive proteins are equivalent to the gain-of-function mutations seen earlier.

thyroid-stimulating hormone (TSH)) are proteins and therefore direct gene products. In either case, the lack of or abnormality in a hormone reflects a gene defect (Figure 12.3).

Mutations in the genes for three hormones have been reported to be the source of non-functional agonists: parathyroid hormone, TSH, and insulin [405]. People with these defects have normal receptors, and they respond well if they are given injections of the normal hormone.

12.1.5 Receptor Mutations

If the receptor gene is mutated so the protein does not form properly, or if the receptor does not reside in the proper place in the cell (cytoplasm for steroid receptors, plasma membrane for cell-surface receptors), then even normal concentrations of hormone will have no function. Other gene mutations might reduce the numbers of receptors synthesized, or might alter the binding site so it does not bind the hormone. Mutations in the epithelial chloride transporter that cause cystic fibrosis (CF) and receptors in the low-density lipoprotein (LDL) belong in this last category (more complete discussions of the roles of these proteins occur later in this chapter).

12.1.6 Post-receptor Mutations

Most types of receptors couple to a membrane or cytoplasmic biochemical system, for example, to a second messenger as in the G-protein-coupled receptors (GPCRs) (Chapter 7), or to a kinase as in the growth receptors (Chapter 8). Thus, post-receptor defects can cause diseases relating to transport, metabolism, development, or cell division.

- In the GPCRs, mutations of the post-receptor type (Figure 12.3C,D) involve alterations in the G proteins or the effector enzymes activated by them, such as adenylyl cyclase, phosphodiesterase, phospholipase, and transcription factors.
- In the receptor tyrosine kinases, mutations may occur in the cytoplasmic domain, source of the kinase function. Such mutated receptors may be present in normal numbers and may exhibit normal hormone binding, but either do not activate biochemical processes in the cell or may cause hyperactivity.

12.2 Receptor Pathologies

12.2.1 Ion Channel Superfamily

The importance of ion channels in neural and muscular activity, as well as their roles in epithelial transport, suggests that abnormalities can lead to serious disorders of sensation, movement, or thought. This discussion covers a small number of ion "**channelopathies**" [406] and receptor disorders.

12.2.1.1 Calcium Channels

Alterations in any of the several types of Ca^{2+} channels are seen in several neurological conditions, the best known of which is epilepsy. This condition is characterized by spontaneous seizures that are not associated with other known neurological damage; it can be caused by hyperexcitability of T-type Ca^{2+} channels, leading to long (>100 ms) depolarizations that cause voltage-gated Na^+ channels to open and generate action potentials. The cause of the increased excitability of the Ca^{2+} channels is not known. Other disorders associated with Ca^{2+} channelopathies are autism, ataxia, and some forms of night blindness and deafness [406].

12.2.1.2 Transient Receptor Protein (TRP) Channels

These are also cation channels, involved in sensation (e.g., taste, warm/cold). They couple to other channels (e.g., Ca^{2+}, K^+, and Na^+ – Figures 6.12 and 6.13); thus, TRP mutations may affect multiple functions in excitable and non-excitable cells. Indeed, mutations in all seven members of this family in humans have been linked to disorders. Some examples are kidney failure, pain, muscular weakness, and impaired night vision [179].

12.2.1.3 Voltage-Gated Na^+ Channels

These channels control neural and muscular depolarization, and are responsible for most sensation and movement. Mutations in their structure will therefore produce major sensory or motor disorders. The main voltage-gated Na^+ channelopathies have been noted in skeletal and cardiac muscle, and in brain neurons. Disorders originating in the skeletal muscle are mainly paralytic; the primary disorder derived from malfunctioning cardiac muscle Na^+ channels is sudden infant death syndrome; in the brain, epilepsy is the primary dysfunction related to altered function of this channel [407].

12.2.1.4 Ligand-Gated Na^+ Channels

These channels are located at neural and neuromuscular synapses; they are activated by the neurotransmitter ACh; channel malfunction thus may produce large alterations in movement. The best-known pathological state of this channel is myasthenia gravis, caused by antibody-induced deactivation of the channel. The antibodies are produced in an autoimmune process of unknown origin; they bind to the channel and have two main effects: they prevent transmitter binding and they reduce channel numbers by causing endocytosis of the receptor–antibody complex [408].

12.2.1.5 Chloride Transporter – Cystic Fibrosis

The cystic fibrosis transmembrane regulator (CFTR) (Figure 12.4a) is a Cl^- channel that opens in response to agonists such as EP, VIP (vasoactive intestinal polypeptide), and ADO (adenosine) (Chapter 7), all of which activate receptors coupling to G_s and cAMP (cyclic adenosine monophosphate). Binding leads to phosphorylation of the regulatory subunit of the CFTR, allowing ATP to bind to the NBD 1 and 2 (Figure 12.4b); these steps promote unblocking of the channel and the diffusion of Cl^- and fluids out of the cell (Figure 12.4c).

In persons with CF, the CFTR is inactive [410] (Figure 12.4d). As a result, the pump is ineffective in transporting chloride, so the fluids secreted are abnormally thick and viscous. The lungs secrete mucus that is too thick for proper elimination of aerosols. CF may be caused by any of several mutations in the amino acid composition of the CFTR, although most occur in the nucleotide-binding domains, primarily NBDs.

12.2.2 G-Protein-Coupled Receptor Superfamily

12.2.2.1 Cholera

The infectious disease cholera has clear connections to G protein function through control of guanosine diphosphate/guanosine triphosphate (GDP/GTP) exchange and coupling to receptor binding (Chapter 7). Study of the G protein malfunction in this disease has not only improved our understanding of cholera but it has also given researchers the tools to better understand normal operation of G protein transduction processes. Cholera is an infectious disease caused by the bacterium Vibrio cholerae. The bacterium is spread by contact with water contaminated by human waste. After ingestion, the bacterium lodges in the intestines and releases a toxin that acts on intestinal cells, causing secretion of large amounts of water and salt that are lost via diarrhea.

The mechanism of action of the toxin is the addition of an ADP-ribose molecule – a portion of the metabolic cofactor flavin adenine dinucleotide (FAD) – to the α subunit of the G protein when it is bound by GTP [411]. This process prevents GTP hydrolysis, fixing the G_s in the "on" position so adenylyl cyclase continuously synthesizes cAMP. If the G protein continues to synthesize cAMP, intestinal cells continue to pump ions into the intestinal lumen. As ion pumping across a cell membrane is always accompanied by the flow of water across the same membrane, cholera causes loss of both ions and fluids.

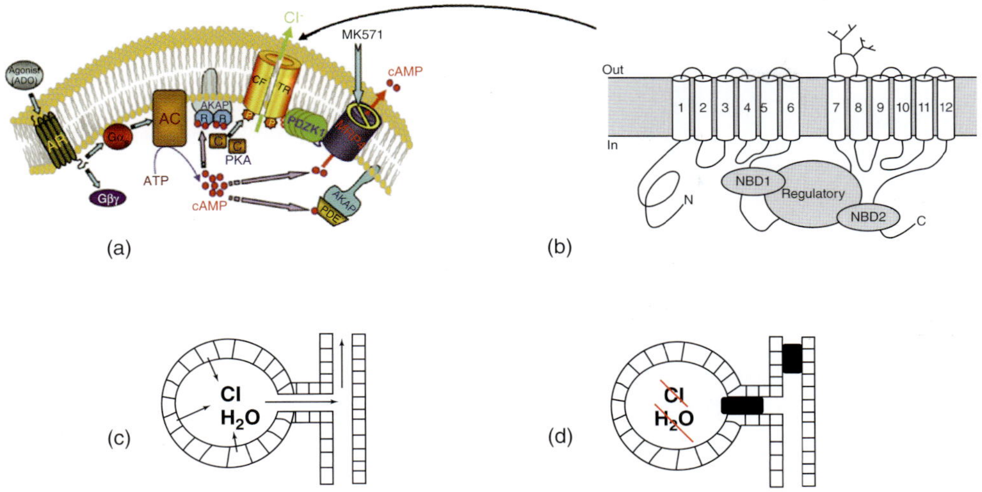

Figure 12.4 Changes in ion and water transport in cystic fibrosis (CF). (a) Intracellular processes responsible for normal transport. Exposure to irritating aerosol in lung causes release of agonists (e.g., adenosine (ADO) in blood. ADO binding produces cAMP, which activates protein kinase A (PKA). This kinase phosphorylates the regulatory domain of the CFTR so the CFTR transports chloride out the luminal side of the cell. (b) CFTR structure. The transporter has 12 transmembrane (TM) domains in two groups plus two nucleotide-binding domains (NBD 1 and 2). The regulatory domain aids channel gating and ATP binding to the NBDs. (c) Normal duct anatomy with epithelial cells that transport Cl. Na, H_2O, and protein flow in normal duct function; in addition, cilia present in ducts aid transport. (d) In CF, ducts become clogged as normal ion and water flows are prevented. Panel (a) modified from Figure 7D in Ref. [409] and panel (b) Figure 5.17 of Ref. [161]. Panel (b) Used with permission.

12.2.2.2 Thyroid Diseases

Changes in thyroid activity may result from a lack of thyroid hormone (see below under nuclear receptors), from a lack of the receptors for the pituitary peptide TSH (Chapter 4) that causes release of thyroid hormone, or from antibodies that increase or decrease TSHR (thryoid stimulating hormone receptor) function. Receptors for TSH are found in the thyroid gland, and belong to the G-protein-coupled superfamily. On binding TSH, the receptors increase cellular cAMP, which activates enzymes involved in the synthesis of thyroid hormone.

One cause of hypothyroidism is mutation of TSHR. In this condition, the receptor does not bind TSH and therefore does not activate cAMP. Patients with a mutated TSH receptor have very low thyroid hormone levels [405]. Another condition leading to hypothyroidism is Hashimoto's thyroiditis, in which inhibitory autoantibodies antagonize either thyroid production in thyroid cells [412] or they antagonize the TSH receptor [413]. Both effects decrease production of thyroid hormone.

Hyperthyroidism, or Graves' disease, is caused by stimulatory autoantibodies to the TSH receptor, causing increased production of thyroid hormone [412].

12.2.2.3 Cardiovascular Disease

Increased blood pressure is an important chronic disease in industrialized countries, and may lead to stroke, heart disease, or damage to other organs. A minority (10%) of cases can be linked to **tumors** or specific kidney pathologies, but 90% of cases have no known cause. Nonetheless, it is clear that two physiological processes, increased sympathetic nervous system stimulation of the heart and vasculature and activation of the renin–angiotensin system, may play important roles in hypertension [414].

Sympathetic Activity: Sympathetic nerves release the transmitter norepinephrine (NE), which activates α- and β-adrenoceptors in heart and vascular tissue (Chapter 7). Both α_1 and α_2 receptors cause constriction of vascular smooth muscle, and excessive release of NE can lead to increased blood pressure. β_1 adrenoceptor stimulation in heart by EP leads to increased heart rate and strength of beat, also causing increased blood pressure. However, EP activates β_2 adrenoceptors in vascular cells, and activation of this receptor causes dilation via inhibition of Ca^{2+} influx [415]. Thus, the GPCRs of the cardiovascular system have differing effects on blood pressure. A summary of adrenoceptor influences on vascular smooth muscle contraction and relaxation is shown in Figure 12.5.

Renal Effects: In addition, the kidney determines fluid balance within the body by regulating salt

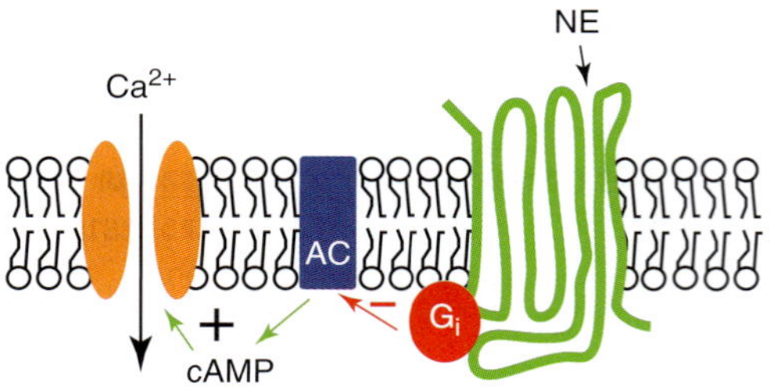

Figure 12.5 Influence of alpha-adrenoceptor signals on vascular smooth muscle contraction. Calcium influx may be stimulated by several sources (here, $alpha_2$ adrenoceptor action is shown). This contraction can be inhibited by interfering with Ca^{2+} channel action or antagonist binding to receptors.

and water excretion. Reduced blood pressure in the vessels supplying the kidney cause secretion of the enzyme renin from the kidney. Renin cleaves a blood peptide, angiotensinogen, thereby producing the peptide, angiotensin I. Another vascular enzyme, converting enzyme, converts angiotensin I to its active form, angiotensin II. This peptide is an agonist of GPCRs in cells of the adrenal cortex, where it causes secretion of the steroid aldosterone (Figure 9.9). This hormone causes uptake of Na^+ from the kidney tubules, and therefore increases salt and water content of blood. Abnormal increases in renin production can cause hypertension [416]. *The mechanisms underlying these changes may involve alterations in the structure of the catalytic unit in adenylyl cyclase or changes in the structure of the G_s subunit of the G proteins* [417].

Because both the kidney and sympathetic nervous systems affect blood pressure, it is not surprising that hypotheses regarding hypertension invoke both systems. The primary therapeutic approaches to hypertension are therefore linked to the processes that inhibit one or more of the steps in the vasoconstrictor response or the activation process of angiotensin.

12.2.2.4 **Obesity**

Body weight is a resultant of several factors involving food intake, metabolism, and activity. Behavior also influences food intake, and both behavior and metabolism are affected by neurotransmitters, brain peptides, and hormone–receptor systems [418].

Several regions of the hypothalamus – the ventromedial nucleus, the lateral hypothalamus, and the paraventricular nucleus – communicate with each other and with other brain regions responsible for eating behavior. NE acting on hypothalamic **β**-adrenoceptors reduces eating behavior; NE acting on α-adrenoceptors stimulates food intake. **β** receptors are more numerous in the hypothalamus of lean animals and alpha receptors are more numerous in obese ones [418]. Changes in neuropeptide balance also affect feeding behavior. Neuropeptide Y (NPY), a peptide that occurs in the sympathetic neurons of brain and adrenal medulla along with NE and EP, is especially important, as it decreases metabolic rate and increases food intake if released within the hypothalamus.

Familial patterns of obesity have long suggested that "obesity genes" regulate eating behavior and metabolism. These genes encode signal peptides or receptors involved in metabolism or satiety. The best-studied genes involved in models of obesity are the Ob and Db genes. The dominant allele Ob encodes a circulating protein called **leptin**. The dominant allele Db encodes the leptin receptor. Leptin is a **satiety factor** (signal to stop eating). It is released during eating, travels to the hypothalamus, binds to its receptor, and decreases the release of NPY. Thus, leptin acts both to reduce food intake and to increase metabolism.

The understanding of how leptin and its receptor work has created a potentially powerful animal model for obesity. Ob/Ob or Ob/ob mice are normal because they produce adequate amounts of leptin. Mice of the ob/ob genotype are obese because they lack leptin; the lack of leptin encourages release of NPY, which reduces metabolic rate and encourages food intake. The administration of leptin to obese ob/ob mice produces weight loss by inhibiting production of NPY [419].

The leptin receptor is a cytokine receptor in the Ig superfamily; it is not itself a kinase, but couples to Janus kinase (JAK) (Chapter 10). Mutant mice with the db/db genotype do not show response to administered leptin. Thus, a clear relationship exists between state of the Ob and Db genes and eating behavior in mice. However, though humans also have these genes, clinical experiments showed that such a simple model does not explain many cases of human obesity, as obese people often have high blood levels of the satiety factor leptin and no evidence of Ob gene mutations. In these people, leptin should signal satiety and prevent eating, but it does not.

The likely reason is that the high levels of triacylglycerides in the blood of obese people lead to impaired leptin transport from the blood to the hypothalamus, decreasing or eliminating the effectiveness of leptin as a satiety signal [420]. As high triacylglycerates are present in people with low fat and high carbohydrate intake [421], increased carbohydrate intake (rather than mutated Ob or Db genes) is the most probable cause of obesity.

12.2.2.5 Depression

Depression is a **mood disorder** involving an extended period of sadness, grief, or disappointment in which a person loses interest in his or her surroundings and is often unable to function normally. It is a complex disease, and many aspects of depression are influenced by brain neurochemistry [422]. Mechanisms responsible operate at any of several levels: alteration in the amount of transmitter available at the synapse, changes in transmitter–receptor interaction, reduction or increase in the number of receptors, or altering the post-receptor processes of cell activation. Some causal factors are stress, accidents causing minor brain damage, and familial inheritance.

Several drugs exert antidepressant effects. The effects of the classical antidepressants fall into two categories: the blockade of uptake of NE and serotonin (Ser) into nerves, and blockade of enzymes that degrade NE and Ser. Thus, these drugs are thought to operate clinically by potentiating the action of NE and Ser at their receptors, by increasing the concentration of the transmitter in the synapse and increasing the residence time of the transmitter at the receptor. The best-known drugs are the NE uptake inhibitors such as imipramine, the Ser uptake inhibitors such as prozac, and the monoamine oxidase inhibitors such as tranylcypromine. The phytochemical hyperforin, from St. John's wort (Hypericum perforatum), is also a monoamine uptake inhibitor, and is likely the source of the plant's antidepressive action [423].

The effectiveness of substances that inhibit uptake and degradation of monoamines led to the transmitter hypothesis of depression, in which the amount of NE or Ser at the receptors was thought to be the deciding factor in mood. However, continued therapeutic advances produced a paradox – the antidepressant drugs do not immediately relieve symptoms of depression, but instead work over a much slower time course (10–20 days) that implies they are not simply blocking uptake of transmitter. In addition, simply removing NE and Ser from normal people does not produce depression. Thus, although the amount of transmitter in the synapse is an important factor in depression, it is not the entire answer. Recent research has been adding to this model of depression by extending it to include cellular mechanisms after receptor binding occurs.

Postsynaptic cells regulate the number of receptors according to the concentration of transmitter that is present (Chapter 7). Antidepressant drugs therefore decrease the number of NE and Ser receptors in spite of increased amounts of transmitter at the synapse [422]. Even though receptor number is reduced, the elevation in transmitter concentration increases activation of G proteins coupled to adenylyl cyclase. This in turn leads to an increased production of cAMP and an increase in cAMP-initiated protein synthesis. cAMP activates the transcription factor cAMP-related binding protein (CREB), which initiates the synthesis of the peptide **brain-derived neurotrophic factor** (BDNF). BDNF is a ligand for a neurotrophic factor receptor (Chapter 8) involved in the strengthening of existing synapses and in the development of new synapses.

When a person is under stress, the BDNF gene is repressed, leading to loss of neurons in the brain that are controlled by BDNF, and in turn leading to depression [422]. BDNF shows antidepressant effects if administered by itself, and levels of BDNF in the blood (as well as its receptor in the brain) are increased by traditional antidepressant treatment. The time course over which BDNF increases in the body is consistent with the time course of decline in clinical depression. This expanded hypothesis involving BDNF helps explain many of the clinical features of depression.

Taken together, the rate at which antidepressants develop their effectiveness and the role of BDNF suggest that antidepressants are effective because they decrease neuronal loss and encourage new cell growth [424].

12.2.2.6 Schizophrenia

This condition is a **psychosis** (more serious than the **mood disorder** depression). It involves alterations in mood and perception, in which hallucinations, distortions of reality, and erratic behavior play major roles. Even though both disorders involve alterations in neurotransmitter processes in the brain, schizophrenia usually requires more extensive drug and institutional treatment than does depression. Symptoms of schizophrenia have been divided into positive and negative types. The negative symptoms include lack of energy and blunting of emotions; positive symptoms include delusions, hallucinations, and unusual behavior. It is not known whether the negative and positive symptoms have the same cause, nor is the detailed cause of the disease known.

The major brain areas involved in schizophrenia are the limbic system, the frontal lobes, and the basal ganglia, all parts of the forebrain. Schizophrenia is associated with increased activity of dopamine (DA)-releasing neuronal pathways in these areas, an idea that has been formalized as the **DA hypothesis**. Two classes of DA receptors exist in the brain, D_1-like and

D_2-like (Chapter 7). The D_1-like receptors contain the D_1 and D_5 subtypes; the D_2-like receptors contain the D_2, D_3, and D_4 subtypes. All are GPCRs; the D_1-like subtypes activate adenylyl cyclase and the D_2-like subtypes inhibit the enzyme. The effectiveness of certain antipsychotic drugs is proportional to their ability to inhibit adenylyl cyclase in the brain, and the most effective drugs, such as haloperidol, decrease cAMP levels by preventing binding of DA to its receptor. Thus, the main line of research in this area is focused on the mechanisms involved in D_2-like receptor behavior.

Knowledge of receptor types implicated in schizophrenia has not yet led to a complete understanding of the disease. So far, it has been determined that schizophrenic patients show an increased amount of DA released and an increase in numbers of D_2 receptors. The increased release of DA combines with the larger receptor concentration to decrease neuronal depolarization and cell activity. The ability of a drug of the haloperidol class to act as an antipsychotic is directly proportional to its affinity for D_2 receptors [425].

12.2.3 Immunoglobulin Superfamily

12.2.3.1 Diabetes Mellitus

This important disease involves loss of insulin or reduction in insulin receptors (Chapter 8). A major function of insulin binding is the transport of glucose into non-neural cells; thus, the main symptom of diabetes is hyperglycemia because glucose uptake is reduced. In addition to its glucose transport functions, insulin activates enzymes for glycogen and fat synthesis; thus, lack of insulin-receptor binding allows glycogen and fat breakdown to occur unabated, causing release of glucose into the blood and increased fatty acid metabolism.

Two main types of diabetes mellitus (DM) are distinguished, depending on whether the condition is caused by lack of insulin (type 1 DM) or by insulin resistance (type 2 DM). In type 1 DM, mutations in the insulin gene cause changes in the insulin amino acid sequence and thus can have two effects: they prevent the proper formation of insulin so it is not processed properly within the pancreatic cells where it is made (for normal processing of insulin, see Chapter 8), or they prevent insulin from binding to its receptor. In addition, autoimmune attack of insulin-producing cells can prevent release of insulin. People with type 1 DM are born with the condition, and require insulin injections for the duration of their lives.

In type 2 DM, normal insulin is produced, but the person's cells do not respond to it because of a lack of receptors. This condition generally develops in midlife and is caused by overeating or lack of exercise. The excess intake of food stimulates insulin production; continued insulin-receptor binding causes sequestration of receptors. Thus, glucose uptake is inadequate, but for reasons different from type 1 DM.

12.2.3.2 Atherosclerosis

Atherosclerosis refers to the buildup of cholesterol and other materials within blood vessels, reducing their diameter so local blood flow is impeded. It seems to be caused largely by improper transport of cholesterol into cells by the LDL receptor (see Chapter 8 for discussion of LDL receptor mechanisms).

The most serious form of this disease – called familial hypercholesterolemia (FH) – is one in which people have seriously abnormal LDL receptors. Those with FH have one or more of several mutations in the gene for the LDL receptor [283].

- The most common mutation (50% of the total) is the complete lack of LDL receptors.
- Another type is a change in the shape of the ligand-binding site so LDL will not bind to the receptor.
- A third type of mutant has no transmembrane region so the receptor is not placed in the membrane after synthesis but is secreted from the cell.
- A final type of mutant is caused by a change in the cytoplasmic portion of the receptors so they no longer can aggregate with other receptors that have bound LDL, and therefore cannot bring cholesterol into the cell.

If the LDL receptor does not bind and internalize LDL, then two important events occur: (i) the LDLs remain in the bloodstream and may become part of atherosclerotic plaques and (ii) the enzyme that controls cholesterol synthesis (HMG COA (3-hydroxy-3-methyl-glutaryl-CoA) reductase) increases in activity, producing more cholesterol than if cells had normal LDL receptors.

12.2.4 Nuclear Receptor Superfamily – Steroid Receptors

12.2.4.1 Alterations in Transcription

As described in Chapter 9, nuclear receptors are cytosolic receptors that bind to ligands within the cell and exert an action, either on downstream transcriptional targets or upon components of existing signaling cascades. The steroids are divided into three groups: the sex steroids, the glucocorticoids, and the mineralocorticoids. Defects in receptors for the sex steroids (testosterone, progesterone, estrogen) promote abnormal development of primary and secondary sexual characteristics. Defects in receptors for the glucocorticoids (mostly cortisol) cause abnormal release of glucose into the bloodstream.

Defects in the testosterone receptor are the most common in this group. The gene for the testosterone receptor is on the X chromosome, so men have only one copy of this gene, and if it is mutated, no second copy exists to produce a normal receptor. People who are chromosomally male (XY) may be phenotypically female if the mutated testosterone receptor does not bind hormone, and therefore does not activate the processes needed for producing primary and secondary male characteristics. The affected individuals are not normal females, however, because they have blind-ending vaginas and they lack ovaries. Conditions of this type are called testicular feminization, and binding studies show the cells of these people do not have an adequate number of intracellular receptors for testosterone [426], although they do have adequate amounts of testosterone.

Defects in estrogen receptors (ERs) produce skeletal abnormalities in males and females. In males, reduced ER number produces greater than normal skeletal growth with delayed bone maturation; in females, it produces delayed bone maturation as well as increased virilism [427].

Defects of progesterone action affect fertility and sexual behavior in females. In the absence of an adequate number of progesterone receptors (PRs), mice show abnormal behavior [428]. Males lacking PRs show better parenting behaviors: they do not kill their own young and they are less aggressive toward young in general [429].

Defects of the receptors for corticosteroids also occur. One condition is called cortisone resistance, and involves a reduction in affinity of the receptors for cortisol, so activation of target cells is reduced, and although increased amounts of cortisol are produced, it is ineffective. Although the condition is rare, it is a source of abnormal potassium metabolism, hypertension, acne, and menstrual disorders [430].

12.2.4.2 Additional Effects

The ERs are examples of nuclear receptors tied to a number of healthy states such as development and reproduction; they are also implicated in many disease states, including breast cancer, ovarian cancer, and obesity.

The large-scale synthesis of chemicals for agricultural and industrial uses has produced some organic molecules that have specific interactions with steroid receptors. Some of these compounds mimic estrogens and bind to ERs, initiating the synthesis of proteins normally activated only by endogenous estrogens.

Three pathologies have been noted:

- "Gender bending," or the appearance in female animals of characteristics normally associated with males, and vice versa. Several poly-chlorinated biphenyls (PCBs) have the ability to bind to estrogen and androgen receptors at the hormone-binding site, thereby mimicking the effects or normal hormone binding; the interactions produce abnormalities of development, including abnormal structural features and reductions in fertility [431]. These compounds bind with high affinity to receptors, so even low environmental concentrations may have significant physiological effects.
- Antagonism of normal steroid action. The DDT (dichlorodiphenyltrichloroethane) metabolite DDE (dichlorodiphenyldichloroethylene) binds the androgen receptor, thereby inhibiting normal androgen binding, but does not activate transcription. Because the steroid-stimulated pathways of growth and development are so important for normal growth and maturation, DDE reduces reproductive success in many species of animals [431].
- Recent increases in breast cancer in women may be a consequence of the interactions of these chemicals with steroid receptors [431].

The remainder of this chapter develops the relationships between signaling processes and cancer.

12.3 Signaling Mutations Leading to Cancer

12.3.1 Pathogenesis of Cancer

Normal development of a eukaryotic multicellular organism requires controlled cell division of somatic cells, a condition that exists only where controlled intra- and intercellular communication exists. This communication provides signals for differentiation, and it also tells migrating cells when they have arrived at their destinations, when they are packed into a proper shape for organ function, and when it is time for them to die. Rather than being a single disease, cancer is a group of 100 or so related diseases that strike humans, animals, and plants. The primary changes occurring in cancer are these [432].

- Certain types of cells divide rapidly and uncontrollably, produce tumors that crowd out normal cells,

take large amounts of energy, and prevent normal functioning of tissues.
- Cells may act as if they have received a signal to divide, even though they have not.
- Cells may act as though they are alone, even though they are already in a crowded state.

The body parts most often affected are epithelial tissues such as skin, lining of the gut, and respiratory system. Other common forms are cancers of the reproductive system and the blood-forming organs. The above cell types are all ones in which continued mitosis throughout life is necessary; tissues that are least likely to form tumors are those of organs such as nerves and muscles where cell division is rare [432].

12.3.2 Cancer as a Disease of Signaling Molecules

The mechanisms of immortalization, transformation, and metastasis involve changes in the genes encoding several types of proteins involved in cellular communication:

- Mitogenic factors and other cell-signaling molecules (e.g., v-sis – homolog of platelet-derived growth factor – PDGF);
- Receptors (e.g., pp60src – homolog of the insulin-like growth factor (IGF1) receptor);
- Transducing proteins of the plasma membrane (e,g., Ras – analog of G protein alpha catalytic unit);
- Transcription regulators, particularly those that regulate checkpoints of the cell cycle.

The process of phosphorylation causes activation of pathways that enhance growth and proliferation of cells in the embryonic state. In that state, rapid cell division is necessary for development of the adult organism. In adult life, however, rapid cell division produces tumors. Thus, any mutation that disassociates normal signaling could be oncogenic.

A mechanism of this sort is shown in Figure 12.6, in which a normal developmental signaling system, Hedgehog–Patched–Smoothened (Figure 11.12), is used by tumor cell–stromal cell signaling to promote mitogenesis of tumor cells.

The process of phosphorylation causes activation of pathways that enhance growth and proliferation of cells in the embryonic state. In that state, rapid cell division is necessary for development of the adult organism. In adult life, however, rapid cell division produces tumors. Thus, any mutation that disassociates normal signaling could be oncogenic.

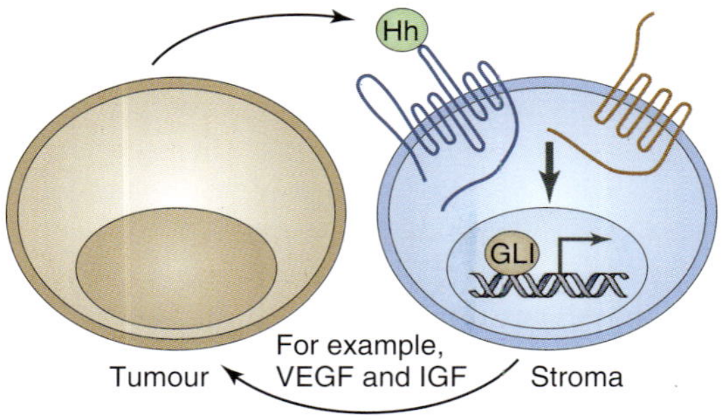

Figure 12.6 Mechanism by which normal developmental mitogenic signaling can lead to inappropriate mitogenesis. Normally developing cells respond to the ligand Hedgehog (Hh) by activating the transcription factor GLI (from glioma tumor), which transcribes genes for mitogenesis. Hh is often produced by tumor cells, and stimulates the stromal cell by autocrine mechanisms, via the receptor Patched (Ptch1) and its associated protein Smoothened (SMOH). Hedgehog stimulates transcription that causes release of other mitogenic ligands such as vascular endothelial growth factor (VEGF) and IGF. These cause further mitogenesis in the tumor cell. Figure 2c in Ref. [433]. Used with permission.

12.3.2.1 Oncogenes that Encode Mutated Transmitters

The receptor link to cancer came to light in the early 1980s when protein sequences from oncogenes were compared to the sequences of other proteins and were found to be similar to certain ones involved in cellular communication. One important such molecule is the peptide, PDGF [434] (see Chapter 8). PDGF is a mitogen released by platelets in response to tissue damage. It binds to cell-surface receptors and stimulates cell division in the area of a wound.

The oncogene in this case is v-sis, the causal agent in **si**mian **s**arcoma; its protein product is nearly identical to a short segment of PDGF, and its chemical properties are identical to those of PDGF. Data suggested that the v-sis oncogene is a simian gene picked up by the virus and then mutated. It mimics PDGF, but in a way that is abnormal and outside the normal control mechanisms of cellular PDGF. This oncogene therefore produces uncontrolled cell division [435].

12.3.2.2 Oncogenes that Encode Mutated RTKs

Some oncogene products are similar to receptors: the protein product of the erbB oncogene of avian erythroblastosis virus is very similar in structure to the epidermal growth factor (EGF) receptor, a cell-surface molecule in the same family as the PDGF receptor [434]. The EGF receptor has extracellular, membrane-spanning, and cytoplasmic domains. The erbB protein is equivalent to the membrane-spanning and cytoplasmic domains, and is missing most of the extracellular EGF-binding domain. This missing domain may partly explain its transforming nature: because of the loss of the binding region, the receptor may be permanently deregulated and stuck in the "on" position, activating the intracellular machinery of cell division. Expression of erbB in cell lines produces changes in epithelial cell structure and changes in cell-adhesion molecules, both of which can explain the cancerous nature of the condition [436].

Mutations in the human EGFR (epidermal growth factor receptor) occur primarily in 4 of the 28 exons (Figure 12.7). These mutations affect ATP binding involved in kinase activity, leading to constitutive activity. Therapies against these cancers involve using kinase inhibitors that target sequences on particular exon-encoded portions of the receptor. The most sensitive regions are shown in Figure 12.7.

Therapies for erbB-induced cancers also involve developing **monoclonal antibodies** against the receptor to stop its function and blockade of dimerization to prevent activation of the intracellular kinase. The same receptors whose binding properties can be altered to form **oncoproteins** have intracellular domains subject to mutation (next section).

12.3.2.3 Oncogenes that Encode Mutated G Proteins

The ras oncogene products are proteins that belong to the guanine-nucleotide-binding (G protein) class [434] (see Chapter 7). These proteins are part of the inner membrane complex of enzymes activated by cell-surface receptors of the G-protein-coupled group. In turn, they activate other enzymes involved in intracellular signaling. In normal cells, G protein hydrolysis of GTP to GDP stops cell activation; the ras oncogene product, however, lacks hydrolytic activity, and therefore remains in the "on" position, maintaining activation of second messenger systems for gene transcription.

One class of G proteins, the G_{12} family (Chapter 7), activates processes of cell division via activation of the Rho GTPases; this G protein group is therefore being studied intensively with respect to cancer mechanisms and therapies [438].

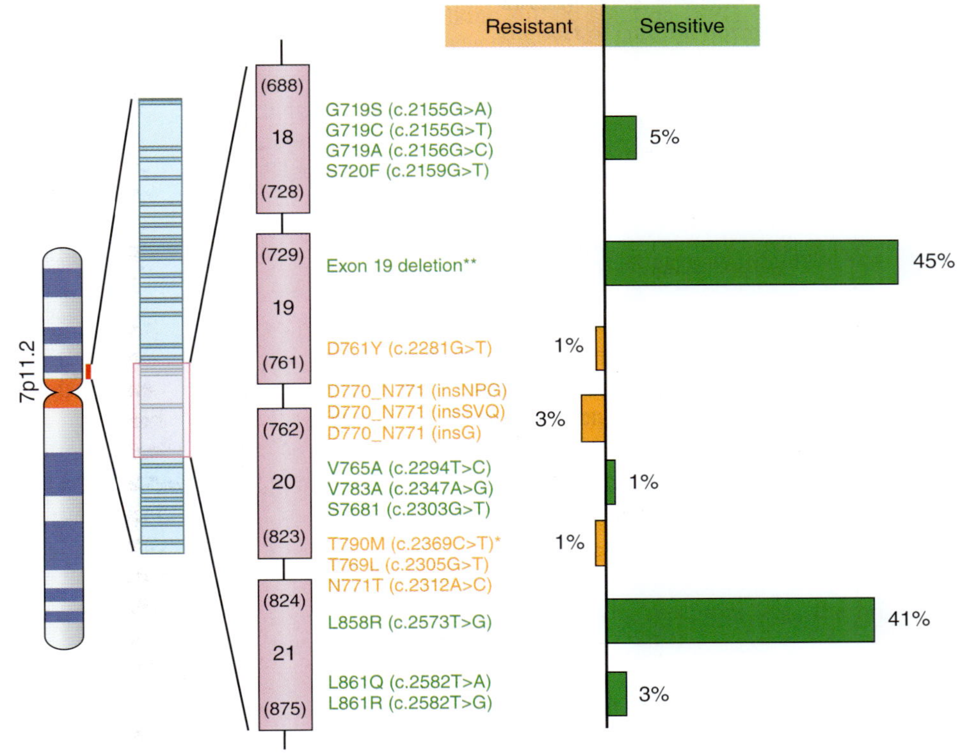

Figure 12.7 Mutations in the epidermal growth factor receptor (EGFR) gene. Mutations in EGFR can lead to cancer. In this study, cancerous mutations in exons 18–21 were analyzed for their resistance (orange) or sensitivity (green) to chemotherapy. Figure 2 in Ref. [437]. Used with permission.

12.3.2.4 Oncogenes that Encode Mutated Transcription Factors – Steroid Receptors

Receptors involved in breast cancer are a receptor tyrosine kinase, the ErbB receptor (Chapter 8), as well as the ER and PR (Chapter 9). Because these are transcription regulators, mutations in these receptors can have serious effects on cell proliferation.

Estrogen and progesterone are transported in blood and diffuse across cell membranes into the cytoplasm. They serve as ligands for the ER and PR: each ligand–receptor complex binds to targeted estrogen response elements on DNA and serves as a transcription factor for new cellular proteins.

Many of the newly synthesized proteins are involved in cell division; thus, one might suppose that an absence of ERs would prevent cell division; however, estrogen likely works in concert with many other signaling pathways to achieve its proliferative effect. Triple negative breast cancer (TNBC) is a disorder that highlights the complex signaling processes that may lead to unchecked cell division.

Receptor-Positive and Receptor-Negative Breast Cancers The primary signaling molecules involved in cancers are transmitters, receptors, and effectors of the types of signaling systems involved in the normal process of mitogenesis (i.e., the RTK and nuclear receptor superfamilies). Receptor-positive cancers grow in the presence of the hormone (e.g., estrogen or progesterone) for the receptor. These cancers respond favorably to administration of selective estrogen receptor modulators (SERMs) (Chapter 9) or inhibitors of the enzyme aromatase (Figure 4.23). Both these therapies inhibit estrogen activation of transcription factors leading to cell proliferation. Cancers caused by RTK mutations can be treated by addition of drugs that inhibit either receptor dimerization or kinase activity [439].

However, about 25% of breast cancers are negative for one or more receptors that promote cell proliferation. This suggests that intracellular proliferative pathways are present and can be activated by other signaling systems. Thus, because prostaglandins may promote MAPKinase signaling, drugs such as aspirin or non-steroidal anti-inflammatory drugs (NSAIDs) that inhibit prostaglandin (Chapter 4) formation are also useful therapies in negative receptor cancers [439].

Triple-Negative Breast Cancer In patients with TNBC, tumor cells are lacking three specific receptors: the ER, the PR, and HER2, a member of the EGF receptor family. Without an ER present, these tumor cells are not sensitive to the proliferative effects of estrogen. However, in these cells, a form of the EGFR is upregulated. It is likely that the metastatic effects are mediated through inappropriate action of this receptor. Cancer cells with this receptor can be extremely aggressive and metastatic, leading to the spread of TNBC cells, often to the bones or brain. While the TNBCs are a heterogeneous group of cancer types, several types can be classified as having upregulated EGFR.

Because of the absence of ERs, PRs, and the HER2 receptors, these TNBCs are not sensitive to the hormone-based chemotherapies that have been successful in other types of breast cancers. This lack of response leads some TNBCs to be classified as having a poor prognosis for patients. However, another ER-like transcription factor, *estrogen-related receptor alpha (ERRα)*, is associated with TNBC. This protein is homologous to the ER, but is active in the absence of ligand binding. As this factor is upregulated in TNBC while estrogen, progesterone, and HER2 receptors are missing, it has become a target for drug discovery [440]. It is our hope that therapies for cancers such as TNBC can develop out of increased knowledge of receptor signaling functions – in this case, a better understanding of the role of the EGFR and ERRα.

13
Receptors and the Mind

> All organisms with complex nervous systems are faced with the moment-by-moment question that is posed by life: What shall I do next?
>
> *S. Savage-Rumbaugh and R. Lewin [441]*

> It is the business of scientists to explain away the magic in the world. The largest coherent body of magic remaining is the behavior of (humans) and animals.
>
> *W. Quinn and J. Gould [442]*

> Behaviour evolved before nervous systems.
>
> *G. Jekely [443]*

13.1
Origins of Behavior

All organisms are conscious of their existence: they respond to external stimuli, take in food, excrete wastes, and reproduce. Vertebrates use their extensive central nervous systems (CNSs) with multiple sensory inputs, interconnections, and motor outputs to coordinate behavior that is rapid and individualized. In contrast, small invertebrates have decentralized nerve networks, and single-celled bacteria and protists have no nervous systems and no cellular differentiation at all. Yet, as we describe below, these non-vertebrates all have a remarkable behavioral repertoire. Thus, behavioral complexity is not a property of complex brains alone. It derives from the abilities of cells to communicate with other cells.

All organisms use the same chemical mechanisms of transmitter–receptor–transducer to promote behavioral integration among parts of the body or between individuals. At the chemical level, little difference exists between the thought of a philosopher and the response of an ameba to an inviting stimulus, and all organisms can think in the sense that a thought represents a pattern of purposeful cellular signaling in response to an earlier cellular or environmental signal.

For these reasons, the evolution of vertebrate neural organization has not depended on evolutionary changes in the fundamental mechanisms of communication. Instead, the evolution of complex brains emerged from alterations over time in the types and numbers of connections made between cells and in the ways that cells are organized into systems. Thus, our understanding of the structure and function of the human nervous system is based in large part on inferences from the remarkable array of signaling processes in the communication systems of other organisms [444].

13.1.1
Bacterial Short-Term Memory

Organisms move by use of pseudopodia, cilia, flagella, or muscular limbs. Each of these effector organs requires ATP hydrolysis, a process that is governed by extracellular signaling. Even the motions of single cells are governed by receptor processes. Bacteria (e.g., *Escherichia coli*) have receptors for nutrients as well as for noxious chemicals, and binding of these chemicals activates sophisticated intracellular pathways, leading to appropriate changes in activity of proteins controlling flagellar direction [445]. Low concentrations of attractant cause flagella to rotate counterclockwise to create a directional movement, while high concentrations of attractant change flagellar motion to clockwise and disrupt directional movement. These receptor–flagellar connections are activated in immediate response to chemical gradients, but the same receptor systems are also capable of longer term learning.

Learning is modification of behavior based on experience, and must have two characteristics: the signaling apparatus (receptor, transducer) must be modifiable, and mechanisms must exist for biochemical storage of the task, at least for an amount of

Receptor Biology, First Edition. Michael F. Roberts and Anne E. Kruchten.

time that assures its retention. Thus, in addition to adaptive attraction or avoidance reflexes, bacteria show memory of their recent movements in relation to nutrient gradients. This memory is not long in human terms, but it is long enough to provide adaptive responses to changes in nutrient gradients [445]. For example, memory in *E. coli* lasts from 1 to 10 s, or enough time for a cell to travel 10–100 bacterial lengths. If memory were much shorter, the cell would not go far before it forgot what it was supposed to look for; if memory were much longer, it would be remembering earlier parts of the gradient, inappropriate for current needs [446].

Bacteria have other receptor-based environmental adaptations. In most aquatic environments, temperature is inversely related to oxygen content. Thus, if bacteria detect a rise in temperature, they correctly predict that anaerobic conditions will occur in the near future; they therefore deactivate aerobic enzymes and activate anaerobic enzymes. However, this response is subject to adaptive modification: if an experimenter alters the usual relation between temperature and oxygen such that a rise in temperature is followed by an *increase* in oxygen content of the solution, the bacteria will, over a number of generations, evolve a new predictive capacity and activate *aerobic* enzymes following a rise in temperature [447].

The above responses of bacteria are sophisticated and adaptive; they are all the more remarkable in involving signaling systems within single cells. More complex learning can occur in multicellular animals.

13.1.2 Animals Without True Neural Organization: The Porifera

The simplest animals are the sponges. They have only a few types of cells (no true nerve cells), and no true organs. They have localized response systems in which the cell receiving the stimulus is also the effector cell: the pore cells on the sponge surface respond to environmental stimuli, and make appropriate local responses that affect water movement into the sponge. Communication from one location to another occurs between contacting cells via cytoplasmic bridges or within large syncitial cells [366].

However, sponges do have well-developed sensory systems: the cilia are associated with photodetectors that alter ciliary beating; sponges also have many genes encoding G-protein-coupled receptors (GPCRs) that are probably chemosensory in nature [443]. No peptide neurotransmitters have been identified in sponges, but these animals do synthesize acetylcholine (ACh), as well as proteins used by eumetazoans for processing peptide transmitters [443].

13.1.3 Animals with Neural Networks: The Cnidaria

The anemones, corals, jellyfish, and hydras have a diploblastic structure with true tissues and a nerve net with true neural cells that communicate via transmitter/receptor interactions. Cnidarians have three types of neural cells: sensory, ganglionic, and **neurosecretory**. They have peptide **neurohumors** as well as other transmitters (catecholamines and ACh), though the cells that release the different transmitters appear to be morphologically identical.

So far, however, cnidarians show no evidence of recognizable tracts of neurons nor do they have centers of integration and organization. Thus, they have **neural networks** rather than **nervous systems**. The term neural network refers to a situation in which sensory cells act directly on effector cells. In a nervous system, in contrast, sensory receptor cells connect first to a central integrating area (the brain or other element of the CNS), then to the effectors (Figure 13.1).

In both types of organization, communication occurs via diffusion of transmitters from one cell to another rather than via direct electrical connections between cells. The neural net pattern of organization provides adequate responses for stationary organisms, organisms that are slow moving, or those that do not possess a unique "direction" to their movements. A nervous system, in contrast, allows rapid

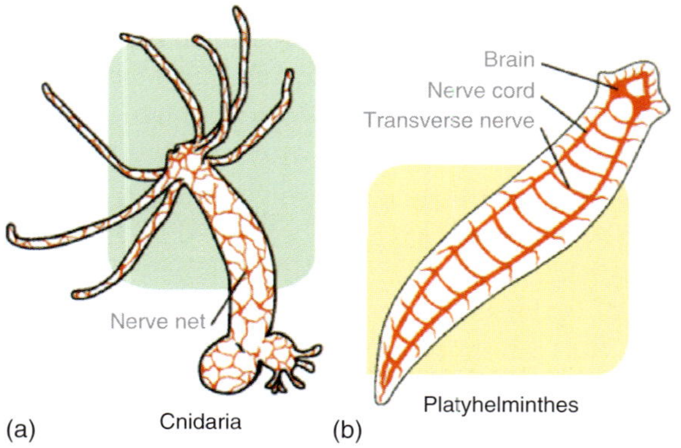

Figure 13.1 Comparison of diffuse nerve network with a central nervous system. (a) Hydra neural net structure, with potential for any neuron to make contact with any other. (b) Flatworm with two segmented longitudinal nerve cords, and all connections occurring within a segment. Figure 17.2 in Ref. [448]. Used with permission.

central processing of sensory information and an integrated motor response that allows rapid motion with a direction.

Cnidarian neurons have intracellular organelles that synthesize and secrete transmitters and humors, yet they have no axon-dendrite organization, nor do they show close synapses. These cells are not permanent as vertebrate nerve cells are, but are constantly renewed. They have receptors and transmitter mechanisms, and two sorts of synapses: one type also seen in bilaterian animals with button-shaped termini and another, unique "en passant" system in which the transmitter may be released along the axon rather than only at the end.

13.1.4 Bilaterally Symmetrical Animals: The Acoela

The members of this flatworm phylum are thought to be similar to the first bilaterians and to be a sister group to all remaining animal phyla [449]. The pattern of neural development reflects their bilateral structure, as they have a well-developed neural network in the epidermis combined with a concentration of cells in the anterior part associated with eye spots. The network consists mostly of orthogonal neurons (neurons that cross at right angles) but no large ventral or dorsal nerves [449]. They also have more types of nerve cells than cnidarians, as well as greater morphological diversity in their neurosecretory cells. Some species have been shown to use ACh and serotonin as transmitters [450].

The remainder of the bilaterian animals form two large groups, the Protostomes and Deuterostomes (Chapter 11). These have well-developed true centralized nervous systems rather than nerve nets, and have complex signaling connections, synapses, and multiple transmitters.

13.2 Nervous Systems

13.2.1 Organization

Up to the eighteenth century, the brain was thought of as a glandular tissue and neurons were thought to be tubes that conveyed brain fluids to the rest of the body. By about 1800, the electrical properties of neural tissue began to be appreciated, though the interactions between neurons were not considered until after Bernard, Langley, and Ehrlich (Chapter 2) proposed the receptor concept.

In the late 1800s, Golgi showed that neurons were individual cells, and proposed that they made direct cytoplasmic connections with each other. Ramon y Cajal argued against Golgi's hypothesis of direct cytoplasmic connections when he stained individual neurons and observed that though they make close contact at the synapse, neurons remain distinct from each other. He proposed three concepts that have become the cornerstone of modern neurophysiology: that neurons signal with a chemical mediator across the short distances that separate them; that within one cell, action potentials are directional, from dendrites to nerve cell body to axon; and that neurons from one area make specific connections to target cells rather than forming an indiscriminate network [451].

In addition, several other novel concepts have arisen in the past few decades: first, a given neuron might release more than one type of transmitter at synapses on different axonal endings. Second, neurotransmitters are released not only from brain cells but also from non-neural cells in many organs of the body. These molecules often act as classical hormones on cells far from their site of origin. Brain cells thus affect, and are affected by, peptide-secreting cells in many organs of the body. Not only does this action give a wider spatial dimension to signaling, it also spreads signals out over time. It is important to think of the mind as a chemical communication system that includes all the body's cells, not just the brain.

The remainder of the chapter deals primarily with neurons and the transmitters permitting the complex forms of nervous signaling involved in memory and mood. Research on both invertebrates and vertebrates has contributed to our understanding of the workings of the mind.

13.2.2 Neurons

13.2.2.1 Cell Structure

Each neuron has four regions important in signaling:

- The cell body containing the nucleus;
- Dendrites, processes that extend from the cell body and receive incoming signals;
- One axon that leaves the cell body and carries impulses to other cells;
- Axon terminals, the sites of transmitter release (Figure 13.2).

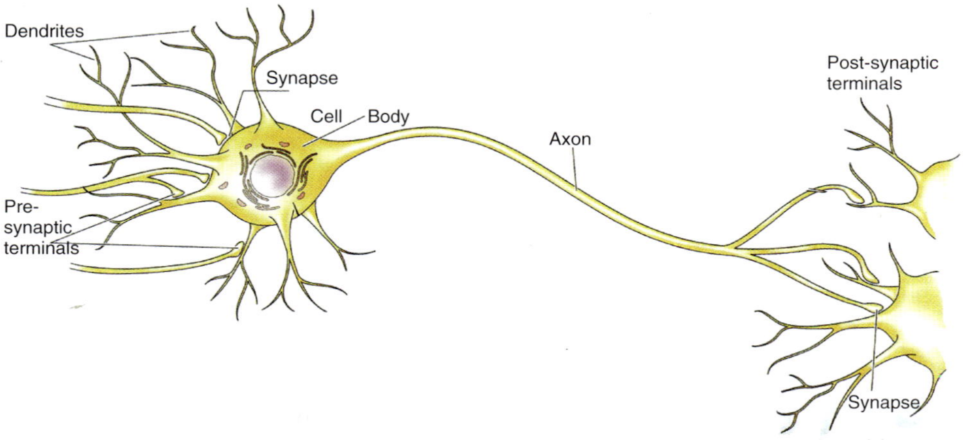

Figure 13.2 Vertebrate neuron structure, showing presynaptic cell, synapse with excitatory and inhibitory axons, postsynaptic cells. Modified from Figure 11.2 in Ref. [87]. Used with permission.

13.2.2.2 Mechanisms

The mechanism of signaling is the same in all neurons:

- A transmitter is released from one neuron and diffuses to a dendrite; it binds to ligand-gated ion channels (Chapter 6) on the dendrite surface.
- The binding opens the ion channel of this postsynaptic cell and allows cations to flow into the cell, creating a localized depolarization.
- If the depolarization is strong enough, it will open enough additional ion channels to form an action potential.
- Each action potential opens cation channels in adjacent parts of the neuron, allowing the depolarization to be conducted along the dendrite to the cell body and out along the axon.
- At the axon terminals of this postsynaptic cell, the action potential will again cause the release of transmitter to a third cell, and the process will continue.

13.2.3 Transmitters

The number of known nervous system transmitters stands at a hundred or so: the best known are ACh, gamma-aminobutyric acid (GABA), glutamic acid, glycine, and the monoamines norepinephrine (NE), serotonin or 5-hydroxytryptamine (5-HT), and dopamine (DA). In addition, steroids and peptides – some newly discovered and others that had already been known from different organ systems [452] – are being added to the list almost daily.

Amino Acids/Amines: Certain molecules known to be neurotransmitters in vertebrates are also produced by other animal phyla [453]. Acetylcholine, catecholamines, and 5-HTs are found in all eukaryotes, even in unicellular protists. It is possible that they are used for other purposes (e.g., in the synthesis of other molecules [453]), and experiments would have to demonstrate their packaging into vesicles, release during stimulation, and effectiveness at synapses before they could be accepted as transmitters. However, receptors are present for many of them, and their wide distribution implies that the receptors and synthetic pathways for these transmitters were established well before the origin of animal phyla about a billion years ago.

Peptides: The first **peptidergic** neurons discovered in the brain were those that released oxytocin and vasopressin (Chapter 4) into the circulation of the posterior pituitary. Next came the **hypothalamic-releasing hormones**. All these peptides are usually considered **neurosecretory** substances rather than transmitters because they are released into the bloodstream rather than into a synapse between two adjacent cells. The discovery of true peptide **neurotransmitters** came later, after immunological methods became available for visualizing them in the brain.

More than 50 peptides have been identified as transmitters in the brain and other neural tissue; in addition, for the most part their receptors have been identified, isolated, and characterized. Peptides with possible roles in communication do not appear to be present in unicellular organisms, though they are found in all metazoans, and are encoded by families of genes that seem to be shared in many phyla. In the

Table 13.1 Phyletic distribution of transmitter families.

Animal taxon	Peptide transmitters	
Cnidaria	Substance P (SP)	
	Gastrin/CCK	
Platyhelminthes	ACTH	
	Somatostatin (SS)	
	Neurophysin	
Annelida	Pancreatic polypeptide (PP)	β-endorphin
	Vasoactive intestinal polypeptide (VIP)	SP
	Enkephalin	ACTH
Mollusca	α-Endorphin	TRH
	Enkephalin	ACTH
	Met-enkephalin	Insulin
	Gastrin/CCK	Vasopressin
Arthropoda	Neurophysin	PP
	Vasopressin	Gastrin/CCK
	Insulin	α-Endorphin
Vertebrata	CCK	Angiotensin II
	SP	Bradykinin
	VIP	ADH
	Neurotensin	Oxytocin
	Met-enkephalin	CRH
	Leu-enkephalin	GHRH
	Insulin	LHRH
	Glucagon	Dynorphins A, B
	TRH	α Neo-endorphin
	GnRH	α melanotropin
	SS	Gastrin
	ACTH	NPY
	β-Endorphin	Calcitonin
	Somatostatin	CGRP
		Bombesin

CCK, cholecystokinin.

invertebrates, peptidergic cells have been found in neurosecretory roles and in transmitter roles.

The peptides so far found in invertebrates (though not all have yet been shown to be transmitters) are listed in Table 13.1 [448].

The peptide transmitters above are agonists of either GPCRs (Chapter 7) or receptor kinases (Chapter 8). Amines, ACh, and amino acids are agonists of GPCRs or ion channel receptors (Chapter 6).

13.2.3.1 Synthesis and Release of Brain Transmitters

The neuronal mechanisms of synthesis and transport differ for peptide and amine transmitters (Figure 13.3).

The majority of cellular proteins are synthesized in the nucleus. Thus (with an exception noted later in the chapter (CPEB, cytoplasmic polyadenylation element binding, proteins)), any proteins located in the axonal endings of neurons must have been moved there by cytological transport systems. Neurons that secrete transmitters make proteins that are either (i) the transmitters themselves or (ii) enzymes for synthesis of transmitters from non-peptide precursors. In both cases, the synthesized proteins are transported down the axon from the cell body to the terminus. The proteins are first packaged in Golgi bodies in the nucleus, and then the vesicles containing the proteins are attached to tubulin "motors" that carry the vesicles at rates of up to 40 cm/day along the axon [454].

Release of both types of transmitters occurs by the same mechanism: depolarization of the nerve ending causes Ca^{2+} influx, which leads to migration of the vesicle to the surface and fusion with the plasma membrane. The contents of the vesicle then become part of the extracellular fluid, and diffuse across the synapse to the nearby postsynaptic cell membrane.

Once released, amine hormones may diffuse away from the junction, but are mainly taken up into the presynaptic cell and repackaged into vesicles for later use. The proteins involved in uptake are the transporters discussed in Chapter 6. Amines may also be degraded by enzymes in presynaptic or postsynaptic cells. Transmitter availability at receptors can be affected by these transporters and degradative enzymes: rapid reuptake or rapid degradation will reduce transmitter concentration, and anything that interferes with reuptake or degradation will increase transmitter concentration. The consequences of these two processes for transmitter disposition will be considered later in the chapter.

Peptides are not taken back up into neurons [452], but diffuse away from the synapse and are degraded by extracellular enzymes. Each peptide molecule that is released during nerve stimulation must be replenished by protein synthesis at the cell body and moved to the nerve ending by axonal transport. These energy-requiring processes may explain why peptidergic neurons have lower concentrations of transmitter and release fewer molecules per impulse, and are also consistent with the observations that peptide receptors show higher affinities for their transmitters than do amine receptors for their transmitters [452].

For the remainder of this chapter, we focus on extensions of these signal transduction mechanisms in three areas of brain function: **Learning and memory** represent alterations in transduction processes of ion channel receptors and GPCRs; **addiction** represents alterations in availability and

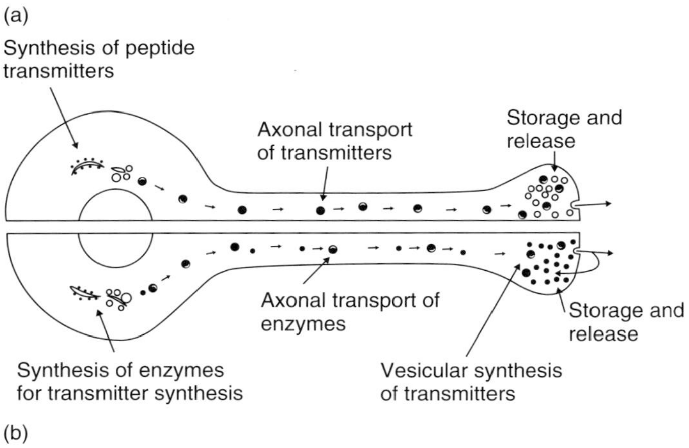

Figure 13.3 Synthesis and transport of brain transmitters. (a) Peptide transmitters are synthesized in the nucleus and transported to the axon terminus for storage and release. (b) Classical (amine/amino acid) transmitters are synthesized in the axon terminus by enzymes that are transported down the axon. Modified Figure 1 of Ref. [452]. Used with permission.

function of receptors in tissues after exposure to transmitter-like drugs; and **mood** is affected by alterations in transmitter availability at receptors.

13.2.3.2 **Converting Short-Term Memory to Long Term**
Above we described the basis for short-term memory in bacteria. A critical part of learning is the conversion of experiences from **short-term memory** into **long-term memory**. For example, people retain the memory of their names for their lifetimes; they retain a merchant's phone number only long enough to dial it (yes, the phone industry still uses the term "dial" for touch tone phones). Certain memories can be converted from short term to long term by continued use: to keep the merchant's number in memory, people have to repeat it many times and/or use it frequently.

In microorganisms, "long-term memory" is encoded in biochemical pathways, such as the relationship between temperature and anaerobiosis in *E. coli* [447]. Individual long-term memory is present in animals with nervous systems, and the mechanisms of memory have recently begun to be approached by biochemical means. It is now clear that the same cellular mechanisms are responsible for learning in invertebrates and vertebrates.

13.3 Animal Memory: Invertebrates

13.3.1 Discovery of the Signaling Contribution to Memory

Ramon y Cajal proposed in the 1890s that learning consisted of alterations in the structure of synapses rather than in the creation of new neurons [451]. This suggests that alterations in signaling are important in learning, and in the 1960s and 1970s the first synaptic mechanism of learning (long-term potentiation) was discovered, based on studies of neurons in the hippocampus of the rabbit brain [455]. These studies implied that a transmitter–receptor interaction was involved, but at the time, the specific cellular mechanisms involved could not be determined.

Kandel reasoned that a simpler system might provide a successful approach, arguing that invertebrates must have memory, and it is likely that the same overall mechanisms of memory are present as in humans. He decided to study learning in the marine mollusc *Aplysia* for several reasons: these animals have only 20 000 neurons, their systems are organized into a small number of **ganglia** that allow identification of individual neurons, and the cells are large enough (1 mm diameter) to be easily studied [444]. This experimental model allowed Kandel and colleagues to develop useful models of the mechanisms for both short-term memory and long-term memory [444, 446].

13.3.2 Receptor Mechanisms of Nerve Cell Interactions

13.3.2.1 **The Gill Withdrawal Reflex of Aplysia**
The *Aplysia* external gill is housed in a cavity covered by a muscular mantle shelf terminating in a siphon (Figure 13.4a). *Aplysia* has a rapid reflex for protective withdrawal of the gill if danger threatens. For example, when an investigator gives the siphon or mantle shelf a light touch, the entire

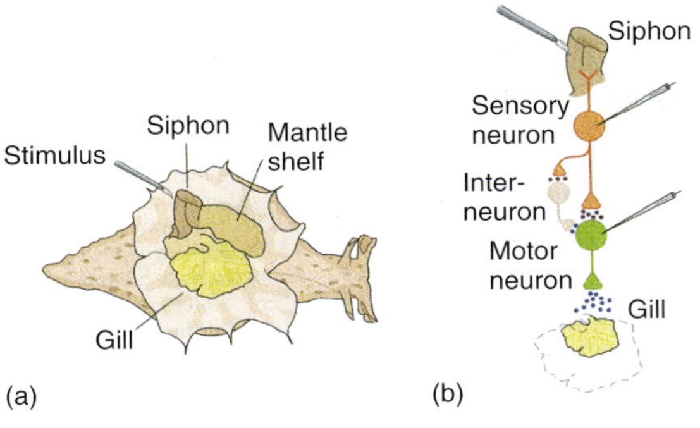

Figure 13.4 (a) Anatomical drawing of *Aplysia* (anterior is to the right), showing gill, siphon, and mantle shelf. (b) Simplified diagram of neural connections in gill withdrawal circuit. Modified from Figure 66.2 in Ref. [448]. Used with permission.

structure contracts and is withdrawn into the cavity (Figure 13.4b).

As in all defensive responses, the gill withdrawal reflex is modifiable by learning; in this case, the animal learns after 10–15 stimuli that repeated light touch is not itself a source of danger, and reduces the intensity of its withdrawal response. This change is called **habituation** , and involves a reduction in transmitter release by the sensory nerve [457].

The opposite can occur as well: the withdrawal reflex can be strengthened (**sensitized**) if one gives an electrical stimulation of the tail while touching the mantle. When these two stimuli are given together, the motor nerve to the mantle releases more transmitter, so the withdrawal reflex lasts longer and is more difficult to habituate.

The neural network involved in **sensitization** is shown in Figure 13.4c. The siphon skin touch receptor is the axonal ending of a sensory neuron (SN). This neuron has axons projecting directly to a motor neuron (MN), as well as indirectly to the same motor neuron via an excitatory interneuron. When the siphon is touched, the transmitter released (from either the sensory neuron or the interneuron) causes contraction of the muscle of the gill. The direct pathway by itself cannot explain sensitization of the reflex to a dual stimulus of electrical shock and touch; thus, the mechanism of sensitization must involve contributions from the excitatory interneuron.

13.3.2.2 Mechanisms Underlying Sensitization and Short-Term Memory

Figure 13.5 shows the entire neural network for activation and facilitation of the gill withdrawal reflex.

The facilitating interneuron (FI) receives sensory input from the tail. In turn, the FI sends axons to the synapse of the sensory neuron on the motor neuron.

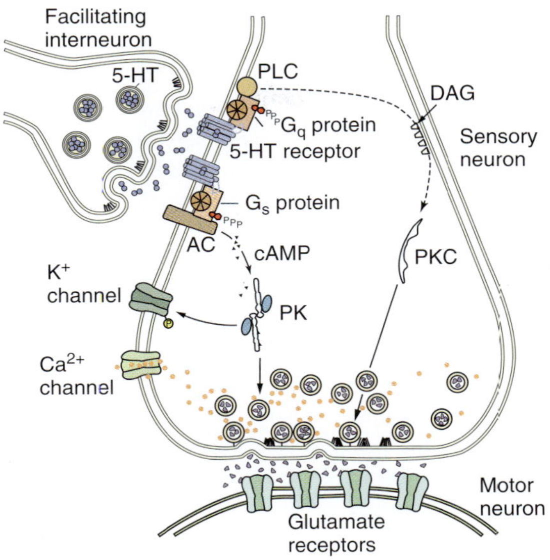

Figure 13.5 *Aplysia* facilitation circuit. Resultant at the sensory neuron of facilitation shown in Figure 13.4. PKA: protein kinase (activated by cAMP) that phosphorylates the K^+ channel. PLC: phospholipase C that synthesizes diacyl glycerate (DAG) to activate Ca^{2+} channels. Modified from Figure 66.4 in Ref. [448]. Used with permission.

Kandel and colleagues discovered the mechanism for sensitization by hypothesizing that FI somehow facilitates transmitter release from the sensory neuron. Because transmitter release is caused by increased intracellular Ca^{2+}, they proposed that facilitation is an increase above normal in intracellular Ca^{2+} brought about by one or more second messenger mechanisms following receptor binding.

13.3.2.3 Ion Flows in Nerve Action Potentials

When an unfacilitated mollusc sensory neuron fires, the depolarization spreads down the axon by the sequential opening and closing of adjacent Ca^{2+} and K^+ channels: Ca^{2+} channel opening allows Ca^{2+} into the cell, driving its membrane potential positive and depolarizing it. Repolarization, or the ending of the action potential, is aided by the opening of K^+ channels, allowing K^+ to leave the cell and reducing the number of positive ions inside (this mechanism of impulse propagation is different from that in vertebrate neurons, where alternate openings of Na^+ and K^+ channels drive propagation). If a K^+ channel could be kept closed, the depolarization would be prolonged and more transmitter would be released from the sensory nerve.

Thus, at the molecular level, the process appears to be as follows (Figure 13.5):

1) Transmitter (5-HT) (left side of sensory neuron) is released from FI following electrical stimulation of the tail.
2) The 5-HT binds to its receptor (a GPCR – Chapter 7) on the presynaptic terminal of SN, activating a G_s protein that in turn activates adenylyl cyclase to produce cyclic adenosine monophosphate (cAMP).
3) cAMP binds to the regulatory subunit of a cAMP-dependent kinase (PK (protein kinase)), causing the dissociation of the regulatory unit from the catalytic unit (Chapter 7).
4) The activated catalytic unit phosphorylates the open K^+ channel, causing it to close.
5) The closure prolongs depolarization and permits greater amounts of Ca^{2+} to enter the cell, increasing the release of transmitter from SN24.
6) An additional action of 5-HT is on another GPCR that activates a G_q protein to produce phospholipase C (PLC), which splits the membrane lipid phosphoinositide bis-phosphate (PIP_2) to produce the second messenger diacylglycerate (DAG). This mobilizes transmitter vesicles, contributing to increased transmitter release.

The end result of sensitization is the increase in activity of SN, causing an increase in the sensitivity of the siphon to light touch. Sensitization is thus a relatively simple event involving receptors and intracellular messengers; yet this simple receptor process has a crucial influence on the behavior of the whole animal, producing the short-term memory (up to an hour) [457] of the initial electrical shock.

13.3.2.4 Consolidation into Long-Term Memory (LTP)

Kandel and Schwartz [456] proposed that the consolidation of long-term memory could involve enhancement of some of the same processes as short-term memory. They showed that with repeated electrical stimulation of the tail, the memory becomes much longer lasting – up to a few weeks [457]. This new sort of memory involves changes in the same receptors as created sensitization, but these changes are consolidated in two ways: into structural changes in the synapses, and in production of new proteins (not just phosphorylated old proteins).

The synaptic changes involve an increased ability to release transmitter by several mechanisms: increasing the number of axonal regions where contact occurs with a motor nerve, increasing the number of locations on the presynaptic membrane where transmitter vesicles form, and increasing the number of transmitter vesicles.

In addition to structural changes, Kandel and Schwartz showed that cAMP activates the MAP-Kinase system (Chapters 7 and 8) to activate a transcription factor that binds to DNA (at the CRE (cyclic nucleotide response element), or cAMP response element – nuclear events in Figure 13.6) to activate the transcription of new kinases and new growth enzymes.

The new kinases then break down the regulatory subunit of Kinase A so the catalytic portion persists in active form. This persistence allows increased release of transmitter.

In addition, LTP involves entirely new mechanisms: synthesis of new proteins occurs in specific synapses rather than in the cell body; thus, mRNAs for new proteins are transported to the synapses rather than completed proteins. This implies that some sort of translation signal must be present to control this synthesis. That signal is the presence of a new type of protein – the CPEB protein – that is somehow activated in particular synapses, and then causes localized translation of mRNAs to occur [444]. CPEB is present in many cellular locations, but it is activated by increased transmitter release at those synapses accentuated during LTP, so its activation is localized (Figure 13.6, lower right).

13.4 Animal Memory: Vertebrates

13.4.1 Intracellular Mechanisms of Potentiation

LTP in vertebrates has a mechanism similar to that seen in molluscs, though the vertebrate brain has more cells than the pathways studied in *Aplysia*. The hippocampus, a part of the forebrain, is a major location for the storage of implicit long-term memories [444]. An important hippocampal pathway involves the synapse between two types of neuron in the hippocampus, the CA3 and the CA1 cells. Though the synaptic structures resemble those in *Aplysia*, the CA3 facilitating neuron releases glutamate rather than 5HT and the channels involved are ion channel receptors (NMDAR and AMPAR – Chapter 6) rather than GPCR metabotropic receptors.

Research on the mouse CA3–CA1 system has provided evidence for both short-term and long-term memory processing. Stimulation of CA3 with one short train of impulses produces early LTP, in which glutamate acts on *N*-methyl-d-aspartate (NMDA) receptors to increase intracellular Ca^{2+}. The Ca^{2+}

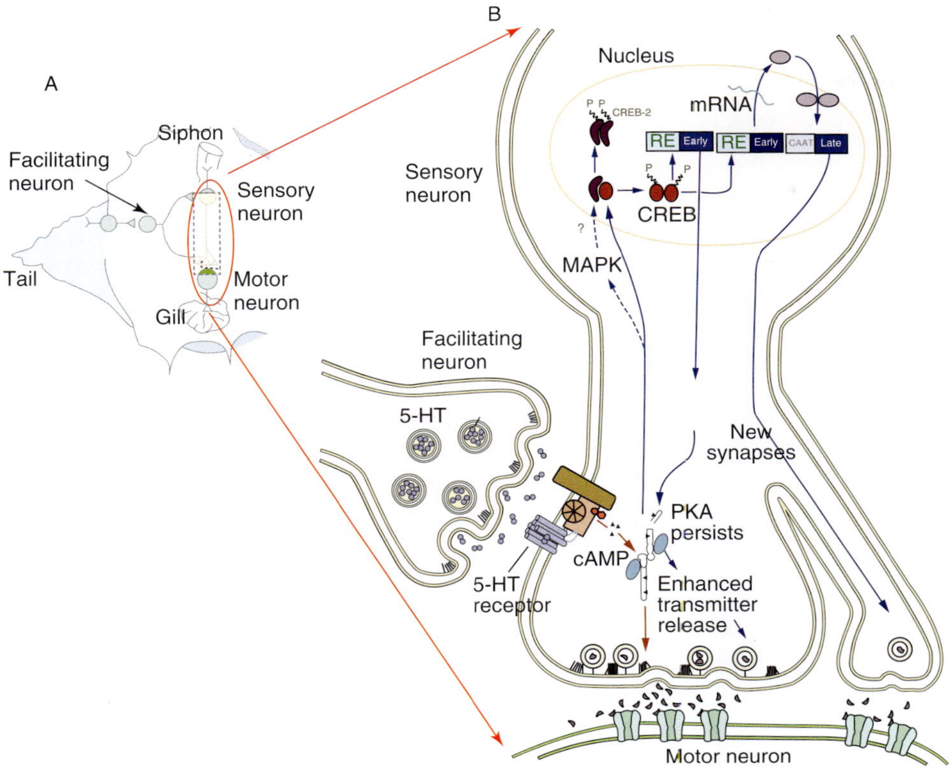

Figure 13.6 *Aplysia* cAMP mechanism of presynaptic facilitation. A: Overview of *Aplysia* sensory, facilitating, and motor nerves; red oval: detail of the memory circuit. B: Detail within red oval of A. CREB: cyclic nucleotide response element binding protein (transcription factor). CRE: cyclic nucleotide response element (promoter on DNA). Blue oval at lower right: new synapses formed by action of CPEB protein. Modified from Figure 66.6 in Ref. [448]. Used with permission.

binds calmodulin, activating three types of kinases (protein kinase A (PKA), protein kinase C (PKC), and calcium–calmodulin kinase), which phosphorylate AMPA receptors and the transcription factor cyclic-AMP-related binding protein (CREBP).

Longer trains of impulses through the CA3 produce late LTP, in which the Ca^{2+}-calmodulin activates PKA to phosphorylate the transcription factor CREBP. Transcription of new proteins is thus the result of LTP. These new proteins are the basis for manufacture of new synapses and additional receptors. The step-by-step process is shown in Figure 13.7:

Before LTP: glutamate released from presynaptic cell (a CA3 neuron synapsing on a CA1). Glutamate binds to either NMDA receptors or AMPA receptors; both are Ca^{2+} channels and promote depolarization of the CA1 cell.

Postsynaptic-LTP (Part A): low-frequency firing of the CA3 neuron causes *postsynaptic LTP* in which new AMPA receptors are added to the CA1 cell membrane, thus increasing the sensitivity of the CA1 cell to incoming transmitter.

Pre- and Postsynaptic LTP (Part B): high-frequency firing of the CA3 neuron causes release of BDNF, which binds to its receptor on both presynaptic and postsynaptic cells, promoting increased glutamate release from CA3, increased numbers of AMPA receptors, and opening of Ca^{2+} channels. All these effects increase sensitivity of the CA1 neuron and contribute to its retention of the memory of the increased activity in the CA3 neuron. The Ca^{2+} channels activated by LTP belong to the transient receptor potential family (Chapter 6), in particular the TRPC1 type [178].

The process of LTP indicates that a "memory" is an enhancement of activity in a pathway that produced a particular thought or action. Short-term memories are increased Ca^{2+} flows into a neuron, leading to longer depolarization times. Long-term memories involve enhancing Ca^{2+} flows by making new receptors or synapses; this latter process requires synthesis of new proteins. The process of LTP thus converts short-term memories that are labile and unstable into longer term memories that are permanent.

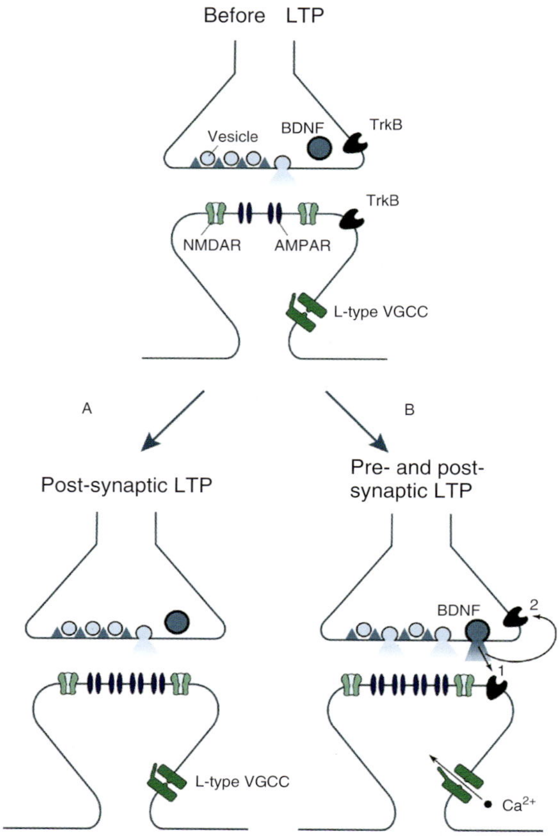

Figure 13.7 A model for the contribution of brain-derived neurotrophic factor (BDNF) to LTP at CA3–CA1 excitatory synapses. Vesicles contain glutamate; NMDAR and AMPAR are glutamate receptors, and before LTP are present in small numbers. TrkB is a receptor for BDNF, which at this stage is not released in large quantities. Voltage-gated Ca^{2+} channels are not opened. A: NMDAR-dependent LTP pathway leads to a purely *post-synaptic* mechanism of LTP via insertion of additional AMPA receptors into the postsynaptic plasma membrane and without participation of BDNF. Ca^{2+} channels are not opened. B: The NMDAR/VGCC (voltage-gated Ca^{2+} channel)-dependent LTP pathway leads to both postsynaptic and presynaptic changes. BDNF acts on TrkB receptors, either on the same neuron (2) or on the postsynaptic cell (1). In addition, additional AMPA receptors are recruited and Ca^{2+} channel function is potentiated. Modified from Figure 9 in Ref. [458]. Used with permission.

13.5 Receptors and Behavior: Addiction, Tolerance, and Dependence

13.5.1 Opioid Receptors

The euphoric, analgesic, and **somniferous** effects of the opium poppy, *Papaver somniferum*, have been known for thousands of years. The plants have long been cultivated in the eastern Mediterranean and Asia Minor, and their chemical components now find their way to all parts of the world.

Box 13.1 Opioids and opiates

The terms referring to the plants and their effects are a large part of Greek and other Mediterranean mythic cultures, as morphine, the purified form of opium, derives its name from that of Morpheus, the Greek god of dreams [459]. The terms "opiate," "opioid," and "euphoria," come from the Greek "opion," meaning poppy juice. "Opiate" refers to substances that are based on the structure of morphine, while "opioid" refers to substances that are not based on that substance, yet still bind the same receptor. Thus, the opioids are natural peptides that are endogenous agonists of opioid receptors; opiates are plant products that happen to bind to the opioid receptors [460].

The analgesic drugs derived from poppies found their first wide medical use in the United States during the Civil War, as they were used to reduce pain from wounds and surgery. Their addictive and toxic effects decreased their usefulness while it increased demand for them. Drug companies attempted to find non-addictive substitutes; one such was heroin, a form of morphine altered by the addition of two acetyl groups. However, all substances that had morphine's analgesic powers also had its addictive properties. It appeared that the two actions were linked in a mechanistic way [459].

Thus, by the early 1970s, it appeared that a receptor existed for the morphine-like drugs, and that natural analgesic ligands (pain-reducing neurotransmitters) for these receptors might also exist. At the time it was impossible to obtain these receptors for study, but Cuatrecasas [461] developed a method of tagging insulin with a radioactive label, incubating this labeled ligand with cellular fragments, and collecting the bound label on filters. Pert and Snyder [253] used radioactive naloxone, an antagonist of morphine, to identify opioid binding sites in the brains and intestines of rats, guinea pigs, and mice. This study is significant for two reasons: for showing the location of action of opiate drugs, and in being one of the first to use radiolabeled substances as probes for the study of receptor distribution.

Pert and Snyder also showed that opiate receptor binding properties parallel their physiological properties, especially their abilities to block pain: those

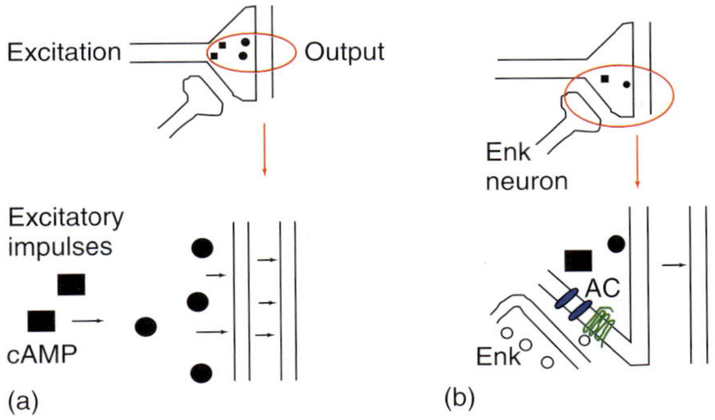

Figure 13.8 Endogenous opioid pathway for reducing pain perception. (a) Excitatory neuron (nociceptor) releases transmitters (glutamate and peptides – closed circles) that provide an alert of the painful stimulus. (b) Enkephalin neuron releases enkephalin (open circles), which binds to opioid receptors on axon ending of excitatory neuron, reducing its output of transmitter, and reducing p. ain perception. Exogenous morphine has the same effects as enkephalin.

substances that had the highest binding affinity were the most likely to be effective analgesics. Other classes of brain-active drugs unrelated to opioid analgesics (such as monoamines, phenobarbital, histamine, and atropine), do not bind to the receptor at all [253].

They also discovered that similar receptors exist in the intestine (associated with its intrinsic nervous system, called Auerbach's plexus) [253]. This work extended the role and significance of peptide transmitters and illustrated the unitary structure of the body: organs other than the brain contain receptors for transmitters of the "mind."

13.5.1.1 Opioid Neuron Pathways in the Brain

Several endogenous peptides are known (Chapter 4). In the brain, the opioid-secreting neurons work in the following way to alter perception of pain [448] (Figure 13.8):

1) A painful stimulus arrives in the brain via sensory (excitatory) nerves from the skin.
2) The excitatory nerve releases an excitatory transmitter at its axonal ending; this stimulates projection neurons to activate the conscious pain reception centers of the cerebrum. Thus, as long as the excitatory transmitters can be released, pain will be perceived.
3) Enkephalin-releasing neurons from the reticular formation of the brainstem have their axonal endings on the endings of the excitatory cells. The excitatory neurons have opioid receptors in their cell membranes.
4) If enkephalin-releasing neurons fire while the pain fibers are firing, transmitter release from sensory cells is inhibited because the binding of endorphins to their receptors reduces the duration of the action potential in the excitatory nerve. This in turn reduces the ability of receiving neurons to transmit pain information to the cerebrum. The receptor mechanism for this effect is discussed below.

The opioid receptors have emerged as the common denominator of the actions of morphine analogs. All substances that cause analgesia or euphoria (or those that specifically block morphine actions) act on one or a few related receptors in the brain tissue. The evidence for specific receptors is as follows:

- The drugs with similar effects all have similar structures (Figure 13.9).
- Some synthetic analogs of morphine are even more powerful than morphine at activating the receptors. The high degree of efficacy and specificity suggests a particular receptor on which the drug works.
- The optical isomers of the drugs have very different abilities to activate the receptors, with only the levorotating isomer being effective [253].
- Euphoria-blocking drugs also have structures similar to those of the morphine analogs, again suggesting that the agonists and antagonists are acting on the same receptor.

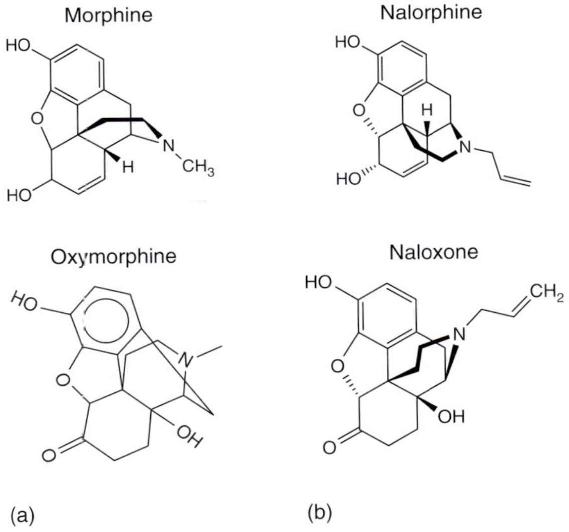

Figure 13.9 Structures of several morphine-like drugs. Agonists of the opioid receptors are shown in (a) and antagonists are shown in (b). Radiolabeled naloxone was the substance used by Pert and Snyder in their important initial study [253].

13.5.1.2 The Opioid Peptides and Receptors

Neuroscientists working on the biology of opiates realized early on that plant products could not be natural ligands for animals' receptors; the high affinity with which analgesic drugs bind to the opioid receptors was considered an accident of shape and electrostatic charge distribution. By the early 1970s, opiate receptors had been localized to the synaptic endings of brain cells [459], and the natural ligands of these receptors were actively sought by adding brain extracts to cells in the absence and presence of specific morphine analogs that would be expected to block the actions of whatever substances might be in the extracts. The first natural ligands were discovered by Hughes and colleagues in 1975 [255]. They found two similar peptides of five amino acids each and named them enkephalins because of their location in the head. The two are met- and leu-enkephalin, differing only in their terminal amino acid.

In the years since, enkephalins have been found in nerve endings of several brain areas, especially the thalamus, limbic system, and central regions of the spinal cord. These areas are the ones most involved in transmission of dull pain [459]. In addition, researchers have found morphine itself in animal tissues, where it acts on μ-opioid receptors (Chapter 7) to modulate immune system activity [462].

Opioid peptides and their receptors are present in the regions of the CNS involved in emotion and reward. Four different opioid receptors – mu (μ), delta (∂), kappa (κ), and opioid receptor like-1 (ORL1) and their genes have been localized and characterized pharmacologically and at molecular levels [460].

13.5.1.3 Mechanisms of Transduction

The opioid receptors are GPCRs that couple to inhibitory G proteins ($G_{i/o}$ – Chapter 7). After agonist binding of artificial or endogenous substances, the $G_{i/o}$ protein subunit activates the **inwardly rectifying** potassium channel, Kir_3 (Chapter 6). This process causes hyperpolarization of the neuron and inhibition of neural activity [460]. The result is a reduction in the Na^+ influx that is normally caused by excitatory neurotransmitters, and a decrease in the amount of transmitter released during nerve stimulation.

13.5.1.4 Characteristics of Responses to Continued Drug Presence

Whenever a drug is taken repeatedly or when a hormone is released continually in much larger amounts than is normal, the body develops a characteristic response to it. The processes of desensitization and downregulation that occur with G-protein-coupled and receptor tyrosine kinsases were discussed in Chapters 7 and 8. **Tolerance** and addiction to a drug, as well as **dependence** on the drug, are three important additional responses.

Tolerance occurs when a drug produces diminishing effects as its use is continued. Dependence is seen by the serious withdrawal symptoms that are manifested when a drug is taken away. One mechanism underlying tolerance and dependence is metabolic: increased amounts of a drug stimulate the synthesis of liver enzymes that break the drug down, so similar tissue levels (and physiological effects) are achieved only by increasing intake [459].

Addiction is the psychological manifestation of tolerance and dependence [459]. An additional plausible mechanism was worked out in reference to the opioids only following the discovery of enkephalin-secreting neurons (Figure 13.10).

In a brain not exposed to excessive amounts of opiates, excitatory pain neurons from the periphery impinge on CNS neurons and release one of the excitatory transmitters (e.g., ACh, NE, or glutamate) (Figure 13.10a). These pain signals can be modified by interaction between excitatory neurons and the enkephalin-releasing inhibitory neurons. Enkephalin neurons appear to release a continuous low level of enkephalins, and can also increase their firing rate if inputs from pain fibers stimulate them. Transmitters from these neurons act on opioid receptors of excitatory neuron axons to decrease the release of excitatory transmitter. The result of this effect is the inhibition of pain signals from peripheral nerves to the CNS (Figure 13.10b).

If a person begins to take opiate drugs (Figure 13.10c), the drugs increase in concentration at excitatory nerve endings. They bind to the enkephalin receptors, inhibiting the excitatory neurons to a greater extent than is seen without the drugs being present. At the same time, the opioids bind to enkephalin receptors on presynaptic sites of the inhibitory neurons and reduce the intrinsic rate of enkephalin release. Continued exposure to opiates causes the cell to make increasing amounts of adenylyl cyclase [459] (Figure 13.10d). This means that excitatory neurons increase the rate of transmitter release in pain pathways. As this release continues, ever higher amounts of opiates are required to achieve the same reduction in cAMP and the same pain relief.

When opiate drugs are withdrawn (Figure 13.10e), the cAMP levels in the excitatory neurons increase because the drugs are no longer inhibiting their production. The increase in cAMP may cause the visible

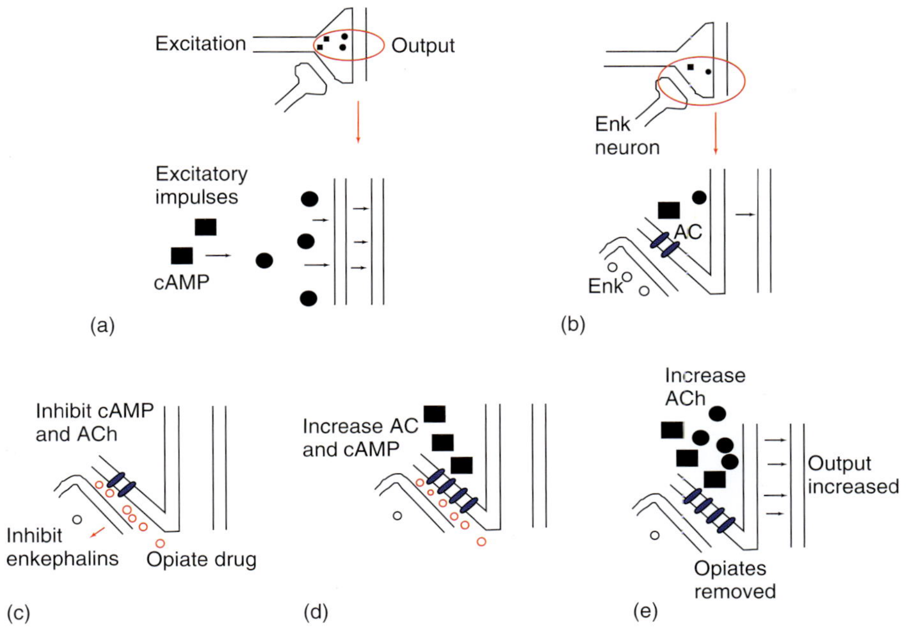

Figure 13.10 Receptor mechanism of addiction. (a) Normal state. Impulses in excitatory neuron cause release of transmitter in pain pathway and (b) release of enkephalins from other (Enk) neuron inhibits cAMP production in excitatory neuron, decreasing release of transmitter in pain pathway. Thus, enkephalins act as analgesics. (c) Initial response to taking opiates. The drugs activate enkephalin receptors on excitatory neurons to reduce pain transmitter release (and thus act as analgesics), but also bind to receptors on enkephalin-releasing neurons to reduce enkephalin release. (d) Tolerance develops after being on drugs for a while. Excitatory neuron makes more cAMP to compensate for loss of signal from enkephalin-releasing neurons. Opiate drugs still present. (e) Withdrawal from chronic exposure to opiates. cAMP now in high concentration in excitatory neuron from earlier response (d), and also because opiates are no longer present. High cAMP levels produce release of more transmitter from excitatory pathway, increasing pain sensations.

symptoms of withdrawal through specific actions of cAMP in enzyme phosphorylation [459].

13.5.2 Individual and Cultural Distributions of Depression

13.5.2.1 Depression

The first sections of this chapter indicate that learning and mood, two important aspects of human behavior, depend on the functions of neuronal proteins such as receptors and ion channels. To this list we add the transmitter uptake transporters and enzymes involved in transmitter destruction. The importance of these latter proteins is attested to by their importance as targets by antidepressive drugs such as the serotonin uptake inhibitor fluoxetine (Prozac) and the monoamine oxidase (MAO) inhibitor imipramine.

Researchers do not generally consider individual differences in their experiments on learning, though polymorphisms for all these proteins are known within populations. Thus, it is not generally considered interesting whether two individual *Aplysia* or mice respond differently during short-term or long-term potentiation. However, human individuality is a source of variation in clinical experiments, and may occur because of individual differences in receptor distribution.

If behavioral differences between people are purely cultural, and not based on biological differences between individuals' proteins, one would expect to see no correlation between a psychological state, for example, depression, and subtype differences in receptors, transporters, or enzymes for mood-related pathways. If, however, individual differences relate in part to biological differences, then behavioral responses may be predictable from an understanding of which of these proteins is possessed by a particular person. In the same way, behavioral differences between societies may be predictable from knowledge of polymorphisms for receptors or the proteins involved in transmitter disposition within the population. This possibility has led to the development of an area of research, "cultural neuroscience," that relates polymorphisms to individual and societal behavior patterns [463].

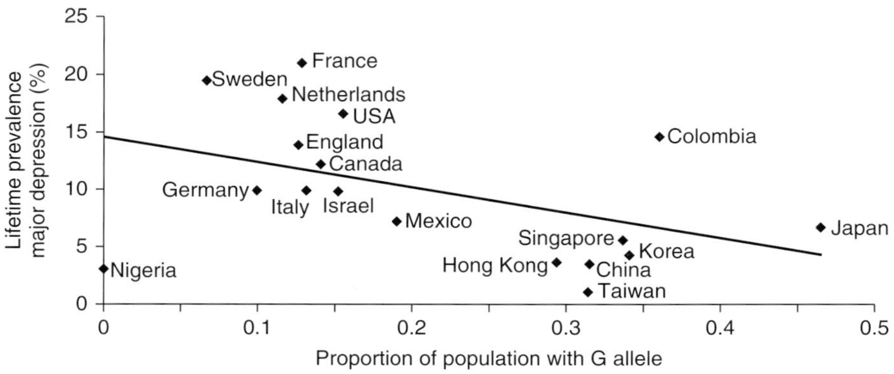

Figure 13.11 Relation between national scores of individualism–collectivism and distribution of the A118G allele for the μ-opioid receptor. Figure 3 of Ref. [463]. Used with permission.

13.5.2.2 Polymorphisms in Neurotransmitter Transporters

The serotonin transporter is responsible for the reuptake of serotonin into brain neurons from the synapse. Efficient uptake decreases the concentration of serotonin at receptors, reducing neurotransmission in this pathway. Any drugs that inhibit the transporter would thus potentiate transmission. The gene for the serotonin transporter is polymorphic, with some alleles having longer DNA sequences than others. The alleles are thus termed "short" and "long" [463].

Any given person could be short/short, short/long, or long/long for these two alleles. People with the short/short genotype have greater risk for depression, anxiety, post-traumatic stress disorder, or suicide than are those of the long/long genotype. The differential effects between the alleles are more statistically significant for people who are exposed to social life stressors such as divorce or the death of a loved one [463].

13.5.2.3 Polymorphisms in Opioid Receptor Subtypes

The endogenous opioid system is involved in more than analgesia against physical pain: infants of several mammalian species show distress when removed from maternal care, but μ-opioid antagonists relieve that distress. This suggests that the level of μ-opioid receptor signaling also indicates how an individual perceives social inclusion.

A polymorphism (A118G) in the μ-opioid receptor gene (produced by a single nucleotide substitution of G for A at position 118) is present in humans, and people with the G allele are more sensitive to perceived rejection than those with the A allele [463].

13.5.2.4 Polymorphisms in Enzymes for Transmitter Disposition

Neurotransmitters such as serotonin, DA, and NE are broken down by enzymes after reuptake into neurons. One such enzyme is MAO. Inhibitors of MAO (e.g., imipramine) also affect mood, and thus are used medically as antidepressants. The A form of monoamine oxidase (MAOA) is specific for serotonin and DA, and is polymorphic. Different alleles are present in humans, and are associated with differences in the amount of MAOA synthesized; those individuals with the low expressing alleles (e.g., the MAOA-uVNTR polymorphism) showed the greatest distress to social exclusion [463].

13.5.2.5 Society-Level Actions

Researchers have also asked whether different human populations have different distributions of alleles for the genes encoding transmitter response and disposition, and whether that distribution might explain societal differences in certain social traits. Few such studies exist at this time, and those that do are focused on the same polymorphisms as have been tied to depression. Peoples' responses to exclusion, as well as their tendencies toward depression, appear to be part of their perceptions of themselves within the spectrum of individualist–collectivist social structures.

Way and Lieberman [463] reviewed studies that related measures of individualism–collectivism to polymorphisms for the above three genes for a large number of nations. The measures of individualism were based on population surveys, linguistic use of pronouns, and sociologists' judgments. European and North American nations tended to score high

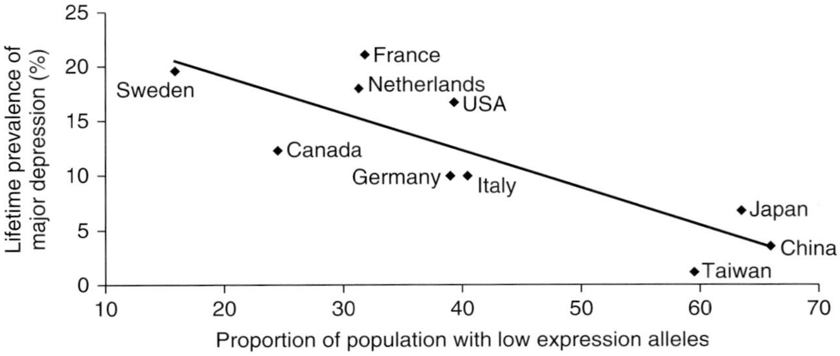

Figure 13.12 Relation between national tendency to major depression and distribution of the MAOA-uVNTR allele for monoamine oxidase. Figure 4 of Ref. [463]. Used with permission.

in individualism, while Asian and Latin American nations scored high in collectivism. This score was correlated with transmitter and receptor variability as well:

- The 5-HTTLPR short protein, a social sensitivity predictor, was much more prevalent in collectivistic populations than in individualistic populations [463].
- The A118G protein, also a social sensitivity predictor, was more prevalent in collectivist than in individualistic societies (Figure 13.11).
- The MAOA-uVNTR allele was also more prevalent in individualist than in collectivist societies, and its frequency was directly proportional to the frequency of major depression (Figure 13.12).

Thus, in all three cases (5-HTTLPR, A118G, MAOA-uVNTR), the alleles that were indicators of sensitivity to rejection in individuals were also more prevalent in collectivistic cultures [463].

13.5.2.6 Possible Mechanisms

It is unclear how causality works in these examples; cultural patterns of behavior might provide directional selective pressure on allele frequencies, or an already existing allelic distribution might influence the sort of cultural norms that emerge in any culture. In any case, the data suggest that individual and cross-cultural psychological differences may be profitably studied by investigating differences in brain neurotransmitter – receptor mechanisms.

14
Evolution of Receptors, Transmitters, and Hormones

> Cell-environment recognition is ... a fundamental prerequisite of evolution
>
> *Csaba [464]*

> Amino acid sequences relentlessly diverge
>
> *Doolittle [465]*

14.1
Introduction

Signaling molecules and receptors are found in all organisms, whether eukaryote or prokaryote, multicellular or unicellular. In addition, all receptor processes have identical mechanisms. Thus, the signaling molecules, their receptors, transduction systems, and intracellular messengers must have had their origins at or near the origin of cellular life. This chapter covers the evolutionary history of each of these molecular types.

14.1.1
Phylogeny of Communication: General Ideas

Hormone signaling systems within a multicellular organism are derived from much older *interindividual* signaling systems of unicellular organisms that originated with pheromones based on sterol or protein structure. The first interindividual communication processes were those that encouraged

- **conjugation** and aggregation in prokaryotes;
- conjugation, cell division, differentiation, and aggregation in simple eukaryotes.

14.1.2
The Receptors

The original receptors were likely proteins that bound amino acids or sugars for transport into the cell. These proteins are widespread in bacteria and seem to be the phylogenetic basis for the transporter and ion channel group of receptors (Chapter 6). The details of evolution in this group are discussed below. The only other type of receptor found in both eukaryotes and prokaryotes is the protein rhodopsin, found in certain Archaea, and its relatives the G-protein-coupled receptors (GPCRs), found only in eukaryotes. Receptor tyrosine kinases (RTKs) and nuclear receptors appear to be more recent in origin, being found only in eukaryotes in their membrane-bound forms.

14.2
Origins of Transmitters and Receptors

14.2.1
Evolution of Signaling Processes

The transmitters used by organisms are diverse, and belong to several different chemical classes:

1) Genetic components, metabolic cofactors, and energy transducers (adenosine, adenosine monophosphate (AMP), and ATP), used by members of all domains; Chapters 6 and 7.
2) Metabolic intermediates (amino acids, acetylcholine (ACh), monoamines); Chapter 7.
3) Peptides (several in gram-positive bacteria, mating factors in yeast; dozens in eukaryotic neuroendocrine signaling); Chapters 7, 8, and 11.
4) Gases (nitric oxide, hydrogen sulfide, carbon monoxide); Chapter 10.

Of these several groups, it is not clear which were the original signaling molecules; it is likely that organic biosynthetic intermediates and amino acids were early signals, because of their ubiquity and metabolic importance. The nucleotides have also been proposed as the earliest transmitters [466], because of the importance of nucleotides in energy transduction and metabolism, and because all eukaryotes have well-developed purine nucleotide signaling systems [467].

Receptor Biology, First Edition. Michael F. Roberts and Anne E. Kruchten.

At about the same time the first cells were able to detect environmental chemicals, these same cells began to coordinate their division by communication with other cells of the same species, or among several unrelated species. The molecules used for these processes in Bacteria are intracellular ligand-operated transcription factors (Chapter 4), but cell-surface proteins are used as mating factor receptors in single-celled eukaryotes such as yeast (Chapter 7).

As indicated in Chapter 1, signals have meaning that is arbitrary but specific. That meaning is associated with the combination of transmitter and receptor, and is expressed as the receptor transduces the binding event into a set of intracellular processes that form a biochemical chain: binding – transduction – behavioral change [466]. That change in behavior can be any of several actions, such as secretion of another signal, cell membrane depolarization or hyperpolarization, contractile activity, or transcription of genes. These are, as individual responses, fairly simple; but they can bring about major changes in organismal biology, such as motion toward food or away from predators, learning and memory, or communication with other cells or organisms. These are the adaptive functions of the "image-processing" receptors [466]. Over time, the genes encoding those receptors that provide successful behavioral responses will survive and serve as the basis for the evolution of additional signaling processes.

14.2.2 Homologous Sequences

If a trait such as a structure or a physiological process is similar in two organisms *because the two received the trait from a common ancestor*, the traits are **homologous.** Homology is thus central to evolutionary thought because all evolutionary change represents descent of organisms from common ancestral species, and the traits present in an ancestor must be present in homologous form in descendants.

Until the 1960s, morphological characters provided the most information about homologous traits; however, homology applies equally well to nucleotide sequences in genes and to amino acid sequences in proteins, and in 1969, Jukes and Cantor [468] inaugurated the study of protein evolution by sequence analysis. This approach is based on the idea that mutations in genetic sequences cause them to change with time, leading as well to changes in the proteins they encode. This analytical framework made it possible to study the evolutionary history of genes from the change over time in their protein products.

14.2.2.1 Orthologous and Paralogous Sequences

The concept of homology extends in two directions: homologous proteins may occur within one organism if duplication produces an extra copy of a gene. Such sequences are called **paralogous** (Figure 14.1). Another type of homologous protein may also exist: if an ancestral species splits into two daughter species, the two retain the genes and proteins of the ancestor; each sequence in the descendant is termed **orthologous** to that in the ancestor (Figure 14.1).

Paralogy occurs by gene duplication and subsequent mutational divergence of the two sequences, whether or not actual speciation occurs. Orthology represents the inheritance of a gene sequence in both daughter species after a speciation event. Orthology thus represents the evolutionary history of the species of organisms themselves.

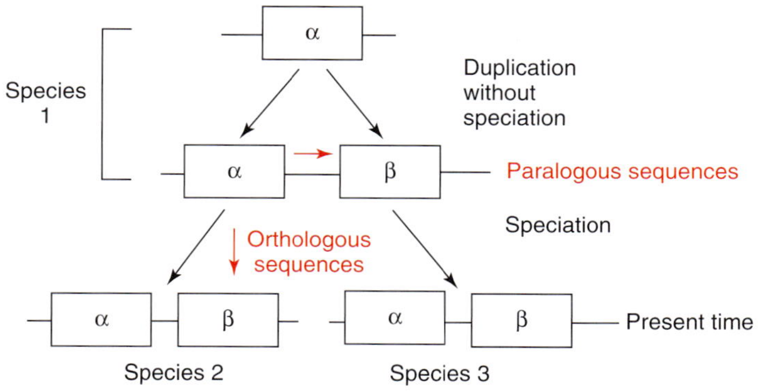

Figure 14.1 Two sequences are **homologous** if they share a common ancestor. This figure describes two types of homology: **Paralogous** sequences result from a duplication and are present in the same organism (thus the α and β proteins of species 1 are paralogs). **Orthologous** sequences are those that are descended from a common ancestor (thus, the α proteins in the three species are orthologs). Modified from Figure 6.6 in Ref. [469]. Used with permission.

The duplication event shown in Figure 14.1 is frequently seen in organismal genomes. Most often, single genes become duplicated by a malfunction in the crossing-over mechanism at meiosis (**homologous recombination**). Because mutation and selection occur in all types of genes, proteins such as hormones and receptors will evolve, acquiring new functions, and spawning related proteins that will function in related ways.

As earlier chapters have indicated, each receptor superfamily has multiple families that reflect evolutionary divergence within the larger unit. The next section illustrates how researchers decide the relationships among these families of proteins.

14.2.3 Phylogenetic Inference

The resemblance between two receptors (or any other pair of proteins) is measured by aligning the nucleotide or amino acid sequences of the two and comparing the amino acids at each position. It is more common to study amino acid sequences than nucleotide sequences because most amino acids are encoded by multiple nucleotide codon triplets, and not all nucleotide mutations alter the amino acid sequence. In any case, the comparisons are always made between two proteins that are already thought from other evidence to be homologous.

Because 20 different amino acids may be used in any protein, a $^1/_{20}$ or 5% chance exists that the same amino acid will be at a given position even in unrelated proteins. However, the probability that a sequence of amino acids will be identical at random decreases rapidly as the number of amino acids increases ($p = [^1/_{20}]^n$, where n is the number of amino acids in the sequence). Thus, two proteins containing even short strings of identical amino acids are very likely to be derived from a common ancestral sequence [465]. For example, oxytocin and vasopressin (nine amino acids each) are related because they differ at only one amino acid and they share identical disulfide bond structure (Figure 4.9). In contrast, the hypothalamic releasing hormones are probably not related because they have no sequence similarity (Table 4.1). This is why oxytocin and vasopressin are thought to be homologous, while the releasing hormones are not homologous.

Long gene and protein sequences are often too complex to be compared by eye, so computer-based algorithms are used in the construction of phylogenies. These approaches are able to make detailed comparisons of large sequences in a short time. To make evolutionary comparisons of DNA sequences, researchers first must determine that the two are worth comparing (i.e., they are likely to code for a homologous protein). Then the two are aligned, for example:

```
A : TCAGACGATTG
      *  ****
B : TCGGAGCTG--                (14.1)
```

Several mismatches are apparent here, indicated by asterisks; in addition, the sequences are of unequal length because of earlier nucleotide deletions or insertions (neither of which is known by the researcher ahead of time). The mismatches can be eliminated by making the unwarranted assumption that all are caused by separate mutational deletion events in the two lineages. The two can thus be realigned as the two "matching" sequences

```
A : TCAG-ACG-ATTG
B : TC-GGA-GC-T-G              (14.2)
```

However, to make the sequences match, researchers had to add in six gaps. It should be clear that one can nearly always improve sequence alignments by arbitrarily adding gaps, but it is not possible to know whether the gaps added represent actual past mutational deletions. An alternative is to minimize both the mismatches (to two) and the gaps (to four) with this alteration:

```
A : TCAG-ACGATTG
            * *
B : TC-GGA-GCTG-               (14.3)
```

Statistical tests are used to determine which is more biologically relevant – allowing mismatches or arbitrarily adding gaps to relieve those mismatches. The result of several statistical approaches is a numerical index of the degree of homology of two sequences. This computed sequence similarity is then used to construct a phylogenetic tree for the two.

Methods such as these are behind the examples of trees that appear in this chapter for sequences of peptide hormones and for receptors.

14.2.4 Phylogenetic Illustration of Protein Relationships

Researchers who perform phylogenetic analysis such as that described in the previous section must then present in graphical form their hypothesis for the relationships among the genes or proteins they studied.

Several types of diagrams are used to illustrate phylogenetic history, all of which illustrate change over time by splitting lineages. We present trees for all the protein hormones and receptors discussed earlier in the book. Each illustration is derived from a publication that based its analysis on the principles described above.

We use a branched structure (e.g., Figure 14.3) in which time goes from bottom to top (occasionally from left to right), and each branch represents an event in which protein sequence is altered by mechanisms shown in Figure 14.1. The diagrams we present are generally simplified versions of the phylogenies shown in the original references; our simplification has two purposes: to present phylogenetic charts with consistent structures and to focus only on the relationships among hormones and receptors that we are discussing in this book. The original references contain more complex diagrams because a given publication tends to focus on a single family containing many members; these members are clearly distinguished and the degree of confidence each researcher has in that branching pattern is shown on the figure. Where we have removed proteins from the original publications, we have done so without altering the original phylogenetic relationships. Readers desiring more complete phylogenetic information than is present in our figures are encouraged to visit the original references.

14.2.5 Whole-Genome Duplication (WGD)

Phylogenetic trees represent genetic relationships at two levels: first, among the organisms themselves: the closer a given protein (e.g., insulin) is in two different species, the more closely related the *two species* are to each other (orthology – Figure 14.1). Second, duplication can occur in a given protein gene in one organism; this produces a second protein similar to the first. Subsequent change in the gene for this protein may produce a protein with a different function, and the tree gives a *family structure to this group of proteins* (paralogy – Figure 14.1).

Mutations causing sequence changes in genes for either receptors or peptide transmitters will most likely be maladaptive and will eliminate the function of either protein. This is because any change in receptor structure must be matched by a corresponding change in the hormone, a very unlikely event. Even if simultaneous mutations were to occur, however, the function exerted by the original hormone–receptor pair may be lost, another maladaptive event.

It is nonetheless the case that families of related receptors and hormones exist in organisms (e.g., the four types of opioid receptors – see Section *14.4.2.2*). The most reasonable hypothesis for this sort of diversity lies in the duplication of genes for receptors and/or for peptide hormones.

In 1970, Ohno [470] proposed that at the time of origin of the vertebrates (about 560 mya), the entire genome became duplicated. According to this idea, the earliest vertebrates were tetraploids as compared to the earliest chordates, and each allele was present in two copies. This hypothesis has been controversial since it was proposed, but evidence suggests it is in general correct, and that multiple WGD duplications have occurred in vertebrate lineages (including two more: one at the step at which jawed vertebrates arose, about 530 mya; and a third about 300 mya within the lineage leading to bony fish) [471].

Duplication of an entire genome produces paralogous copies of all genes (Figure 14.1) but most of the new copies mutate to nonfunctional genes. The lineage will retain only those that mutate and incidentally develop a new function [471]. Genes α and β in Figure 14.1 are examples: β is a paralog of α and originated by duplication; in this example, β is retained because its mutation happens to encode a functional protein. As the original lineage splits to form new daughter species (Figure 14.1), the two functional paralogs are retained in the new species.

Hittinger and Carroll [472] also argue that most duplicated genes do not obtain entirely new functions, but become "**subfunctionalized**" such that the paralog and its duplicate share related functions. Though the genes Hittinger and Carroll describe (GAL1 and GAL3) do not encode receptors, they do encode DNA-binding proteins, suggesting that receptor-binding modifications might also occur by similar subfunctionalization following duplication.

Receptor families provide good evidence for the earlier paralogous duplication of genes; as becomes clear in later sections, all the superfamilies of receptors contain multiple families that most likely arose via large-scale duplications. Subsequent mutation and selection allowed them to be retained as receptors for different hormones.

The remainder of the theoretical sections of this chapter discusses the evolution of protein domains and the simultaneous evolution of receptors and hormones.

14.2.6 Origins of Novel Domains

Protein domains were discussed in Chapter 3. They are short segments within proteins with characteristic shape and function, for example, α helix or β pleated sheet, whose forms are determined by the primary amino acid sequence.

Even a small protein has a large number of ways to order its amino acids (20^n, where n is the number of amino acids in a protein). Thus, because the 3-D structure and function of a protein derive from its primary structure, one might suppose that the number of domains in the proteins of all organisms might be huge. In fact, only a few fundamental domains exist in proteins (Chapter 3), and the question arises as to why proteins do not fold into additional fundamental ways.

The answer lies in natural selection and the fact that only certain configurations of proteins have catalytic, structural, or binding functions. However, it is still valid to ask whether natural selection "chose" the existing primary sequences because they produced proteins that fold into stable functional domains, or whether only a few stable domains are possible, and each of them might be able to be encoded by more than one primary sequence.

Li *et al.* [473] have developed a simple lattice model of protein structure that implies that the second idea is correct: that existing protein domains exist because they are thermodynamically the most stable and can be encoded by the most number of different primary sequences. The model uses a 27-amino acid "protein" randomly assembled into a primary sequence from two amino acids (each embodying one of the two fundamental properties of amino acids – being either hydrophobic or polar) and filling the lattice of a $3 \times 3 \times 3$ cube. Li *et al.* used standard thermodynamic rules to determine what 3-D structures could be formed from each sequence based on the interactions of adjacent amino acids. They determined the structures built by every one of the 2^{27} (134 million) sequences.

The results were surprising: only a few (0.1% of the number possible) stable structures exist, and the most stable domains resemble those in existing proteins. Further, many different primary sequences can encode each of the most stable arrangements. An example is shown by Figure 14.2, in which a hypothetical protein folds into a stable state because of interactions among its constituent amino acids.

This work implies that α helices, β pleated sheets, and other common domains occur most often in natural proteins simply because they are some of the few stable structures that *can be made* by sequences of amino acids. In addition, these structures are the most resistant to change by mutations in primary sequence. An example of the resistance of proteins to change is in the various families of GPCRs, in which all share the same seven transmembrane (TM) structure though few of the families have homologous amino acid structures. Thus, protein evolution operates by two principles: that thermodynamically stable protein structures must be able to be encoded by many different primary sequences, and that these structures will be relatively insensitive to changes in amino acid sequence.

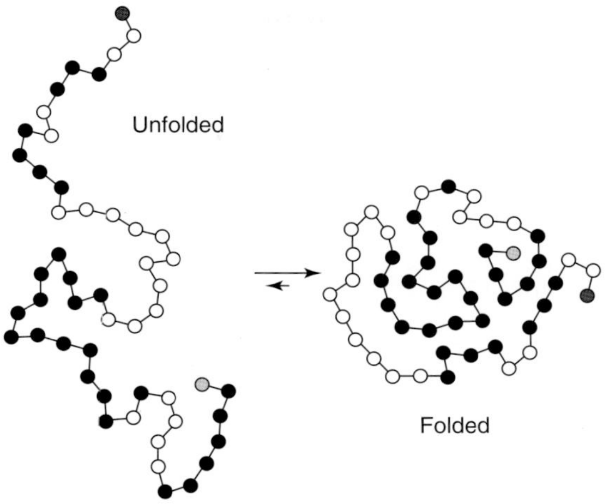

Figure 14.2 The transition between the newly formed unfolded protein and the folded functional shape. The connected circles are amino acids; open symbols are polar amino acids and closed symbols are nonpolar. This particular arrangement is thermodynamically stable, as interactions among the nonpolar amino acids largely determines the final shape. Source: Wikipedia article on Protein folding: http://en.wikipedia.org/wiki/Protein_folding. Accessed 13 January 2015.

14.2.7 Adaptation of Receptor Systems

Two mechanisms operate to promote change and adaptation of receptor systems:

- Gene sequences change over time through duplication and accumulation of mutations. However, all changes in hormones and receptors are under the control of natural selection.
- Therefore, even mutated receptors and their hormones must continue to show specificity, high affinity binding, and high efficacy for cell activation.

This model allows us to predict that duplication and mutation will generate families of genes coding for related receptors. As we show in later sections, this prediction holds for most proteins and groups of organisms, and has been a powerful stimulus to the study of receptor and hormone evolution.

14.2.8 Coevolution of Components of Signaling Pathways

Mutation alters signaling systems at three levels:

1) Genes for receptors themselves;
2) Genes for peptide transmitters;
3) Genes for biochemical pathway enzymes involved in synthesis of non-peptide transmitters.

14.2.9 Peptide Hormones and Their Receptors

Once a gene is duplicated, the two copies continue to make their products. One of three events may then occur [474]:

- **Null mutation** in and the subsequent inactivation of its sequence.
- **Subfunctionalization** of the original and its copy; a subfunctional process involves mutations in both the original and its copy such that both still function, but their new function involves (i) alteration of the original function and (ii) complementation such that the two gene versions work together in a new function rather than working separately [474]. Subfunctionalization involves changes in regulatory regions rather than coding regions, so we will not consider it further.
- **Neofunctionalization** of the copy alone. A neofunctional receptor is one with a new function (e.g., the β gene in Figure 14.1). Later sections of this chapter provide many examples of neofunctional hormone genes and receptor genes.

These three processes do not necessarily coordinate evolution in hormones and their receptors. But in some cases, all members of a family of peptides are agonists of a single family of receptors. For example, the various peptides in the secretin hormone family (Chapter 4) are agonists of the Family B (secretin-like) receptors (Chapter 7), suggesting that selection has somehow coordinated changes in the genes for both types of peptide [103].

14.2.10 Receptors and Their Non-Peptide Hormones

The evolution of non-peptide ligands (such as nucleotides, amines, or steroids) is different from the preceding in that the structure of the ligand is fixed, unlike the example of protein ligands. For fixed-structure ligands, ligand evolution and receptor evolution are separate, and ligand evolution likely occurs via enzymatic alteration of secreted substances.

For example, the nucleotide ATP is a widespread intracellular and extracellular mediator; it acts as an agonist for a group of ligand-gated ion channels (LGICs) and for a group of GPCRs. ATP is also modified by extracellular enzymes, and the breakdown products (ADP, AMP, and adenosine) are also ligands for GPCRs [467].

In the case of steroid hormone action, the steroid pathway synthesizes a variety of molecules, any of which might fortuitously activate a particular receptor; changes in receptor gene sequences can create new steroid receptors, and new receptors may be activated by existing steroid ligands for which the receptors did not previously exist.

This is the scenario envisioned by Thornton [475] and called **ligand exploitation** . We later examine its application to the evolution of nuclear receptors, but for now we discuss it as a possible mechanism for the coordinated evolution of receptors and non-peptide hormones. As earlier described (Figure 4.23), estrogen is the *final* product in the steroid biosynthetic pathway. The estrogen receptor, however, was the *first* steroid receptor to evolve. Thus, once estrogen was available, most other steroids (e.g., progesterone, androgens, corticosteroids) must have already been present. At the time of their origin, they did not have receptors to activate; however, duplication of the estrogen receptor allowed new receptor types to evolve from the duplicates; by this time, steroids were already available for use by the new receptor types. Any new receptor could thus "exploit" the ligands already present. This mechanism could operate for any receptor type that uses non-peptide ligands [475].

14.3 Evolution of Hormones

14.3.1 Peptide Hormones for G Protein-Coupled Receptors

These peptides are found in all eukaryotes. Some examples are the yeast mating factors and vertebrate hypothalamic releasing hormones, the anterior and posterior pituitary peptides, as well as additional brain and gut peptides discussed in Chapter 4.

14.3.1.1 The Yeast Mating Pheromones

Yeast cells are divided into two mating types, called **a** and **α** on the basis of the peptide pheromones that each secretes. The pheromones act on GPCRs in yeast plasma membranes. They influence behavior and conjugation of cells, leading to cell fusion, and a process similar to fertilization in multicellular organisms. Certain of these peptides and their receptors show close relationships with those in vertebrates: the amino acid sequence of the yeast **α** peptide is homologous with that of mammalian gonadotropin releasing hormone (GnRH), and the **α** peptide is able to stimulate release of gonadotropins from the mammalian pituitary. The **a** mating peptide, in contrast, has no sequence homology with the α peptide, nor with other known peptides.

14.3.1.2 The Anterior Pituitary Trophic Hormones

Six peptide trophic hormones (Chapter 4) come from the anterior pituitary: follitropin (FSH), lutropin (LH), thyrotropin (TSH), adrenocorticotropin (ACTH), somatotropin or growth hormone (GH), and prolactin (Pr). The last two bind to members of the RTK superfamily and are discussed in a later section. The other four bind to GPCRs. The closest relationships among them are shown in the FSH, LH, and TSH peptides. As pointed out in Chapter 4, they consist of two subunits, α and β; the α subunit of 92 amino acids is identical in all three, and the β subunit shows two large domains in which all three are clearly related. ACTH is not homologous to these three, but is transcribed from the same gene as is β-endorphin, a peptide that also binds to a GPCR (one of the opioid receptors – see Section *14.4.2.2*).

14.3.1.3 The Hypothalamic Releasing Hormones

Most of the six hormones above are released from the anterior pituitary by **releasing hormones** that are actually hypothalamic neurosecretory peptides. In one case, release is influenced by a non-peptide neurotransmitter. With the possible exception of TRH (thyrotropin releasing hormone) and GnRH, the releasing hormones show no sequence homologies with each other. The complex pattern is shown in Table 14.1.

Corticotrophin -Releasing Hormon e: CRH is a 41-amino acid peptide that shares homology with the urocortins, brain peptides found in a variety of vertebrates. Urocortins are able to cause release of pituitary hormones, and, in addition, are involved in osmoregulatory and cardiovascular functions [477].

Table 14.1 Releasing molecules for the anterior pituitary (AP) hormones.

AP hormone	Releasing hormone	Effect	Source
ACTH	CRH	Stim	[107]
TSH	TRH	Stim	[107]
GH	GHRH	Stim	[107]
GH	Somatostatin	Inhib	[107]
LH	GnRH	Stim	[107]
FSH	GnRH	Stim	[107]
Prolactin	TRH, NT, OT	Stim	[476]
Prolactin	DA	Inhib	[476]

Thyrotropin-Releasing Hormone: This three-amino acid peptide (sequence: EHP) is also found in non-neural tissue where it serves cardiovascular and reproductive roles [478].

Growth-Hormone-Releasing Hormone: This 44-amino acid peptide is homologous to secretin and glucagon (see below).

Gonadotropin-Releasing Hormone: This decapeptide is widespread among animals. Most phyla express a neuropeptide similar to GnRH, but the molecule plays a reproductive role only in bilaterians. The reproductive function in chordates may have originated after the genome-wide duplication event that occurred at about the time chordates evolved [470].

Prolactin-Releasing Hormones: Three peptides are known to stimulate release of prolactin from the anterior pituitary: TRH, oxytocin (OT), and neurotensin (NT) [476]. These peptides show no sequence homology, though TRH and GnRH have the same two initial amino acids. Prolactin release is inhibited by dopamine (DA), which is a monoamine and not a peptide.

14.3.1.4 The Posterior Pituitary Hormones

Oxytocin and vasopressin consist of nine amino acids and differ only at two sites. Each also has a single disulfide bond, giving folding patterns that generate very similar three-dimensional structures. They are found in all vertebrate groups, so the duplication event that generated the pair must have occurred early in vertebrate evolution. They are homologous with the neurophysins, proteins that transport them in the axons of hypothalamic neurons. Oxytocin is unusual in that it also serves as a releasing hormone for prolactin.

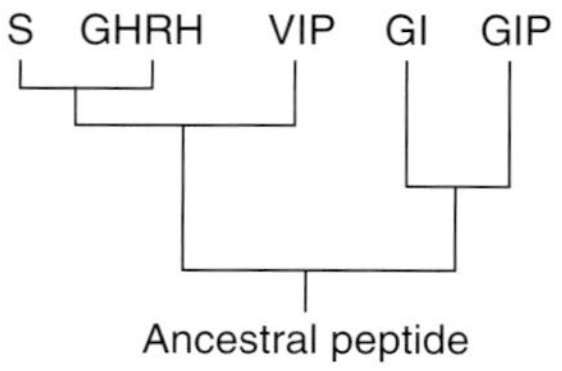

Figure 14.3 Homologies of secretin (S), growth-hormone-releasing hormone (GHRH), vasoactive intestinal peptide (VIP), glucagon (Gl), and gastric inhibitory peptide (GIP). Modified from Figure 1 in Ref. [103]. Used with permission.

14.3.1.5 Miscellaneous Peptide Hormones

Secretin, GHRH, VIP, Glucagon, and GIP: These hormones form a group of related peptides, all of which are agonists of the family B GPCR – secretin receptors – though they do not have similar functions. Their phylogenetic tree is given in Figure 14.3.

Gastrin and CCK: Gastrin and cholecystokinin (CCK), peptides from the gut, also form a related family. CCK has 58 amino acids, while gastrin exists in three forms – with 14, 17, and 34 amino acids. The relationships among the peptides are shown by the identity of the five amino acids at the C-terminal ends, and the existence of sulfur-containing groups attached to tyrosine residue (Chapter 7). So far, no other peptide hormones appear to have related sequences.

Bombesin, Neuromedin-B, and Gastrin-Releasing Peptide: Both bombesin and gastrin-releasing peptide cause release of the hormone gastrin from the cells of the stomach and duodenum; they also inhibit feeding. A related peptide, neuromedin-B, appears to have similar effects; however, because each of these three peptides can activate the receptors of the other peptides, it is difficult to distinguish the roles of the three receptors. Relationships among the three receptors are shown in Figure 14.4.

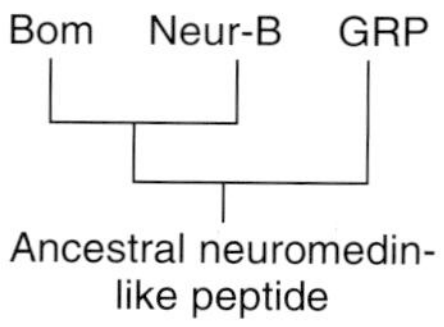

Figure 14.4 Phylogenetic relationships among bombesin, neuromedin-B, and gastrin-releasing peptide. Modified from Figure 3 of Ref. [479]. Used with permission.

14.3.2 Hormones of the Receptor Tyrosine Kinases

All ligands of this receptor superfamily are peptides. The several families of ligands are diverse and appear not to be related to each other. Within each family, however, sequence similarities indicate phylogenetic relationships of the genes encoding these hormone groups.

14.3.2.1 The Insulin Family

Insulin has been detected in all organisms [480]. Its effect is similar wherever it is found – to facilitate the penetration of nutrients into cells. It is not yet known how close the primary sequences of the peptides and their receptors are in vertebrates and unicellular organisms, though the binding site for insulin in the protist *Tetrahymena* is a peptide similar to insulin itself [481]. The fungus *Neurospora* has an insulin receptor (of unknown structure) that possesses tyrosine kinase activity and that activates carbohydrate-storage pathways similar to those of vertebrates [480].

Insulin-like growth factor (*IGF*) shows an organization, disulfide linkages, and sequence that is one-third identical to insulin. *Relaxin* is related to insulin, but is an agonist of GPCRs. The phylogenies of these three peptides [482] are shown in Figure 14.5.

14.3.2.2 The Neurotrophins

These peptides show no sequence similarity to insulin, but do have disulfide linkages. The related sequences include brain-derived neurotrophic factor (BDNF), nerve growth factor (NGF) and the neurotrophins (NTr) (Figure 14.6).

14.3.2.3 The Growth Hormone Family

This family includes three peptides, GH, chorionic somato-mammotropin (also called Placental Lactogen – PL), and Pr. The three are transcribed from genes on the same region of chromosome 17 in humans, and thus are likely paralogous. GH and Pr are released from the anterior pituitary, while PL is released from the placenta during pregnancy. Their relationships are shown in Figure 14.7.

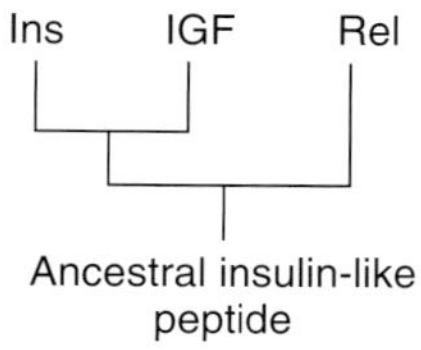

Figure 14.5 Relationships of insulin (Ins), insulin-like growth factor (IGF), and relaxin (Rel).

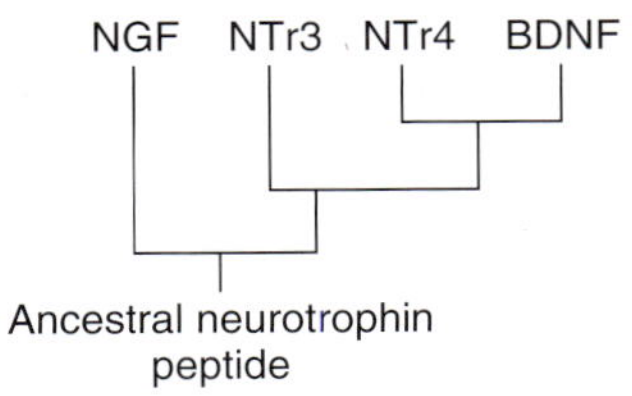

Figure 14.6 Phylogenetic relationships among NGF, BDNF, and NTr proteins. Modified from Figure 5 of Ref. [483]. Used with permission.

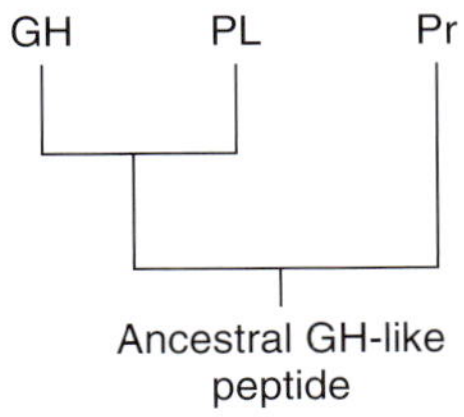

Figure 14.7 Relationship of mammalian growth hormone (GH), placental lactogen (PL), and prolactin. Modified from Figure 1 of Ref. [484]. Used with permission.

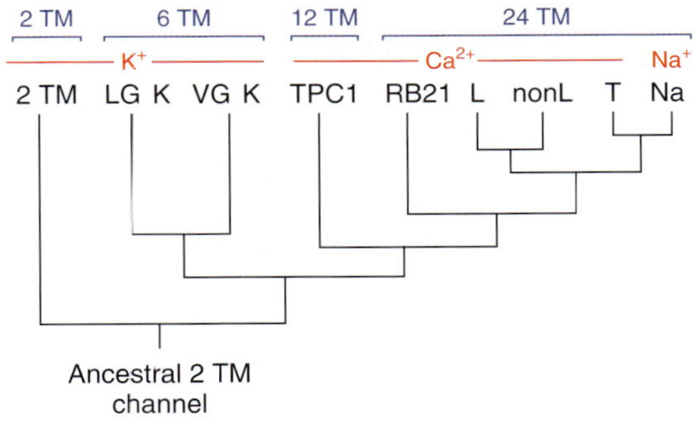

Figure 14.8 Phylogeny of voltage-gated ion channels. 2 TM–24 TM: the number of transmembrane domains; LG K: ligand-gated K^+ channel; VG K: voltage-gated K^+ channel; TPC1: two-pore segment Ca^{2+} channel; RB21: rat brain voltage-gated Ca^{2+} channel; L: L-type Ca^{2+} channel; T: T-type Ca^{2+} channel; and Na: voltage-gated Na^+ channel. Modified from Figure 2 of Ref. [485]. Used with permission.

14.4 Evolution of Receptor Superfamilies

14.4.1 Ion Channels

The ion channel superfamily contains several families (stretch-gated, voltage-gated and ligand-gated channels, transporters, and pumps). Even though the 3-D structures of transporters, voltage-gated channels, and ligand-gated channels are similar, analysis of their amino acid sequences does not allow a firm idea of the evolutionary process that led from one to the other. The most we can say is the following [161]:

- Vertebrate K^+ channels share sequences with some vertebrate transporters (e.g., the cystic fibrosis transmembrane regulator) and some bacterial transporters.
- Bacterial K^+-selective glutamate channels resemble voltage-gated K^+ channels.

The stretch- and osmotic shock-gated channels are considered to be possible precursors of this superfamily (Chapter 6), and the cyclic nucleotide-gated channels and voltage-gated channels likely evolved next. The ligand-gated channels are clearly related to the voltage-gated channels because of their similar 3-D structure, but the two share no common amino acid sequences, so they cannot be placed into a single phylogenetic system. This observation, and the lack of ligand-gated channels in prokaryotes, indicates that the two channel groups became distinct near the beginning of the origin of eukaryotes.

14.4.1.1 Voltage-Gated Channels

Significant similarities of overall structure exist between members of this group: the Na^+ and K^+ channels have the same number of transmembrane domains, similar voltage sensing domains, and have 29% identical residues, even though the Na^+ channel is a single molecule and the K^+ channel consists of four separate identical peptides (Chapter 6).

K^+ channels are the most diverse (at least 12 types), suggesting that they are the most ancient of the standard voltage-gated channels. The voltage-gated K^+ channels consist of four identical peptides, each with six TM domains. *Na^+ channels* evolved from the low-voltage Ca^{2+} channels.

Figure 14.8 shows a probable phylogenetic system for the cation channels in prokaryotes and eukaryotes (all but the LG K^+ (ligand-gated K^+) channel are voltage gated).

14.4.1.2 Ligand-Gated Channels

The LGICs channels exist only in animals, with the exception that bacteria have a K^+-selective glutamate receptor [161]. The lack of other types of neuronal- or muscular-type receptors is consistent with the observation that only animals have large-scale body movements requiring the interplay of neural and muscular elements. The derivation of these channels from bacterial precursors is shown in Figure 14.9.

Researchers can also identify three families within the ligand-gated group:

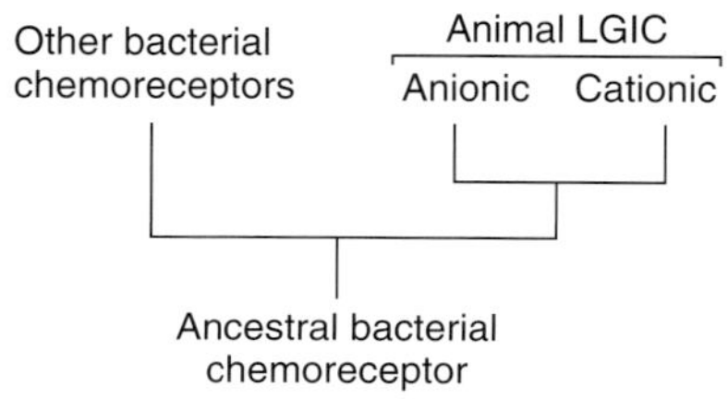

Figure 14.9 Origin of animal LGICs as descendants of bacterial chemoreceptors for amino acids. The anionic (Cl^-) channels are sister receptors to the cation (Na^+, K^+) channels. Modified from Figure 3 of Ref. [189]. Used with permission.

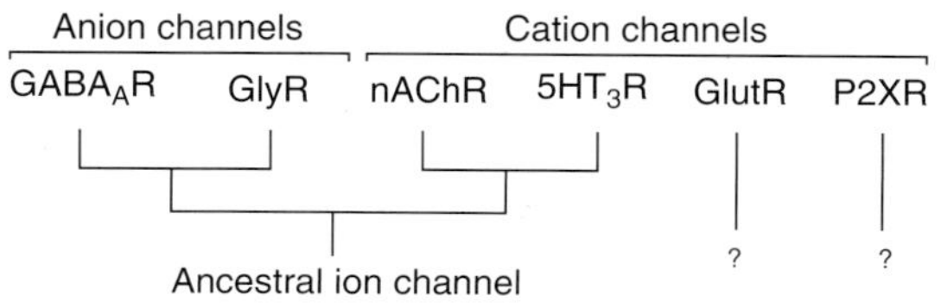

Figure 14.10 Possible origins of the ligand-gated ion channel families. The P2X receptors for ATP are presumed to be the oldest (Ref. [466]), but have not yet been identified in bacteria, and they share no sequence homology with the rest. Glutamate receptors (GlutRs) originated in prokaryotes, but the remainder of the ligand-gated channels are much more recent. Transporters appear to be the initial bacterial proteins from which the voltage-gated and ligand-gated channels evolved. Modified from Figure 22.13 in Ref. [161]. Used with permission.

- The purinergic receptors (P2X) for ATP;
- The glutamate (NMDA (*N*-methyl-d-aspartate), KA (kainate)) receptor group;
- The nicotinic ACh receptor, the ionotropic 5-hydroxytryptamine (5-HT) receptor, the glycine receptor, and the GABA$_A$ (gamma amino butyric acid) receptor group.

These families are distinct from each other and their origins date to the same time period as the split of the voltage-gated and ligand-gated channels; their evolutionary origins are unclear (Figure 14.10), but their possession of an even number of TM domains and use of multimeric structures suggests that they are homologous.

14.4.2 G Protein-Coupled Receptors

The ancient origin of these receptors is suggested by the presence of the protein rhodopsin in the Archaea, the prokaryotes that are the closest relatives of the Eukaryotes. Bacterial rhodopsin is very similar in three-dimensional structure to the rhodopsin that acts as the visual pigment in animal eyes. Bacterial rhodopsin is a transducer of light energy, but not for vision – it generates a transmembrane proton gradient that promotes ATP synthesis for cellular processes [486]. Both types of rhodopsin have seven transmembrane domains, though only eukaryotic rhodopsin couples to a G protein effector (transducin, homologous to G_o).

Amino acid sequences of Archaeal and Eukaryotic rhodopsin do not show significant identity, but the overall structural resemblance is sufficient to indicate evolutionary relationship (see earlier Section *14.2.6*). All the true GPCRs have seven membrane-spanning regions, homologous TM domains, and homologous domains for ligand binding and G protein interaction. As indicated in Chapter 7, these receptors, as found in eukaryotes, belong to six families [213]. Even within the eukaryotes, sequence comparisons do not allow estimation of the ancestral sequence or the relationships among the families. That situation also argues for the great antiquity of the GPCR superfamily.

14.4.2.1 G-Protein-Coupled Receptor Types

The six major families are as follows:

- *Family A:* Rhodopsin-like, the largest group, containing the rhodopsins, the amine receptors, odorant receptors, bitter taste receptors, nucleotide receptors, receptors for some peptides such as the opioids, and three hypothalamic releasing hormones;
- *Family B:* Secretin-like, containing receptors for secretin, glucagon, VIP (vasoactive intestinal peptide), GIP (gastric inhibitory peptide), and the remaining hypothalamic releasing hormones – these are closest to the Adhesion family and found only in Bilateria [487];
- *Family C:* The metabotropic Glutamate receptors;
- *Family D:* Fungal pheromone receptors;
- *Family E:* Cyclic adenosine monophosphate (cAMP) receptors;
- *Family F:* Frizzled/smoothened receptors.

Five of the eukaryote families are also found in mammals. These are often described by the acronym GRAFS [206] for glutamate (G), rhodopsin (R), adhesion (A), frizzled/smoothened (F), and secretin (S) families.

Relationships among the families of GPCRs are shown in Figure 14.11.

14.4.2.2 Family A Receptors – Rhodopsin Family

These receptors have been divided into four classes, α, β, γ, and δ.

The α Class This group contains amine receptors and the opsin-like light receptors; melatonin receptors; a

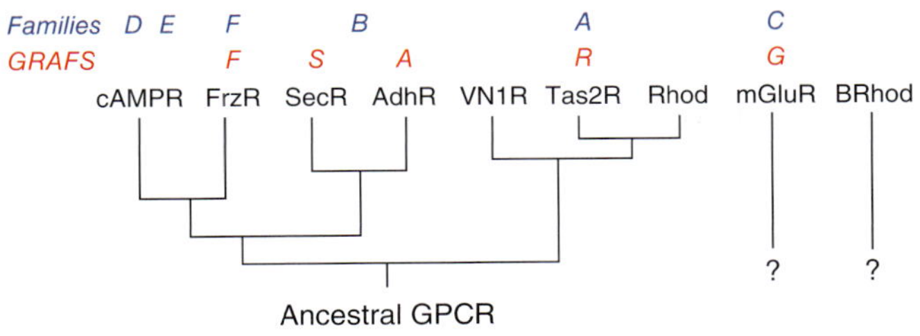

Figure 14.11 Nomenclature and possible relationships among the six clades of the G-protein-coupled receptor superfamily. The families are designated by two lettering systems: A–E (blue) for all eukaryotic families and GRAFS (red) for mammalian families. cAMP: cyclic AMP receptors; Frz: frizzled/smoothened receptors; Sec: secretin receptors; Adh: adhesion receptors; VN1: vomeronasal receptors type 1; Tas2: bitter taste receptors; Rhod: rhodopsin; and mGluR: metabotropic glutamate group. The exact phylogenetic positions of BRhod and the mGluR are not known. Modified from Figure 1 of Ref. [213]. Used with permission.

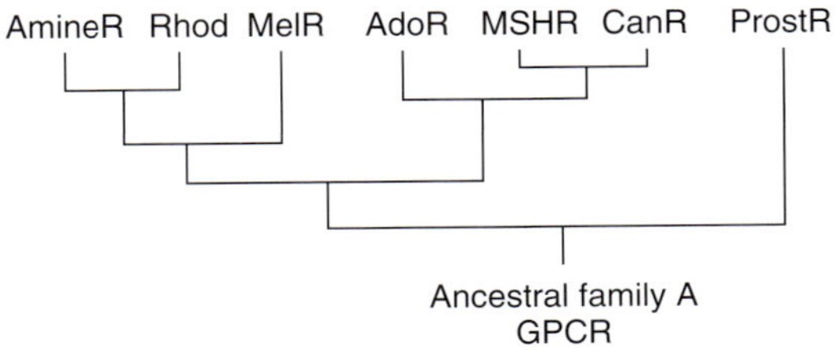

Figure 14.12 Sequence homologies between the family A, group α, G-protein-coupled receptors. The separate but related lineages are amine: monoamines, Rhod: rhodopsin and related opsins; Mel: melatonin receptor; Ado: adenosine receptor; MSH: melanocyte-stimulating hormone receptor; Can: cannabinoid receptors; and Prost: prostanoid receptors. Amine receptors are expanded in Figure 14.13 and prostanoid receptors are expanded in Figure 14.14; Modified from Figure 3 of Ref. [263]. Used with permission.

lineage that contains adenosine, melanocortin, and cannabinoid receptors; and the prostanoid receptors. Their relationships are shown in Figure 14.12.

We provide short descriptions of two of the members of this class:

Amine Receptors: These are the classical metabotropic receptors mediating physiological roles of the sympathetic nervous system (Figure 14.13).

The signaling functions of the amine receptors were covered in Chapter 7. The receptors are named for their agonists, in order from left to right: muscarinic cholinergic (ACh), DA, α_2-adrenoceptor (NE (norepinephrine)), Serotonin, α_1-adrenoceptor (NE), β-adrenoceptor (EP (epinephrine)), and histamine. It is interesting that several have multiple subtypes and these do not always cluster in single phylogenetic clades (e.g., DA, Ser, and α). ACh, Serotonin (5-hydroxy tryptamine), and Histamine are also agonists of LGICs (Chapter 6).

Prostanoid Receptors: Molecular sequences of prostanoid receptors suggest that the first to evolve was able to bind PGE_2. Sequence analysis also suggests that two major branches exist: one group Gs-coupled (DP1, EP2, EP4, and IP) and the other group G_i- or G_q-coupled [488] (Figure 14.14).

The β Class This group contains receptors for neuropeptides; their relations are shown in Figure 14.15.

The γ Class These are also peptide receptors, but belong to a different lineage from those in group β. Figure 14.16 shows the relations among the receptors for these peptides.

Opioid receptors illustrate the WGD scenario described earlier in this chapter. The four receptors in the group are termed ORL (opioid-receptor-like), KOR (kappa opioid receptor), DOR (delta opioid receptor), and MOR (mu opioid receptor). The ORL is likely the original opioid receptor. Its gene was duplicated (at the time of the initial vertebrate duplication event) and the KOR emerged as a mutated version of the ORL; a later genome-wide duplication allowed the DOR and MOR to form. In each duplication, one of the two (ORL and DOR) evolves more slowly while another (KOR and MOR) evolves more rapidly (Figure 14.17) [489].

The δ Class These are the receptors for olfactory molecules, three anterior pituitary hormones, and relaxin. They are shown in Figure 14.18.

14.4.2.3 Family B – Secretin and Adhesion Receptors

The secretin group of receptors includes receptors for VIP, secretin, calcitonin, and parathyroid hormone. They form one clade of a large receptor group that also includes the adhesion receptors (Figure 14.19).

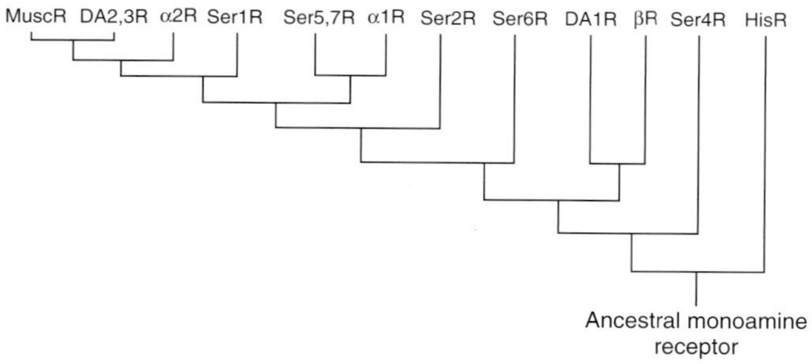

Figure 14.13 Amine receptors MUSC: muscarinic cholinergic receptor; DA: dopamine receptor; ALPHA: α-adrenoceptors; SERs: serotonin receptors; BETA: β-adrenoceptors; The several subtypes of serotonin and dopamine receptors are not monophyletic. Modified from Figure 3 of Ref. [263]. Used with permission.

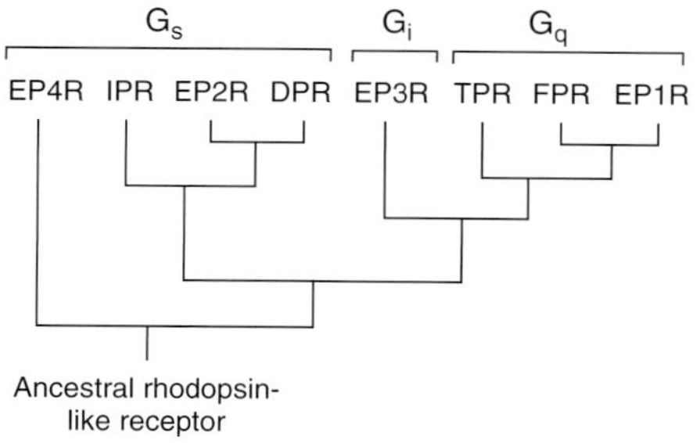

Figure 14.14 Phylogenetic relationships of the prostanoid receptors and their G protein partners.

Family C – Metabotropic Glutamate Receptors The receptors in this diverse family have a large N-terminal region of several hundred to several thousand amino acids; this region binds the ligands, which are also diverse: calcium and sweet/umami taste molecules, as well as two neurotransmitters that also bind to ion channel receptors (Figure 14.20).

14.4.2.4 Family F – Frizzled and Smoothened Receptors

This family contains two receptors for important developmental signaling processes (discussed in Chapter 9). Both Frizzled and Smoothened receptors are involved in setting up the body axis during early development (Chapter 11). Some subtypes of Frizzled (1, 2, 3, 6, and 7) are more closely related to Smoothened than are the others (Figure 14.21).

14.4.2.5 Elements of the GPCR Transduction Pathway

Nearly a dozen proteins are involved the G-protein-coupled transduction system of a given cell:

- The cell surface receptor itself;
- The G_α, G_β, and G_γ subunits of the G protein;
- Adenylyl cyclase;
- Phospholipase C;
- Protein kinases A and C;
- The receptor for IP_3 on intracellular membranes;
- Ion channels and additional kinases stimulated by the second messenger systems.

The following evolutionary discussion focuses on some of these molecules.

We focus on the G proteins and the G-protein-like molecules. The G proteins originated as GTPases or **elongation factors** for protein translation in bacteria. Eukaryotes have retained genes for those proteins and have modified others to encode G proteins as transduction units for the GPCR. G proteins are found in the cell membranes of unicellular protists, plants, Fungi, and Animals [493]. One of the animal G proteins, the recently described G_v, is not found in vertebrates [494]. The molecular weights of G protein subunits in invertebrates are roughly the same as in vertebrates, and their ability to act as substrates for cholera toxin and pertussis toxin indicate that the invertebrate G proteins are phylogenetically and functionally similar to those from vertebrates [480].

The vertebrate and invertebrate lineages diverged about 600 mya. Even so, vertebrate receptors can couple functionally with invertebrate G proteins. This ability is an indication that this class of receptors has highly conserved protein structures and transduction interactions. Evidence of conservation of sequences and structure in these receptors comes from experiments in which artificial membranes have been reconstituted with β-adrenoceptors from mammals and G proteins from birds, amphibians [486], or even insects [480]. These artificial combinations produce functional complexes that activate cAMP in response to epinephrine. Further, sequence analysis of G protein genes [493] has shown that the rates of change of the α chain in protists, fungi, and animals have been very slow, only a few changes per amino acid site per billion years. These are among the most highly conserved proteins found in eukaryotes, as

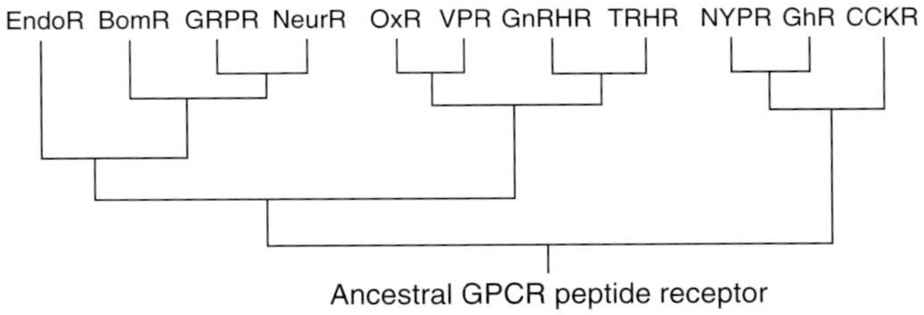

Figure 14.15 Sequence homologies between the family A, group β, G-protein-coupled receptors for neuropeptides, Endo: endothelin receptor; Bom: bombesin receptor; GRP: gastrin-related peptide; Neur: neuromedin-B receptor; Ox: oxytocin receptor; VP: vasopressin receptor; GnRH: gonadotropin-releasing hormone receptor; TRH: thyrotropin-releasing hormone receptor; NPY: neuropeptide Y receptor; GhR: ghrelin receptor; and CCK: cholecystokinin receptor. Modified from Figure 3 of Ref. [263]. Used with permission.

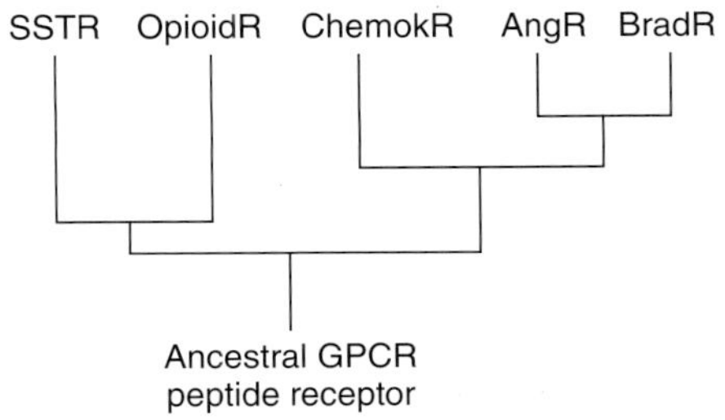

Figure 14.16 Sequence homologies between the family A (group γ) G-protein-coupled receptors for peptides. SST: somatostatin receptor; ORL: opioid receptors; CHEMOK: chemokine receptors; ANG: angiotensin II receptor; and BRAD: bradykinin receptor. The opioid receptors are expanded in Figure 14.17. Modified from Figure 3 of Ref. [263]. Used with permission.

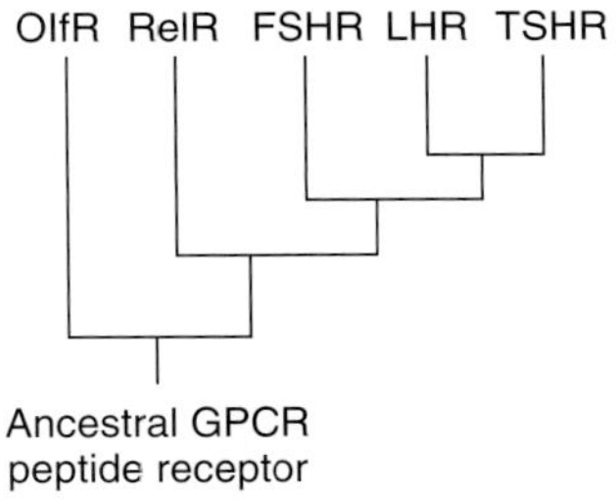

Figure 14.18 Evolution of family A, group δ, G-protein-coupled receptors (GPCR): OLF: olfactory receptors; FSH: receptors for follicle-stimulating hormone; LH: luteinizing hormone receptor; and TSH: thyroid-stimulating hormone receptor. Modified from Figure 3 of Ref. [263]. Used with permission.

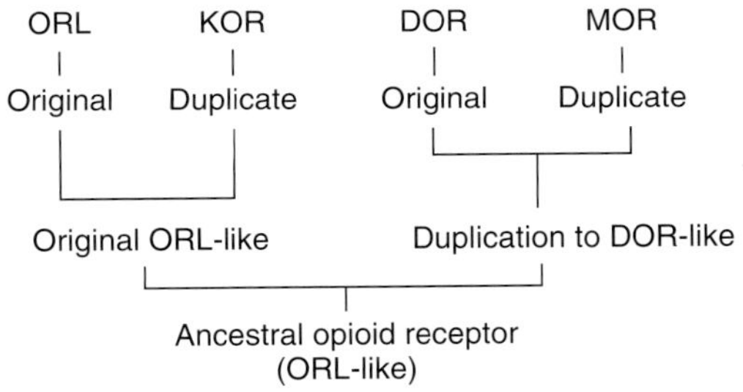

Figure 14.17 Evolution of opioid receptors in group γ. The original was similar to the ORL; it was duplicated, and the duplicate evolved to become the KOR. After this initial duplication, a further duplication produced copies of the ORL and KOR, and these mutated to become the DOR and MOR, respectively.

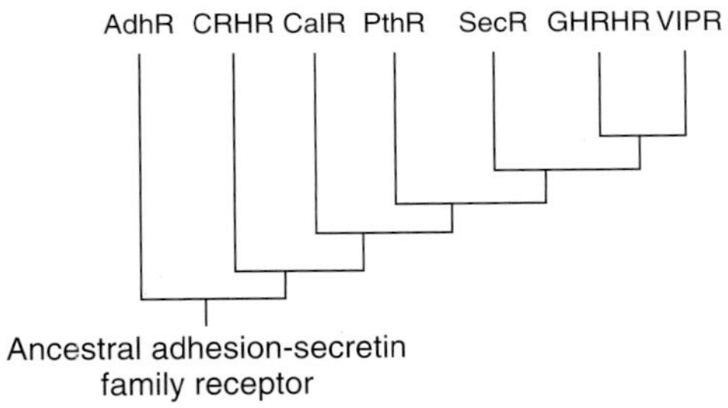

Figure 14.19 Evolution of G-protein-coupled receptors from family B: Adhesion receptors (AdhRs), corticotropin releasing hormone receptor (CRHR), calcitonin receptor (CalR), parathyroid hormone receptor (PthR), secretin receptor (SecR), growth hormone releasing hormone receptor (GHRHR), and vasoactive intestinal peptide receptor (VIPR[2]). Modified from Figure 2 of Ref. [490]. Used with permission.

all the α subunits show at least 40% amino acid identity (out of nearly 400 amino acids per peptide). Their conservation indicates that G protein function depends on a precisely defined structure.

The α chains for those amino acid sequences that have been determined allow the relationships among the several types to be shown [224, 492, 494] (Figure 14.22).

Similarities between Ras Oncogene and G Proteins The G_γ subunit is related to the Ras oncogene protein, whose function was described in Chapter 7. These relationships suggest that the ras and G_γ genes share a common ancestor.

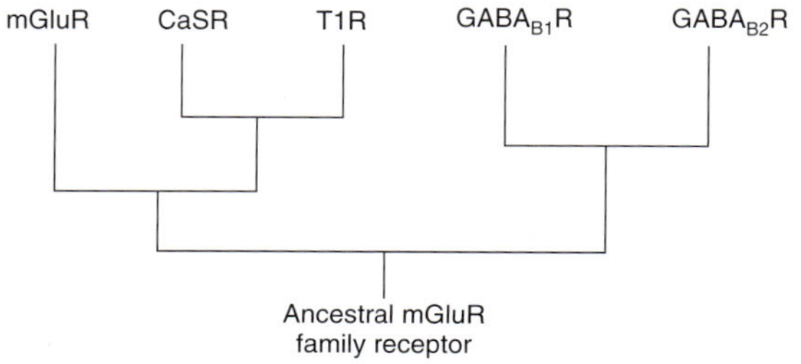

Figure 14.20 Evolution of G-protein-coupled receptors of family C. The family consists of metabotropic glutamate receptors (mGluRs), Calcium-sensing receptors (CaSRs), type 1 taste receptors (T1Rs) for sweet and umami, and the gamma amino butyric acid receptors ($GABA_B$). Modified from Figure 2 of Ref. [491]. Used with permission.

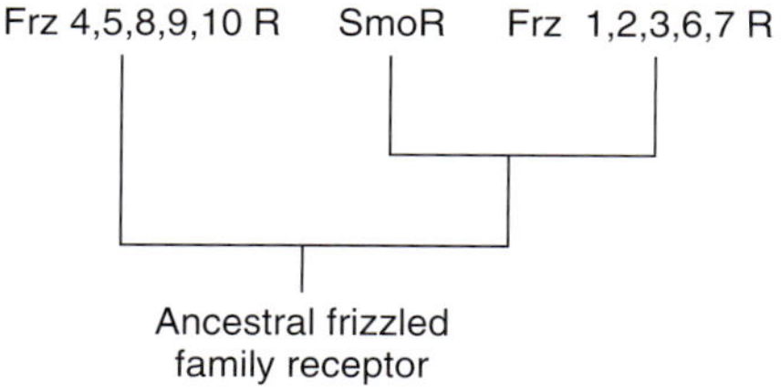

Figure 14.21 Evolution of family F G-protein-coupled receptors. Frz: frizzled receptor and Smo: smoothened receptor. Smoothened evolved from within the Frizzled group. Modified from Figure 1 of Ref. [492]. Used with permission.

14.4.3 The Immunoglobulin Superfamily

The members of this large group of receptors and other cell-surface proteins all have a single TM domain; most have an intracellular kinase domain and extracellular cysteine-rich regions that form disulfide bonds linking portions of the extracellular domain. Not all are involved in immune system responses, but the various molecules – receptor and intracellular kinases, and cell adhesion molecules – show an overall homology that makes their common ancestry likely.

All organisms possess kinases that regulate enzyme action via phosphorylation. Most belong to two types, serine/threonine (S/T) kinases and tyrosine (Y) kinases; a few atypical kinases phosphorylate other molecules such as lipids or choline [495]. Both the S/T kinases and the Y kinases have soluble members and membrane-located members. The membrane-bound Y kinases belong to the large family of **RTKs,** and the membrane-bound S/T kinases belong to a single family, the transforming growth factor receptor group. Though these two types of receptor kinases operate on different substrates, they all appear to belong to a single clade [495], and are all found only in the eukaryotes.

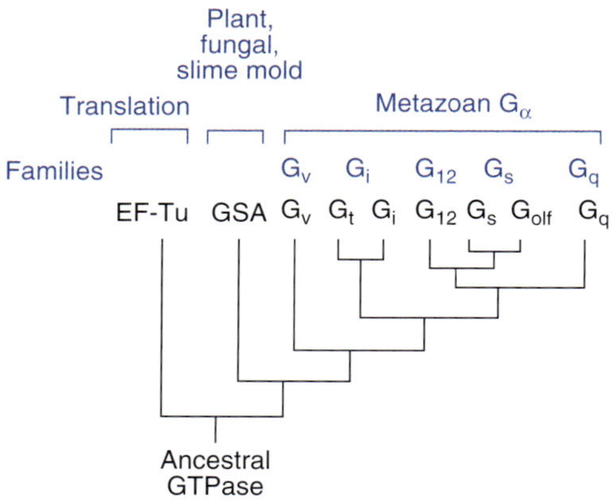

Figure 14.22 Sequence relationships of G protein alpha subunits. Information obtained from Refs. [224, 493, 494]. **EF-Tu:** universal protein translation elongation factor and GSA: alpha subunit as found in plants, fungi, and slime molds. G_{olf}, G_s, G_q, G_{12}, G_o, G_t, and G_i are described in Chapter 7 and are found only in metazoans.

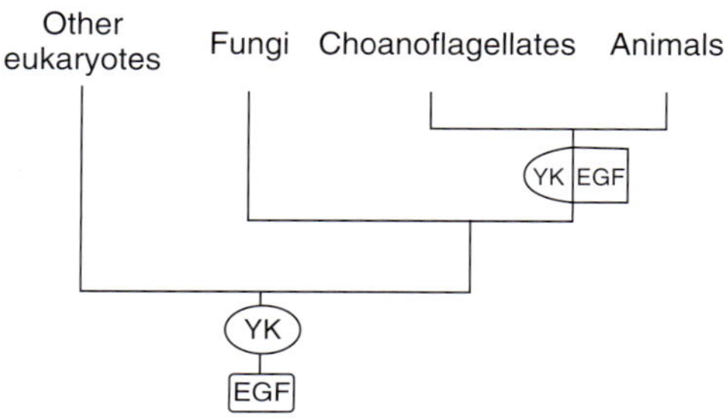

Figure 14.23 Phylogeny of the multicellular eukaryotes showing origins of receptor tyrosine kinases following fusion of the originally separate tyrosine kinase (YK) domain with the epidermal growth factor (EGF) binding domain. Plants and fungi both have intracellular tyrosine kinase proteins and extracellular EGF proteins. Genes for the two became fused only in the clade including the choanoflagellates and animals, leading to the family of receptor tyrosine kinases. Modified from Figure 1 of Ref. [496]. Used with permission.

14.4.3.1 The Receptor Tyrosine Kinases

The RTKs evolved from non-receptor kinases by a fusion between an early eukaryotic kinase gene and the gene for the epidermal growth factor (EGF) domain [496] (Figure 14.23). This original EGF is present as an extracellular protein in most eukaryotes [497]. The RTK family thus descended from the intracellular kinases at the time the animal-choanoflagellate clade evolved.

A likely phylogeny of the major enzyme groups of the family [495, 498] is shown in Figure 14.24.

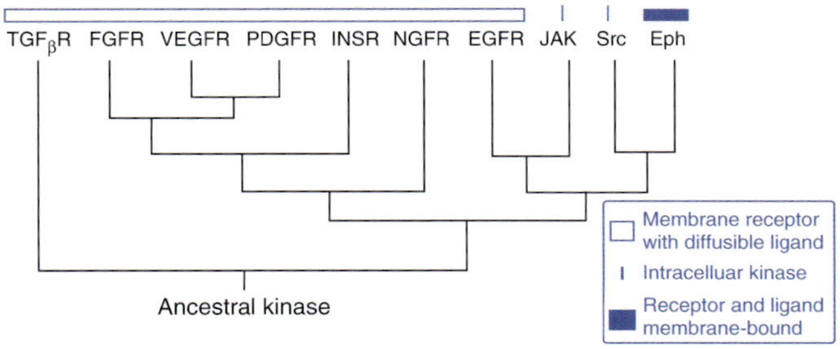

Figure 14.24 Phylogeny of the receptor kinases, intracellular, and membrane-bound. Relationships are based on sequences of the kinase domains. TGF-beta: Transforming growth factor-beta receptor (receptor serine/threonine kinase); FGF: fibroblast growth factor receptor; VEGF: vascular endothelial growth factor; PDGF: platelet-derived growth factor receptors; INS: insulin and insulin-like receptors; NGF: nerve growth factor; EGF: epidermal growth factor receptors. JAK: Janus kinase; Src: sarcoma kinase; and Eph: ephrin receptors. Src and JAK are non-receptor kinases; ephrin receptor ligands are all bound to cell membranes, and are not diffusible. All but TGF-beta are tyrosine kinases.

14.4.3.2 Molecules of the Adaptive Immune System

The cell-surface molecules involved in adaptive immunity are of three types:

- Receptors of the major histocompatibility complex (MHC) that first identify an invader as non-self;
- Receptors on the surfaces of T lymphocytes that bind to MHC and signal other cells to produce antibodies;
- The antibodies themselves (immunoglobulins) that bind invaders and target them for destruction.

This constellation of molecules is found only in the vertebrates. The proteins are homologous with the receptor kinases and thus form a diverse but related group.

14.4.4 Steroid, Vitamin A/D, and Thyroid Hormone Receptors

14.4.4.1 Origin of Nuclear Receptors: The Role of Ligands

The origin of the steroid receptors was proposed by Csaba [341] to lie in a migration of cell-surface receptors to the inside of the cell. However, steroids and their receptors exist only in eukaryotes; further, these "receptors" are actually transcription factors, very similar to those that already exist in cells as unliganded transcription factors: both have Zn finger domains and function in the same way. Thus, the nuclear receptors probably originated – at around the time of the origin of animals – from intracellular transcription factors. Though Csaba's specific proposal is no longer tenable, debate continues as to whether the nuclear receptors began as unliganded or liganded proteins [499].

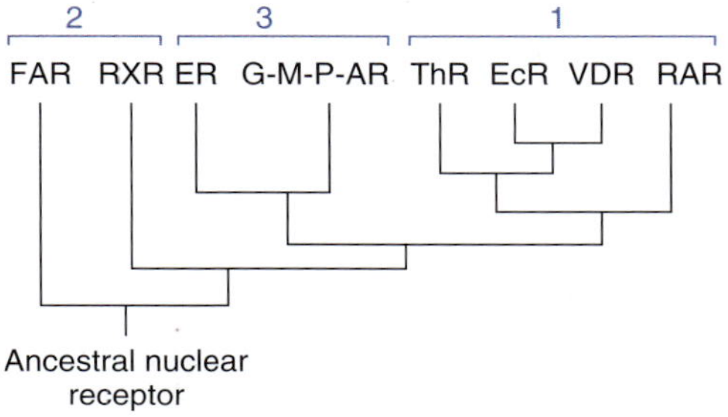

Figure 14.25 Phylogeny of the nuclear receptors. FA: fatty acid receptor; RXR: retinoid-X-receptor; ER: estrogen receptor; G-M-P-A: glucocorticoid, mineralocorticoid, progesterone, and androgen receptors; TH: thyroid hormone receptor; EC: ecdysone receptor; VD: vitamin D receptor; and RA: retinoic acid (vitamin A) receptor. The ER and G-M-P-A complex is amplified in Figure 14.26. Numbers above receptor names indicate the nuclear receptor families (Chapter 9). Modified from Figure 1 of Ref. [499]. Used with permission.

14.4.4.2 The Nuclear Receptor Families

First, it appears that all nuclear receptors are homologous and have a common ancestor in the early metazoans (Figure 14.25).

The oldest nuclear-receptor-like molecule is present in sponges as a fatty-acid-binding transcription factor (AqNR1 and AqNR2 (nuclear receptors 1 and 2 of the sponge *Amphimedon queenslandica*)) [499]. Vertebrates also have nearly identical fatty-acid-binding receptors (e.g., the hepatocyte nuclear factor 4 (HNF4) receptors of mammals), but the fatty acids are permanently bound to the active site, making the "receptor" constitutively active [499]. The likeliest hypothesis for this type of receptor at the base of the animal line is that these receptors evolved as sensors for food or environmental toxins, and binding of such molecules caused transcription of enzymes for food or toxin disposition. Later

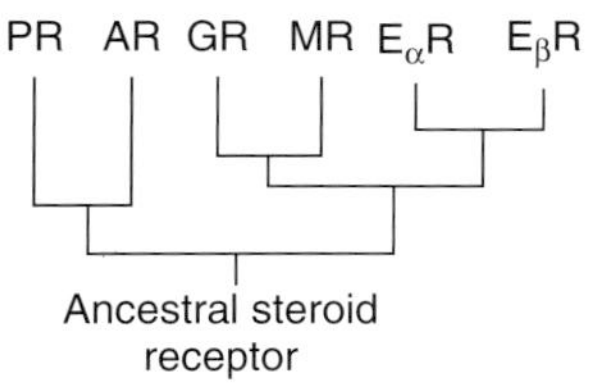

Figure 14.26 Evolution of steroid receptors. PR: progesterone receptor; AR: androgen receptor; GR: glucocorticoid receptor; MR: mineralocorticoid receptor; and Eα, Eβ: estrogen receptors. The ancestral steroid receptor was estrogen receptor-like. PR-like and GR-like receptors evolved by duplication; later duplication produced the MR and the AR. Modified from Figure 1 of Ref. [475]. Used with permission.

evolution of nuclear receptor binding to different ligands may have followed a process similar to "ligand exploitation" (next section). These receptors were discussed in Chapter 9 along with the steroid- and thyroid-hormone-liganded receptors.

14.4.4.3 **Later Evolution of Nuclear Receptors – Ligand Exploitation**

Thornton [475] studied evolutionary changes in six steroid receptors (estrogen α and β, progesterone, androgen, glucocorticoid, and mineralocorticoid) and proposed that the estrogen receptor was the first steroid receptor. From that was derived the progesterone receptor and corticoid receptors (CRs), which in turn split to form the progesterone/androgen line and the mineralocorticoid/glucocorticoid lines (Figure 14.26).

This suggestion is an interesting one, as estrogen is the final steroid in the synthetic pathway, with androgens and progesterone as intermediates (Figure 4.23). Thus, the intermediates would have been produced during a time when no receptor was available for them. Thornton suggests that receptors for progesterone and androgens appeared by a process of "ligand exploitation," in which duplication of the gene for the estrogen receptor (followed by mutational changes in receptor structure) produced redundant receptors that were fortuitously able to bind these intermediates. The great variety of potential steroid structures means they are much like peptides in their "mutability." Thus, mutations in receptor structure are more likely to produce binding sites that are compatible with one of the steroids than with other small molecules (e.g., ACh) that are invariant in structure.

Thornton collaborated with Bridgham and Carroll to develop an extension of the "exploitation" concept as applied to the coevolution of the mineralocorticoid receptor (MR) and its agonist aldosterone [500] (Figure 14.27).

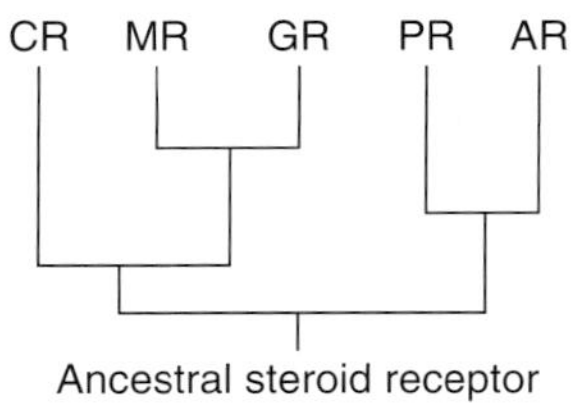

Figure 14.27 Proposed evolutionary sequence of vertebrate steroid receptors other than the estrogen receptor. CR: ancestral corticoid receptors present in Agnatha (lampreys, hagfish). MR: mineralocorticoid receptor; GR: glucocorticoid receptor; PR: progesterone receptor; and AR: androgen receptor. MR is present only in jawed vertebrates, and its ligand aldosterone is present only in tetrapods, though most vertebrate corticoid receptors (CR and MR but not GR) will bind exogenous aldosterone.

Neither the MR nor its agonist aldosterone is present in the jawless vertebrates. The receptor evolved by gene duplication (Figure 14.27) in the first fish with jaws – the sharks – but long before the origin of its ligand. Thus, in the earliest jawed vertebrates, an MR existed without aldosterone. Using phylogenetic analysis, the authors determined the most probable nucleotide sequence of the ancestral CR and expressed it in cell culture. Like the CR of lampreys, this receptor is able to bind aldosterone (which no current vertebrate glucocorticoid receptors (GRs) can). This suggests that the original MR had a broad binding ability and was not specific for any particular steroid [500].

As aldosterone was not present until the evolution of bony fish, the ability of aldosterone to bind the vertebrate CR and the reconstructed ancestral receptor is fortuitous. Aldosterone is synthesized from corticosterone by the enzyme cytochrome P450 11β hydroxylase (CP11H). Both corticosterone and CP11H are present in all vertebrates. However, the form of the enzyme present in lampreys cannot make aldosterone, and that ability only evolved via a mutation in the enzyme gene after the origin of bony fish. The conclusions of this complex pattern of receptor–ligand evolution are as follows:

1) A receptor may evolve from a preexisting protein following gene duplication; that receptor most likely has no specific ligand, but already-existing molecules may activate it; if such a ligand–receptor pair is adaptive, it will be retained.
2) The MR originated in a gene duplication event from the GR in the bony fish.
3) Two subsequent mutations permitted the origin of MR function in vertebrates:

a. Mutation in CP11H permitted aldosterone to be formed.
b. Mutation in the GR copy to MR permitted mineral regulation by the MR receptor.

4) Loss of aldosterone-binding ability occurred only in the GR of fish and tetrapods, separating glucose regulation from salt regulation so one could be activated without activating the other.

In summary, though the parts of a reaction sequence must be adaptive at the time of origin, they may also participate in more than one cellular interaction; thus, as new molecules evolve, they may recruit existing molecules to new functions. The authors call this process "molecular exploitation," and propose that it is a major theme in the evolution of complex interactions [500].

Roles of Steroid Receptors in Vertebrate Evolution The nuclear steroid receptor family evolved only in the vertebrates. Because of their unique signaling mechanisms, the steroid receptors are suggested to be responsible for much of the regulatory complexity within the vertebrate lineage [501].

14.5 Evolution of Receptor Antagonism

Most of the receptor antagonists known are produced in the laboratory for research or therapeutic purposes. For example, α- and β-adrenoceptor antagonists such as prazosin and propranolol have clinical value as antihypertensive drugs. These drugs and others like them work because they bind the receptor without activating its transduction systems and in the process prevent the binding of endogenous agonists. In addition to these artificial compounds, plants produce a number of substances that act as receptor antagonists in animals. Curare, a competitive inhibitor of the nicotinic ACh receptor, is one example.

In contrast, very few receptor antagonists are produced by animals for inhibition of their own receptors or for antagonism of receptors of other individuals. The best known are the cone snail and snake venoms that act in the same way as curare, as well as other venoms that target voltage-gated ion channels. These toxins are analogous to the plant poisons in that they are meant for use on (killing) a *different organism.*

It is also an interesting question why animals don't use receptor antagonism as a mechanism for stopping their own receptor action. Most likely, it is because of the way antagonists must act: in order to oppose the action of a high-efficacy agonist such as epinephrine, an antagonist must have a much higher affinity for that receptor than the agonist. Thus, once an animal has released such an antagonist into its circulation or at a synapse, the antagonist will remain bound to the receptors for a long time, and/or the animal will have to produce large amounts of agonist to displace the antagonist. This clearly will blunt the animal's responses and will slow response times. A far easier way to stop receptor action is simply to stop release of the transmitter.

Even though natural endogenous antagonists are unlikely to be widespread, recent research has uncovered a few examples of naturally occurring receptor antagonists produced within the same animal as the receptor on which it acts:

- The dark hair color of mammals is determined by the action of melanocyte stimulating hormone (MSH) on its GPCR in melanin-producing cells of the hair follicle. Receptor binding elevates cAMP to activate an enzyme that produces dark melanin pigment. This action is, however, antagonized for a short period of time during hair growth: the *agouti* gene temporarily produces a protein that binds as an antagonist to the MSH receptor and prevents the binding of MSH [502]. This impedes dark pigment formation, resulting in a narrow band of yellow on a dark background; when all hairs show this pattern, the result is the flecked "wild type" or "agouti" coat color of mammals.
- The luteinizing hormone (LH) receptor gene produces, by alternate splicing of two exons, some short peptides that bind to the LH receptor at its ligand-binding site, thus preventing LH binding when they are present. The physiological role of these peptides and the control of their production are obscure, but their presence suggests they may be important natural regulators of LH receptor responsiveness.

14.6 A Final Note

An important feature of receptors is that they initiate powerful and complex processes following simple interactions with simple substances of "low information content." This is the essence of the receptor mechanism, and is a surprising discovery, considering early hypotheses that complex developmental and biochemical pathways must be initiated by large, information-rich signaling molecules [385].

Instead, a satisfying simplicity seems to be at work, in which small molecules on both sides of the cell membrane are linked with cell membrane proteins to form a signal pathway. From their beginnings as metabolic intermediates, these molecules first became *intra*cellular second messengers and then *inter*cellular first messengers or hormones. The many problems of physiology, biochemistry, and development are therefore solved by a very few kind of molecules, and the complex patterns of life are painted by using only a very few fundamental colors.

No book can provide a complete and perennially up-to-date discussion of such a rapidly evolving field. It is thus appropriate to end with a return to Plato's dialog *Meno,* to which the translator Jowett has provided an introduction. After discussing the difficulties of defining and teaching "virtue," he notes, "The answer ... given by Plato is paradoxical enough, and seems rather intended to stimulate than to satisfy enquiry [153]." That is the intent of this book as well.

Glossary

A

Acrosome tip of the sperm; a membrane-bound packet of enzymes that is released when sperm binds to the extracellular coat around the egg. The enzymes dissolve the coat and allow the sperm access to the egg cell membrane.

Active site the molecular region of an enzyme or receptor involved in catalytic or binding activity.

Active transport movement of ion up its concentration gradient; requires energy from ATP.

Adaptive immune response production of T cells, B cells, and antibodies by lymphocytes following invasion by foreign molecules or cells. Produces long-term response with memory of previous invasions.

Addiction compulsive drug use with serious dependence and withdrawal symptoms.

Address selection when growing neurons reach their target, they recognize it by characteristic ligand–receptor interactions between target and neuron.

Adenylyl cyclase enzyme bound to the cytoplasmic side of the plasma membrane, activated by G proteins following hormone binding, and promoting the conversion of ATP to cAMP.

Adrenal medulla central portion of the adrenal gland, composed of sympathetic nervous system cells. Releases EP and small amounts of NE into the bloodstream.

Adrenergic neural cells that release epinephrine or norepinephrine.

Adrenoceptor G-protein-coupled receptors whose primary transmitters are epinephrine and norepinephrine.
α-Adrenoceptor – receptor for norepinephrine; causes vasoconstriction.
β-Adrenoceptor – receptor for norepinephrine and epinephrine; causes vasodilation, bronchial dilation, and cardiac excitation.

Affinity measure of the strength of binding of ligand and receptor. High-affinity binding is strong; low-affinity binding is weak.

Agonist substance that activates cellular processes by binding to activated state of receptor.
Partial – agonist that is effective only when most or all the receptors are bound.
Full – agonist that is effective with only a small portion of the receptors bound.

Alkaloid nitrogen-containing organic compound, often important as a drug or as a metabolite in cells.

Allosteric binding site on receptor or enzyme different from the site to which ligand or substrate normally binds. Binding of another ligand to the allosteric site modifies receptor action.

Alpha helix helical shape of a protein caused by hydrogen bonds between adjacent amino acids.

Analgesic pain-killing.

Androgen steroid hormone responsible for male growth and development.

Antagonist substance that inhibits cell response by binding to a receptor without activating it.
Competitive – antagonist that binds to the same site as the agonist.

Anterior pituitary hormone-secreting organ adjacent to hypothalamus and connected to it by a vascular system that delivers releasing hormones.

Antibody protein produced by organism attacked by invader. Specifically binds antigen and neutralizes it.

Antigen protein of an invader that serves to activate host defenses.

Receptor Biology, First Edition. Michael F. Roberts and Anne E. Kruchten.

Antigen receptor protein in surface of host cell that recognizes invading antigen as nonself.

Antiporter transport protein that moves one substance into the cell and another simultaneously out of the cell.

Apoptosis programmed cell death, in which a group of cells is destroyed during development. Examples are the webbing between toes and fingers in humans, and the tails of tadpoles. Comes from Greek – "falling off."

Asymmetry in membrane, the presence of different types of phospholipid on the cytoplasmic and extracellular sides of the membrane. In organisms, a lack of uniformity of structure.

Atherosclerosis buildup of fats and cholesterol in the blood vessels; associated with increased cholesterol content of blood.

Atropine antagonist of **metabotropic** ACh receptors.

Autoinducer chemical signaling molecules that are produced and used by bacteria participating in quorum sensing.

autonomic nervous system nerves to internal organs, controls visceral functions such as heart rate, digestion, and circulation.

Autophosphorylation self-phosphorylation of a receptor peptide

B

B cell lymphocyte of the immune system involved in antibody-mediated immunity.

Basal lamina layer of protein secreted by epithelial cells.

Benzodiazepines group of tranquilizer drugs that stimulates the GABA receptor.

Beta sheet flat sheet-like structure of protein caused by hydrogen bonding between amino acids of nearby regions of a protein.

Bilateria animals with bilateral symmetry.

Bindin sperm protein that binds to the egg plasma membrane in sea urchins.

Blastoderm surface cells of the early embryo.

Blastomere cell early in development: the two cells resulting from division of the zygote are the first blastomeres.

Blastopore organizing region of the gastrula; cells migrate inward to form mesoderm and nervous system.

Body plan a group of anatomical features shared by members of a taxon.

Brain-derived neurotrophic factor (BDNF) peptide ligand for growth receptor involved in strengthening synapse and maintaining neuron integrity.

C

Cadherins calcium-dependent **cell adhesion molecules**.

Catecholamine hormone/transmitter class based on the structure of the amino acid tyrosine. Epinephrine, norepinephrine, and dopamine are members. Related compounds are serotonin and melatonin.

Cell adhesion molecules receptors allowing adhesion of similar cells into tissues.

Cell-surface receptor protein embedded in the cell membrane; binds a transmitter or hormone to activate the cell.

Channelopathy disease caused by impairment in function of ion channels, resulting from mutations or autoimmune responses.

Chemoaffinity hypothesis idea of Sperry that the surfaces of neurons have chemical codes that serve as identifiers to guide them in growth and development.

Chemotaxic stimulus chemical that binds to cell-surface receptor and influences movement of the cell.

Cholesterol fatty substance that is important in membrane formation. Is situated between lipid molecules, and affects fluidity of the membrane.

Chromophore compound such as retinal that is responsible for the light absorption properties of opsin.

Clathrin intracellular protein, bound to plasma membrane, involved in control of cell shape and movement.

Cleavage cell division. The initial period of development in deuterostomes is called *cleavage* because cell division proceeds without visible differentiation, and the embryo often appears as a mass of identical cells.

Cluster of differentiation (CD) molecules proteins that act as agonists or receptors in communication between immune system cells.

Coactivator protein cellular protein that promotes binding and activation of nuclear receptors to genetic elements in the nucleus.

Coated pit membrane aggregate of receptors before they are internalized.

Collagen linear extracellular protein that serves as an anchor for cells and a guide for developing neurons.

Complement a group of blood proteins, activated by antigen–antibody binding that cause the rupture of invading cells.

Conformation physical shape and molecular charge distribution of the binding site of a receptor or other protein.

Conjugation binding of one unicellular organism to another for the purpose of gene exchange.

Constant region portion of antibody molecule (Ig) that is the same from one Ig to the next.

Constitutive expression being in a continuously active state; for a gene, this means always producing protein. For a receptor it means always activating effectors.

Coordination bond linkage between a metal ion, such as Zn^{2+}, and another molecule, such as a protein.

Corticosteroids steroid hormones promoting glucose metabolism.

Cotransport active transport of a sugar or an amino acid into a cell, accompanied by the transport of Na or H in the same or opposite direction.

Cyclic AMP (cAMP) intracellular messenger (second messenger) that consists of adenine + ribose + phosphate, with the phosphate attached both at the 3′ and the 5′ carbons.

Cyclic GMP (cGMP) intracellular messenger (second messenger) that consists of guanine + ribose + phosphate, with the phosphate attached both at the 3′ and the 5′ carbons.

Cystic fibrosis disease characterized by thick mucus and blocked exocrine ducts; caused by mutated chloride transporter.

Cytokine protein hormones that act as ligands for tyrosine kinase receptors; these hormones activate receptors that control growth and development and those that mediate the immune response.

D

DDE pollutant, metabolite of DDT, that binds to testosterone receptor and inhibits testosterone binding.

Dependence physical and physiological symptoms that occur when a drug has been taken for a long time and is suddenly withdrawn.

Depolarize change of cellular membrane potential from the normal value near −70 mV to a value near zero.

Desensitization decrease in the responsiveness of a receptor, either by reducing its affinity for hormone or by uncoupling it from cellular processes.

Deuterostomes group of animals (Echinoderms, Chordates, and related phyla) that shows regulative development in which cell–cell interactions are more important than cell lineages in producing the differentiated state.

Diabetes disease of excess blood glucose caused by inability of cells to take glucose up from blood.

Differentiation origin of different cell types in a developing organism. Caused by differential expression of genes in different cell types under the control of DNA-binding proteins.

Dimer a compound protein consisting of two separate peptide chains.

Homodimer – dimer in which the two peptide chains are identical.

Heterodimer – two unlike proteins forming a functional association.

Diplobastic having two cell layers – ectoderm and endoderm. Characteristic of Cnidaria alone.

Disulfide bridge two cysteines on a peptide attached by sulfur–sulfur link.

Dopamine hypothesis idea that schizophrenia is caused by increased activity of dopamine pathways in the brain.

Domain functional region of a protein molecule, for example, a catalytic portion, a portion that binds an agonist, or a region that spans the plasma membrane and acts independently of neighboring sequences.

Dominance extent to which genes are expressed.

Incomplete – allele in which expression of heterozygotes is less strong than that of homozygotes.

Complete – allele that is expressed the same in heterozygotes (Aa) as in homozygous dominants.

Dominant-negative unusual expression of a particular protein caused by a mutation in a gene for a different protein that had been inhibiting the first.

Downregulation decrease in response to an agonist.

Drug chemical substance that alters cellular function.

E

EF-Tu protein that controls elongation of a peptide chain during protein synthesis.

Effector intracellular protein (e.g., adenylyl cyclase) that converts the binding signal into a biochemical process such as ion movement, vasoconstriction, or secretion.

Efficacy ability of receptor binding to activate cellular response. A high-efficacy agonist is effective at low levels of receptor binding.

Electroplax modified muscle cells in certain fish that allow them to generate high voltages.

Elongation factor protein (resembling G protein) that promotes translation of mRNA into a new protein chain.

Endocrine refers to the process in which a substance is released from a cell and travels through the blood to a target cell or organ.

Endocytosis movement of substances bound to receptors into a cell by invagination of the plasma membrane following binding.

Endogenous originating in the organism being studied.

Endosome intracellular vesicle containing internalized receptor and ligand.

Epidermal growth factor peptide hormone involved in stimulation of mitotic division

Epinephrine hormone released from adrenal medulla; involved in cardiac excitation by binding to cardiac β-adrenoceptors.

Epitope peptide fragment of an invading antigen that is placed on the macrophage cell surface.

Estrogen steroid hormone responsible for egg maturation and yolk production.

Eumetazoa animals with two or three cell layers (includes all animals but Placozoa and Porifera).

Excitatory producing depolarization of cell membrane.

Exocrine gland that secretes a substance into a duct.

F

Facilitated diffusion transport of small molecules across cell membranes by a carrier molecule that does not use energy.

Fertilization membrane or “fertilization envelope”; membrane that resists sperm binding; it is formed just after fertilization to prevent multiple fertilization events.

First messenger hormone, transmitter, or other extracellular signaling molecule.

G

G protein-coupled receptor receptor class with seven membrane-spanning regions and connection to a Guanine-nucleotide binding protein as part of the activation sequence.

Gain-of-function mutation mutation that causes increased activity or expression of a protein over that seen with the usual “dominant” allele.

Ganglion collection of nerve cell bodies outside the CNS.

Gastrulation developmental phase in which cell movements occur and multiple germ layers form, allowing organ development.

Genome total DNA complement of a cell.

Genomic action of steroids initiation of transcription leading to synthesis of new proteins.

Germ layers three general regions (ectoderm, mesoderm, and endoderm) of early embryos. Each germ layer produces particular tissues and organs in the adult.

Glucocorticoid steroid hormone, such as cortisol, which is responsible for increasing glucose release into the blood.

Glycoprotein protein containing chains of carbohydrates linked to particular amino acids in the sequence.

Glycosylation addition of a carbohydrate molecule or chain to a protein.

Grade a taxon united by a level of morphological complexity but not intended as a phylogenetic unit.

Gray crescent region of fertilized zygote that will later form the blastopore.

Growth cone leading edge of a migrating neuron, following chemical cues that determine its movement.

G(uanine-nucleotide binding) Protein membrane-bound protein that binds GDP and GTP and controls intracellular enzymes after interaction with a receptor.

Guanosine diphosphate (GDP) Guanine + ribose + diphosphate. When bound to a G protein, it leaves the G protein in the inactive state.

Guanosine triphosphate (GTP) Guanine + ribose + triphosphate. When it replaces GDP on a G protein, it allows the G protein to activate cellular enzymes.

H

Habituation reduction in intensity of response following repeated stimulation.

Haploinsufficiency inability of one dominant allele to produce normal phenotype. Action similar to *recessive*.

Heat shock proteins proteins of the "chaperone" class; in their role as protectors, they bind enzymes and prevent their denaturation at high temperatures. In the case of nuclear receptors, the heat shock proteins are bound to the receptor in the absence of agonist.

Helper T cell lymphocyte involved in early recognition of foreign antigen – activates B cells

Heterodimer protein consisting of two nonidentical peptides.

Hippocampus region of forebrain associated with long-term learning and emotion.

HMG co-A reductase enzyme that controls synthesis of cholesterol.

Homologous recombination mistaken matching of DNA sequences in meiosis such that one sequence is moved to a position different from the one it occupied earlier.

Homodimer protein consisting of two identical peptides.

Homophilic binding association between two molecules that are identical, for example, cadherins.

Hormone small molecule (protein, amino acid, steroid, or other substance) that binds to a receptor and causes a cellular response.

Hydrogen bond a chemical bond in which hydrogen is bound to oxygen or nitrogen on one molecule and comes close to an oxygen or nitrogen on an adjacent molecule. The electropositive hydrogen becomes shared between the other electronegative atoms.

Hydrophilic polar.

Hydropathy plot assignment of numerical values (from +4.5 for very nonpolar to −4.5 for very polar) for amino acids in a protein.

Hydrophobic nonpolar.

Hyperglycemia excess glucose in the blood, related to lack of insulin or to lack of insulin receptors.

Hyperpolarize change of cellular membrane potential from a normal value about −70 mV to a more negative level (e.g., −90 mV).

Hypertension abnormally high blood pressure.

Hypophyseal portal circulation blood vessels taking blood from hypothalamus to anterior pituitary; transports releasing hormones.

Hypothalamic releasing hormones small peptides released from the hypothalamus that travel via bloodstream to the anterior pituitary and cause or inhibit release of other peptide hormones.

I

Imaginal discs blocks of tissue in the insect larva that will later differentiate into epidermal structures in the adult.

Immortalization ability of a cell to divide indefinitely; characteristic of cancer cells.

Immune response ability of organism to counteract invading tissues and molecules.

Immune system receptors cell- surface antibodies or other proteins involved in self- and nonself-recognition.

Innate immune response activation of cells and chemicals to fight invaders. Operates in the short term and does not produce antibodies or memory of invasion.

Induction process of differentiation in which one cell releases a signaling molecule that acts on an adjacent cell, causing it to alter its pattern of gene expression.

Inhibitory producing hyperpolarization of cell membrane.

Integral membrane protein receptors, channels, and other molecules that are embedded in the cell membrane and often span the distance from extracellular to cytoplasmic faces.

Integrins extracellular proteins that serve to bind cells together and to guide axon growth.

Interactome interactions among proteins of a cell.

Interleukin cytokines involved in stimulation of immune response.

Internalization removal of receptors from the cell surface to an intracellular site where they do not bind hormone.

Intracellular receptor receptor that resides within the cell rather than on its surface.

Inverse agonism preferential binding of ligand to inactive receptor (rather than to active receptor as does a full agonist).

Inwardly rectifying K^+ channel K channel that selectively allows ions out rather than in because it closes when the membrane depolarizes. The outward movement of K ions hyperpolarizes the cell.

Ion channels membrane proteins that serve as regulatable pores for ion movement into and out of cells.

Ionic bond a chemical bond in which one atom donates electrons to another, leaving the former with a net positive charge and the latter with a net negative charge.

Ionotropic (not inotropic) referring to rapid-response ion channel function – inward flux of ions that depolarizes the cell.

K

kDa abbreviation for kilodaltons, a measure of molecular weight. 1 kDa = 1000 g/mol.

Kinases enzymes that catalyze the transfer of a phosphoryl group from trinucleotides (mainly ATP) to other molecules, generally to enzymes.

L

Laminin extracellular linear protein associated with **basal lamina** of cells.

Learning modification of behavior based on experience.

Ligand substance that binds to and activates or deactivates a receptor.

Ligand exploitation the terminal ligand in a biosynthetic pathway is the first for which a receptor evolves; selection for this hormone also selects for the synthesis of intermediates despite the absence of receptors, and duplicated receptors then evolve affinity for these substances.

Ligand-gated ion channel (LGIC) membrane protein that acts to control the movement of ions into and out of a cell; binding of a transmitter to the channel opens or closes it.

Lineage-specific proteins products of genes in particular cells that allow those cells to differentiate in specific ways.

Lipid bilayer two layers of phospholipids in a membrane.

Local modulator hormone or other biochemical substance that acts to change cell activity; refers to action on adjacent or nearby cells.

Loss of function mutation mutation that alters a normal protein to an inactive or ineffective one. Also called *recessive*.

Low-density lipoprotein (LDL) receptor growth factor receptor required for normal uptake and metabolism of cholesterol.

Lymphocyte white blood cell involved in immune response.

Lymphokine see cytokines and interleukin.

Lysosome intracellular organelle containing digestive enzymes.

M

Macrophage large white blood cell involved in immune response; is often the first cell to interact with invading antigen; displays epitopes on its surface for recognition by T and B cells.

Magnocellular neurosecretory cells hypothalamic cells that release oxytocin and vasopressin in posterior pituitary.

Major histocompatibility complex (MHC) cell-surface proteins that identify "self" in an animal.

MAP (mitogen-activated protein) kinase system cascade of enzymes that stimulate cell division after activation of growth factor receptor by its agonist.

Memory retention of the knowledge of previous thoughts, impressions, or events. In the immune system, memory involves production of cell lineages able to recognize pathogens for long periods after the initial invasion.

Memory, long-term recall of information that remains with the person for long periods and cannot easily be disrupted; mechanism involves modification of synapses and cell biochemistry.

Memory, short-term recall of information that is short-lived and easily disrupted; mechanism involves continues activation of a circuit without modification.

Metabolome metabolic products and processes within a cell.

Metabotropic relatively slow (minutes to hours) metabolic effect of receptor activation that is characteristic of G-protein-coupled receptors. The ion channels, in contrast, operate at millisecond time scales.

Metastasis invasive migration of cancer cells into body regions different from their origin.

Metazoa animals, including all multicellular descendants of the choanoflagellates.

Mind–body dualism concept that the mind and the body are separate entities; originated by Plato and developed by Descartes.

Mineralocorticoid steroid hormone, such as aldosterone, which is responsible for Na uptake from kidney tubules.

Mitogenesis cell division.

Mitogenic factors hormones or signaling molecules that promote cell division.

Monoclonal antibodies antibodies produced by a cell that is a hybrid between a normal antibody-producing cell and a cancer cell. The cancer cell confers immortality on the combination so continuous production of antibody can be maintained.

Mood state of mind or feeling.

Morphogen substance that activates genes controlling development.

Morphogenesis cellular processes in development that allow organisms to develop the structures proper for that species.

Mosaic development pattern of development shown by protostomes; cell differentiation is most affected by intracellular substances rather than by cellular interactions.

Muscarinic acetylcholine receptor G-protein-coupled receptor that controls vasoconstriction in certain blood vessels and airways, as well as fluid secretion in certain glands.

N

Neofunctionalization genes achieve a new function following duplication.

Neural net system of interconnections between receptor cells and effectors with no integrative interneurons between them.

Neuroglia cells that support and nourish neural cells.

Neurohumor (neurosecretory substance) hormone or transmitter released into the bloodstream from a neural cell.

Neurosecretory cell neural cell that releases a transmitter or hormone (called a *neurosecretory substance*) into the blood stream rather than into a synapse.

Neurotransmitter substance that is released from a neural cell into a synapse; it immediately acts on a receptor of the adjacent neural, muscle, or gland cell.

Nicotine alkaloid that is an agonist on ACh ion channel receptors of nerves and skeletal muscles.

Nicotinic the action of acetylcholine on ion channels in nerves and skeletal muscles.

Nonpolar containing fat-soluble components.

Nonsteroidal anti-inflammatory drugs (NSAIDS) substances such as aspirin that achieve analgesia by inhibition of prostaglandin synthesis.

Norepinephrine transmitter released from sympathetic nerve terminals, largely in blood vessels; causes vasoconstriction by binding to α-adrenoceptors.

Null mutation mutation that causes a protein to no longer function as the protein of the normal gene did.

Nuclear receptor protein within the cell that binds a signal molecule in the cytoplasm or nucleus; the binding event allows the receptor–agonist combination to become an active transcription factor.

O

Odorant receptor G-protein-coupled receptor that binds odorant molecules and activates neural pathways of smell.

Oligosaccharide carbohydrate with few (oligo) monosaccharide components

Ommatidia sections of the compound eye of insects; each ommatidium contains photoreceptors.

Oncogene gene sequence encoding a protein involved in cell division; mutated versions can cause cancer by permanently turning on the cell division mechanism.

Oncoprotein protein encoded by an oncogene. Generally acts as a growth factor, receptor for a growth factor, or part of the cellular transduction mechanism that promotes cell division.

Opiate compound containing the fundamental morphine or thebaine structure possessing some affinity to any, or all, of the opioid receptor subtypes. Examples are heroin, buprenorphine, and naltrexone.

Opioid naturally occurring peptide transmitter that operates in pain pathways in the brain.

Organ set of tissues that make up a single structure and function together.

Organelle eukaryotic intracellular structure that carries on a particular function. For example, mitochondria, lysosomes, chloroplasts.

Organizational/activational hypothesis the concept that steroids present during early development set up permanent cellular structures and functions for the adult (operational), while steroids that may be added exogenously to an adult have only temporary (activational) effects.

Organizer blastopore region in frog embryo; releases signal molecules that induce mesoderm and neural tissue in migrating cells.

Orthologous gene sequences in different species that are related because of direct descent between the two species.

Osmotic detectors channels that sense changes in ion composition of fluid.

P

Pathway selection growing neurons choose correct pathway based on receptor–ligand interactions between neuron and proteins making up the path.

Paralogous gene sequences in a single organism that are related because one is a copy of the other.

Peptidergic neuron that releases peptide transmitter(s).

Peptide bond covalent bond between the carboxyl group of one amino acid and the amino group of an adjacent amino acid.

Phagocytosis movement of particles into a cell by invagination of the plasma membrane.

Pharmacopeia drug manual listing physiological effects of various natural and man-made substances.

Pheromone substance released from one individual that binds to a receptor in or on another individual and elicits a response.

Phosphatase enzyme that removes a phosphate ion (PO_4^-) from a molecule.

Phosphatidyl inositol 4, 5-bisphosphate [PIP_2] membrane phospholipid that produces two intracellular messengers (IP_3 and DAG) when it is split by phospholipase C.

Phosphodiesterase membrane-bound enzyme that converts ATP into cyclic AMP. In the eye it converts GTP into cyclic GMP.

Phospholipase C enzyme that breaks membrane phospholipid into diacyl glycerate (DAG) and inositol trisphosphate (IP_3).

Phospholipids membrane lipid with long fatty acids attached to a glycerol through an ester linkage; also contain a phosphorylated sugar or other group attached to the glycerol. The polar portion allows association with aqueous solutions and the nonpolar portion allows association with nonpolar tails of other phospholipids to form a lipid bilayer.

Phosphorylase enzyme that adds a phosphate group to a molecule.

Pilocarpine **muscarinic** agonist, mimicking effects of ACh.

Pinocytosis movement of fluid into a cell by invagination of the plasma membrane.

Plasma cell B cell that has been activated to recognize antigen.

Platelets noncellular blood elements, formed from the breakdown of macrophages.

Polar containing water-soluble components.

Polarity in chemistry, the presence of electrical charge on an atom or molecule; in physiology, the direction in a nerve or at a synapse. A polar neural system is one in which transmission is unidirectional; a nonpolar system is one in which transmission may go in either direction.

Post-synaptic cell nerve, muscle, or gland cell that is innervated.

Posterior pituitary hormone-secreting region that is a part of and immediately below the hypothalamus.

Pre-synaptic cell nerve that innervates a muscle, gland, or other nerve.

Primary structure linear sequence of amino acids in a protein.

Product substance that is the result of enzyme action on a substrate.

Progesterone steroid hormone responsible for maturation of the egg-containing follicle and for hormone production during pregnancy.

Protean agonism a given agonist may produce either agonism or antagonism on a given receptor, depending on the effector to which the receptor couples.

Proteome total protein composition of a cell.

Protostomes group of animals (all phyla aside from the echinoderms, chordates, and a few chordate-like phyla) in which embryonic differentiation occurs primarily by cell lineages and localization of intracellular substances rather than by cellular interactions.

Purinoceptor cell-surface receptor that binds purine compounds such as adenine, adenosine, or ATP.

Q

Quaternary nitrogen nitrogen with four groups attached to it. This gives it a net positive charge.

Quaternary structure two or more polypeptide chains held together to make a functional protein.

Quorum sensing communication process in unicellular organisms, giving them the ability to determine population density and to regulate expression of genes for cell division and other biochemical processes.

R

RAMP (receptor-activity-modifying-protein) membrane protein required for proper function of certain receptors.

Ras intracellular protein related to G proteins; binds GDP and GTP and activates cellular kinases. It is encoded by the gene *ras*.

Receptor protein that is bound by a ligand and, in the process, activates the cell.

Receptor cell cell with specialized surface for receiving environmental stimuli (e.g., touch, light, chemical, heat).

Receptor internalization movement of receptor to the inside of a cell following hormone binding. This is one way receptors (e.g., the low-density lipoprotein receptor) transport their ligands into cells.

Receptor kinase in one class of receptors, the cytoplasmic end of the receptor is an enzyme. It is activated by ligand binding to the extracellular portion of the receptor.

Receptor serine/threonine kinase enzymatic intracellular portion of the receptor that phosphorylates cytoplasmic proteins on serine or threonine residues following hormone binding.

Receptor tyrosine kinase enzymatic intracellular portion of the receptor that phosphorylates cytoplasmic proteins on tyrosine residues following hormone binding.

Receptor-mediated endocytosis movement of extracellular substances into cells after being bound by cell-surface receptor.

Recessive allele that is not expressed because the protein it makes is nonfunctional.

Regulative development pattern of development shown by deuterostomes; cell differentiation is most affected by cell–cell interactions rather than by cytoplasmic determinants.

Releasing hormone peptide hormone from the hypothalamus that causes release of an anterior pituitary hormone.

Repression Prevention of transcription of certain genes by steroid hormone–receptor complex. The competition between complex and a transcription protein for the promoter site prevents normal activation.

Response element short DNA sequence to which receptor–hormone complex binds to initiate transcription.

Retinal molecule similar to vitamin A, linked to opsin in the eye. Retinal changes shape when struck by light, allowing the opsin to activate a G protein.

Rhodopsin G-protein-coupled protein in the discs of the rods and cones of the eye. When struck by light, activates a G protein that initiates the process of vision.

S

Satiety factor signaling molecule (e.g., leptin from gut) that alerts feeding centers in the brain that a person has eaten enough food.

Saturated fatty acid chain with no double bonds between carbon atoms.

Second messenger intracellular substance that is produced following receptor activation. It acts within the cell to cause a cellular response.

Secondary structure alpha helix or pleated sheet that forms as a result of interactions between amino acids in a protein.

Segment polarity genes genes encoding proteins that control development of segments in early development.

Selectivity (also called *specificity*) ability of a receptor to bind only a particular type of molecule.

Selectivity filter constriction in membrane-spanning region of ion channel, lined with polar amino acids, that determines the species of ion allowed through the channel.

Sensitization increase in intensity of response following repeated stimulation.

Sequestration process of receptor inactivation caused by bringing the receptor into the cell on endocytotic vesicles.

Sialic acid carbohydrate that forms chains attached to cell-surface proteins. Confers binding specificity for cellular interactions.

Signal peptide hydrophobic leader at the N-terminus of a new protein; governs the placement of the protein into cell membrane.

Signal transduction cell–cell communication by use of hormones and intracellular biochemical mechanisms.

Signaling molecule small chemical substance (also called *hormone, pheromone, agonist, transmitter, local modulator*) that binds to a receptor to initiate a cellular process.

Smads intracellular proteins activated by serine-threonine kinase receptors. They migrate to the nucleus after activation and cause transcription of genes involved in differentiation.

Smooth muscle contractile tissue of glands, blood vessels, gut, and airways.

Somniferous causing sleep.

Sphingolipids lipids in which fatty acids attach to a carbon core through a nonester linkage.

SRM Selective receptor modulators. Agents that function as steroid agonists in some tissues (such as bone, brain, cardiovascular system), but that also function as antagonists in sites where estrogen may promote carcinogenesis, such as breast, uterus, or ovary. Tamoxifen and Raloxifene are examples of selective estrogen receptor modulators (SERMs).

Subfunctionaliztion change in function by genes after duplication.

Substrate substance that binds to an enzyme and is modified by enzyme action.

Subtypes of receptors receptors with similar structures and functions, though with small differences that alter their sensitivity to particular hormones or that alter their intracellular actions.

Superficial membrane protein molecule attached only to extracellular or cytoplasmic face of cell membrane.

Sympathetic nervous system motor nerves that control actions of glands, cardiac muscle, and smooth muscle of blood vessels and gut.

Synapse narrow junction between cells, especially between nerves and muscles; transmitter released from pre-synaptic cell diffuses across the synapse to bind to receptor on post-synaptic cell.

Syncytial blastoderm in insects, the embryo goes through several rounds of nuclear division without cell membrane formation; the result is a syncytium with many nuclei but no cell membranes separating them.

T

Target selection growing neuron nearing its target is guided by diffusible chemical signals released by the target; these interact with receptors on the neuron.

T cell lymphocyte of the immune system involved in cell-mediated immunity.

Tertiary structure folded pattern of secondary structures.

Testicular feminization male with mutated testosterone receptor so male primary and secondary sexual characteristics are not produced.

Testosterone male sex steroid.

Thyroid-stimulating hormone (TSH) peptide from the anterior pituitary that acts on G- protein-coupled receptors to cause secretion of thyroid hormone from the thyroid gland.

Tissue a set of cells that form a structure and function together; several tissues make up an organ.

Tissue-specific transcription factor protein involved in promoting transcription; found only in one or a few tissue types.

Tolerance after repeated doses of a drug, increased amounts are needed to elicit the same response as before the higher exposure.

Transcription production of a messenger RNA (mRNA) as a complement to one strand of the DNA. The mRNA is later used to produce a new protein.

Transcription factor protein that binds to a gene to activate it to make an mRNA for protein synthesis.

Transducer membrane-bound proteins (e.g., G proteins) that transmit the signal from the hormone–receptor complex to an intracellular effector such as adenylyl cyclase.

Transduction transfer of a hormone–receptor binding event to the interior of the cell; activation of the cell.

Transformation conversion of normal cell to one that no longer is affected by contact inhibition, nor by infecting a cell with altered forms of genes for growth factors or receptors.

Transmitter substance such as a hormone that is released specifically from neurons.

Triploblastic having three cell layers – ectoderm, mesoderm, and endoderm; characteristic only of the bilaterian animals.

Trophic (or tropic) hormone peptide from the anterior pituitary that causes release of hormones from other glands.

Tumor mass of cells created by proliferation or uncontrolled cell division.

Tyrosine kinase enzyme (intracellular portion of receptor) that **autophosphorylates** or phosphorylates a soluble enzyme.

U

Uncoupling disconnecting a receptor from its intracellular actions. Binding of hormone is not altered, but the activation of further steps, for example, G proteins, is prevented.

Unsaturated fatty acid chain containing one or more double bonds between carbon atoms.

V

van der Waals bond a chemical bond that exists between any two neutral atoms; its strength is proportional to the seventh power of the distance between the atoms.

Vasoconstriction causing contraction of blood vessel so the flow is reduced.

Vectorial discharge placement of a protein onto the cell membrane by coalescence of a protein-containing vesicle with the outer cell membrane.

Vitelline layer complex protein surface of the egg.

Voltage-gated ion channel ion channel that opens following voltage changes; does not bind transmitter.

Voltage-sensing domain transmembrane domains of voltage-gated ion channels that sense voltage changes across the membrane and control channel opening.

W

Whole-genome duplication replication within the cell of entire chromosome(s), causing polyploidy.

Z

Zinc finger portion of steroid hormone receptor that is responsible for binding to the DNA and for dimerization; contains a Zn atom bound loosely to four cysteines in such a way that a finger-shaped projection of the protein is formed.

Zona pellucida glycoprotein matrix surrounding the oocyte in mammals.

ZP3 zona pellucida protein serving as sperm receptor.

References

1. Jones, S. (1995) *Why is there so much genetic diversity?* Ch. 5 in (ed J. Brockman), The third culture, Simon & Schuster, New York, 413 p.
2. Koshland, D.E. (1958) Application of a theory of enzyme specificity to protein synthesis. *Proc. Natl. Acad. Sci.* U.S.A., **44** (2), 98–104.
3. Pattee, H. (1969) *Communication in Development*, Academic Press, New York, pp. 1–16.
4. Agosta, W.C. (1992) *Chemical Communication: The Language of Pheromones*, Macmillan.
5. Kandel, E.R., Schwartz, J.H., and Jessell, T.M. (2000) *Principles of Neural Science*, 4th edn, McGraw-Hill, Health Professions Division, New York, xli, 1414 p.
6. Hancock, J. (2005) *Cell Signaling*, 2nd edn, Oxford University Press.
7. Kokkola, T. (2005) G Protein coupled receptors, in *Physiology and maintenance, Encyclopedia of Life Support Systems*, UNESCO Editor.
8. Sayle, R.A. and Milnerwhite, E.J. (1995) Rasmol – biomolecular graphics for all. *Trends Biochem. Sci.*, **20** (9), 374–376.
9. Unwin, N. (2005) PDB ID 2BG9: refined structure of the nicotinic acetylcholine receptor at 4A resolution. *J. Mol. Biol.*, **346** (4), 967–989.
10. Palczewski, K. *et al.* (2000) PDB ID 1F88: crystal structure of rhodopsin: a G protein-coupled receptor. *Science*, **289** (5480), 739–745.
11. Li, Q., Wong, Y.L., and Kang, C. (2014) PDB ID 2MFR: solution structure of the transmembrane domain of the insulin receptor in detergent micelles. *Biochim. Biophys. Acta*, **1838** (5), 1313–1321.
12. Schwabe, J.W. *et al.* (1993) PDB ID 1HCQ: the crystal structure of the estrogen receptor DNA-binding domain bound to DNA: how receptors discriminate between their response elements. *Cell*, **75** (3), 567–578.
13. Reiter, R.J. (1991) Melatonin: the chemical expression of darkness. *Mol. Cell. Endocrinol.*, **79** (1-3), C153–C158.
14. Dennett, D.C. (1991) *Consciousness Explained*, 1st edn, Little, Brown and Co, Boston, MA, xiii, 511 p.
15. Mayr, E. (1982) *The Growth of Biological Thought: Diversity, Evolution, and Inheritance*, Belknap Press, Cambridge, MA, ix, 974 p.
16. Plato, B.J. and Tileston, M.W. (1882) *Socrates. The Phædo of Plato*, Roberts Brothers, Boston, MA, 141 p.
17. Ryle, G. (1949) *The Concept of Mind*, Senior Series, Hutchinson's University Library, London, New York, 334 p.
18. Walsh, C.T. and Schwartz-Bloom, R.D. (2010) *Levine's Pharmacology: Drug Actions and Reactions*, 7th edn, Informa Healthcare, London.
19. Magendie, F. (1822) *A Summary of Physiology*, Edward Cole, Baltimore, MD.
20. Devereux, S.M.A., Donahoe, H.B., and Kimura, K.K. (1956) A physiological basis for the grouping of the elements—James Blake (1815–1893). *J. Chem. Educ.*, **33** (7), 340.
21. Bernard, C. (1864) Etudes physiologiques sur quelques poisons americains. *Rev. Deux Mondes*, **53**, 164–190.
22. Langley, J.N. (1875) The action of jaborandi on the heart. *J. Anat. Physiol.*, **10**(Pt. 1), 187–201.
23. Parascandola, J. and Jasensky, R. (1974) Origins of the receptor theory of drug action. *Bull. Hist. Med.*, **48** (2), 199–220.
24. Silverstein, A.M. and Ehrlich, P. (2002) *Paul Ehrlich's Receptor Immunology: The Magnificent Obsession*, Academic Press, San Diego, CA, xix, 202 p.
25. Muir, R. (1915) Paul Ehrich. 1854–1915. *J. Pathol. Bacteriol.*, **20** (2), 350–360.
26. Ada, G.L. and Nossal, G. (1987) The clonal-selection theory. *Sci. Am.*, **257** (2), 62–69.
27. Fischer, E. (1894) Einfluss der Configuration auf die Wirkung der Enzyme. *Ber. Dtsch. Chem. Ges.*, **27** (3), 2985–2993.
28. Langley, J.N. (1901) On the stimulation and paralysis of nerve-cells and of nerve-endings: part I. *J. Physiol.*, **27** (3), 224–236.
29. Langley, J.N. (1905) On the reaction of cells and of nerve-endings to certain poisons, chiefly as regards the reaction of striated muscle to nicotine and to curari. *J. Physiol.*, **33** (4-5), 374–413.
30. Langley, J.N. (1906) Croonian Lecture, 1906: on nerve endings and on special excitable substances in cells. *Proc. R. Soc. London, Ser. B* (Containing Papers of a Biological Character), **78**, 170–194.
31. Michaelis, L. and Menten, M.L. (1913) Die kinetik der invertinwirkung. *Biochem. Z.*, **49** (333–369), 352.
32. Elliott, T.R. (1905) The action of adrenalin. *J. Physiol.*, **32** (5-6), 401–467.
33. Loewi, O. (1921) Über humorale übertragbarkeit der Herznervenwirkung. *Pflügers Arch. Eur. J. Physiol.*, **189** (1), 239–242.
34. Bonner, J.T. and Savage, L. (1947) Evidence for the formation of cell aggregates by chemotaxis in the development of the slime mold Dictyostelium discoideum. *J. Exp. Zool.*, **106** (1), 1–26.

Receptor Biology, First Edition. Michael F. Roberts and Anne E. Kruchten.

35. Vodopich, D.M. and Moore, R. (2002) *Biology Laboratory Manual*, 6th edn, McGraw Hill Higher Education.
36. Overton, E. (1899) Ueber die allgemeinen osmotischen Eigenschaften der Zelle, ihre vermutlichen Ursachen und ihre Bedeuting fir die Physiologie. *Vierteljahresschr. Naturforsch. Ges.*, **44**, 88–135.
37. Weissmann, G. and Claiborne, R. (1975) *Cell Membranes: Biochemistry, Cell Biology, & Pathology*, HP Publishing Co, New York, xvi, 283 p.
38. Gorter, E. and Grendel, F. (1925) On bimolecular layers of lipoids on the chromocytes of the blood. *J. Exp. Med.*, **41** (4), 439–443.
39. Frye, L.D. and Edidin, M. (1970) The rapid intermixing of cell surface antigens after formation of mouse-human heterokaryons. *J. Cell Sci.*, **7** (2), 319–335.
40. Danielli, J.F. and Davson, H. (1935) A contribution to the theory of permeability of thin films. *J. Cell. Comp. Physiol.*, **5** (4), 495–508.
41. Randall, D.J. *et al.* (2002) *Eckert Animal Physiology: Mechanisms and Adaptations*, 5 edn, W.H. Freeman and Co, New York, xvii, 736, 56 p.
42. Lenard, J. and Singer, S.J. (1966) Protein conformation in cell membrane preparations as studied by optical rotatory dispersion and circular dichroism. *Proc. Natl. Acad. Sci.* U.S.A., **56** (6), 1828–1835.
43. Singer, S.J. and Nicolson, G.L. (1972) The fluid mosaic model of the structure of cell membranes. *Science*, **175** (4023), 720–731.
44. Heung, L.J., Luberto, C., and Del Poeta, M. (2006) Role of sphingolipids in microbial pathogenesis. *Infect. Immun.*, **74** (1), 28–39.
45. Hannun, Y.A. and Obeid, L.M. (2002) The ceramide-centric universe of lipid-mediated cell regulation: stress encounters of the lipid kind. *J. Biol. Chem.*, **277** (29), 25847–25850.
46. Rimon, G., Hanski, E., and Levitzki, A. (1980) Temperature dependence of beta receptor, adenosine receptor, and sodium fluoride stimulated adenylate cyclase from turkey erythrocytes. *Biochemistry*, **19** (19), 4451–4460.
47. Rothman, J.E. and Lodish, H.F. (1977) Synchronised transmembrane insertion and glycosylation of a nascent membrane protein. *Nature*, **269** (5631), 775–780.
48. Anfinsen, C.B. (1973) Principles that govern the folding of protein chains. *Science*, **181** (4096), 223–230.
49. Bella, J., Brodsky, B., and Berman, H.M. (1995) Hydration structure of a collagen peptide. *Structure*, **3** (9), 893–906.
50. Brucker, E.A. (2000) Genetically crosslinked hemoglobin: a structural study. *Acta Crystallogr. D Biol. Crystallogr.*, **56** (Pt. 7), 812–816.
51. Kyte, J. and Doolittle, R.F. (1982) A simple method for displaying the hydropathic character of a protein. *J. Mol. Biol.*, **157** (1), 105–132.
52. Brändén, C.-I. and Tooze, J. (1991) *Introduction to Protein Structure*, Garland Publishing, New York, xv, 302 p.
53. Reece, J.B. and Campbell, N.A. (2011) *Campbell Biology*, 9th edn, Benjamin Cummings, Boston, MA, p. xlvi, 1263 p.
54. Wilkins, M.R. *et al.* (1996) Progress with proteome projects: why all proteins expressed by a genome should be identified and how to do it. *Biotechnol. Genet. Eng. Rev.*, **13**, 19–50.
55. Lederberg, J. and Mccray, A. (2001) The scientist:'ome sweet' omics--a genealogical treasury of words. *Scientist*, **17** (7), 8–8.
56. Oliver, S.G. *et al.* (1998) Systematic functional analysis of the yeast genome. *Trends Biotechnol.*, **16** (9), 373–378.
57. Cohen, J. (2001) The proteomics payoff. *Technol. Rev. (Manchester)*, **104** (8), 54–61.
58. Humphery-Smith, I. (2004) A human proteome project with a beginning and an end. *Proteomics*, **4** (9), 2519–2521.
59. The Signaling Gateway (2014) Alliance for Cellular Signaling, February 26, 2014. http://www.signaling-gateway.org/aboutus/afcs.html (accessed 9 Junly 2015).
60. Robinson, G.E. (1996) Chemical communication in honeybees. *Science*, **271** (5257), 1824–1825.
61. Roshchina, V. (2010) in *Microbial Endocrinology: Interkingdom Signaling in Infectious Disease and Health*, Chapter 2 (eds M. Lyte and P.P.E. Freestone), Springer, New York, xvi, 316 p.
62. Hardie, D.G. (1991) *Biochemical Messengers: Hormones, Neurotransmitters and Growth Factors*, Springer.
63. Parascandola, J. (2010) Abel, Takamine, and the isolation of epinephrine. *J. Allergy Clin. Immunol.*, **125** (2), 514–517.
64. Bayliss, W. and Starling, E. (1902) Preliminary communication on the causation of the so-called "peripheral reflex secretion" of the pancreas. *Lancet*, **159** (4099), 813.
65. Starling, E. (1905) The chemical correlation of the functions of the body. Croonian lecture II. *Lancet*, **2**, 423–425.
66. Grobstein, C. (1956) Trans-filter induction of tubules in mouse metanephrogenic mesenchyme. *Exp. Cell Res.*, **10** (2), 424–440.
67. Hampl, V. *et al.* (2009) Phylogenomic analyses support the monophyly of Excavata and resolve relationships among eukaryotic "supergroups". *Proc. Natl. Acad. Sci.* U.S.A., **106** (10), 3859–3864.
68. Bassler, B.L. (2002) Small talk. Cell-to-cell communication in bacteria. *Cell*, **109** (4), 421–424.
69. Kolenbrander, P.E. *et al.* (2002) Communication among oral bacteria. *Microbiol. Mol. Biol. Rev.*, **66** (3), 486–505.
70. Clarke, M.B. *et al.* (2006) The QseC sensor kinase: a bacterial adrenergic receptor. *Proc. Natl. Acad. Sci.* U.S.A., **103** (27), 10420–10425.
71. Boland, W. (1995) The chemistry of gamete attraction: chemical structures, biosynthesis, and (a)biotic degradation of algal pheromones. *Proc. Natl. Acad. Sci.* U.S.A., **92** (1), 37–43.
72. Poulson, K.L. (2013) Sublethal interactions between the harmful alga karenia brevis and its competitors. PhD Dissertation, Georgia Institute of Technology.
73. Shimomura, O., Suthers, H.L., and Bonner, J.T. (1982) Chemical identity of the acrasin of the cellular slime mold Polysphondylium violaceum. *Proc. Natl. Acad. Sci.* U.S.A., **79** (23), 7376–7379.
74. Astell, C.R. *et al.* (1981) The sequence of the DNAs coding for the mating-type loci of Saccharomyces cerevisiae. *Cell*, **27**(1, Pt. 2), 15–23.

75. Singh, A. *et al.* (1983) Saccharomyces cerevisiae contains two discrete genes coding for the alpha-factor pheromone. *Nucleic Acids Res.*, **11** (12), 4049–4063.
76. Betz, R. *et al.* (1987) Amino acid sequences of a-factor mating peptides from Saccharomyces cerevisiae. *J. Biol. Chem.*, **262** (2), 546–548.
77. Doty, R.L. (2010) *The Great Pheromone Myth*, Johns Hopkins University Press, Baltimore, MD, xi, 278 p.
78. Butenandt, A., Beckmann, R., and Hecker, E. (1961) On the sexattractant of silk-moths. I. The biological test and the isolation of the pure sex-attractant bombykol. *Hoppe Seylers Z. Physiol. Chem.*, **324**, 71–83.
79. Stern, K. and McClintock, M.K. (1998) Regulation of ovulation by human pheromones. *Nature*, **392** (6672), 177–179.
80. Wyatt, T.D. (2009) Fifty years of pheromones. *Nature*, **457** (7227), 262–263.
81. Vale, W. *et al.* (1981) Characterization of a 41-residue ovine hypothalamic peptide that stimulates secretion of corticotropin and beta-endorphin. *Science*, **213** (4514), 1394–1397.
82. Lovejoy, D.A. *et al.* (1992) Distinct sequence of gonadotropin-releasing hormone (GnRH) in dogfish brain provides insight into GnRH evolution. *Proc. Natl. Acad. Sci.* U.S.A., **89** (14), 6373–6377.
83. Kastin, A.J. (2006) *Handbook of Biologically Active Peptides*, Academic Press, Amsterdam, Boston, MA, xliii, 1595 p., 40 p. of col. plates..
84. Shen, L.P., Pictet, R.L., and Rutter, W.J. (1982) Human somatostatin I: sequence of the cDNA. *Proc. Natl. Acad. Sci.* U.S.A., **79** (15), 4575–4579.
85. Mueller, G.P. *et al.* (1980) Thyrotropin-releasing hormone and serotonin secretion from frog skin are stimulated by norepinephrine. *Endocrinology*, **106** (1), 1–4.
86. Ben-Jonathan, N. and Hnasko, R. (2001) Dopamine as a prolactin (PRL) inhibitor. *Endocr. Rev.*, **22** (6), 724–763.
87. Hill, R.W., Wyse, G.A., and Anderson, M. (2008) *Animal Physiology*, 2nd edn, Sinauer Associates, Sunderland, MA.
88. de Herder, W.W. *et al.* (2003) Somatostatin receptors in gastroentero-pancreatic neuroendocrine tumours. *Endocr. Relat. Cancer*, **10** (4), 451–458.
89. Chauvigne, F. *et al.* (2010) Functional and evolutionary analysis of flatfish gonadotropin receptors reveals cladal- and lineage-level divergence of the teleost glycoprotein receptor family. *Biol. Reprod.*, **82** (6), 1088–1102.
90. Raffin-Sanson, M.L., de Keyzer, Y., and Bertagna, X. (2003) Proopiomelanocortin, a polypeptide precursor with multiple functions: from physiology to pathological conditions. *Eur. J. Endocrinol.*, **149** (2), 79–90.
91. Nicholls, D.G. (1994) *Proteins, Transmitters, and Synapses*, Blackwell Scientific Publications, Oxford, Boston, MA, xii, 253 p., 4 p. of plates.
92. Li, C.H. *et al.* (1977) Isolation, structure, synthesis and morphine-like activity of beta-endorphin from human pituitary glands. *Ann. N. Y. Acad. Sci.*, **297**, 158–166.
93. Chavkin, C. and Goldstein, A. (1981) Specific receptor for the opioid peptide dynorphin: structure--activity relationships. *Proc. Natl. Acad. Sci.* U.S.A., **78** (10), 6543–6547.
94. Zadina, J.E. *et al.* (1997) A potent and selective endogenous agonist for the mu-opiate receptor. *Nature*, **386** (6624), 499–502.
95. Curatolo, A. *et al.* (1965) Distribution of N-acetylaspartic and N-acetylaspartylglutamic acids in the neuraxis of birds. *Boll. Soc. Ital. Biol. Sper.*, **41** (11), 591–593.
96. Tatemoto, K. (1982) Neuropeptide Y: complete amino acid sequence of the brain peptide. *Proc. Natl. Acad. Sci.* U.S.A., **79** (18), 5485–5489.
97. Lundkvist, J. *et al.* (1995) cDNA sequence, ligand binding, and regulation of galanin/GMAP in mouse brain. *Neurosci. Lett.*, **200** (2), 121–124.
98. Harmar, A.J. *et al.* (2012) Pharmacology and functions of receptors for vasoactive intestinal peptide and pituitary adenylate cyclase-activating polypeptide: IUPHAR review 1. *Br. J. Pharmacol.*, **166** (1), 4–17.
99. Carraway, R. and Leeman, S.E. (1975) The amino acid sequence of a hypothalamic peptide, neurotensin. *J. Biol. Chem.*, **250** (5), 1907–1911.
100. de Lecea, L., Sutcliffe, J.G., and Fabre, V. (2002) Hypocretins/orexins as integrators of physiological information: lessons from mutant animals. *Neuropeptides*, **36** (2–3), 85–95.
101. Kojima, M. and Kangawa, K. (2005) Ghrelin: structure and function. *Physiol. Rev.*, **85** (2), 495–522.
102. White, A., Handler, P., and Smith, E.L. (1973) *Principles of Biochemistry*, 5th edn, McGraw-Hill, New York, xv, 1295 p.
103. Park, C.R. *et al.* (2013) A novel glucagon-related peptide (GCRP) and its receptor GCRPR account for coevolution of their family members in vertebrates. *PLoS One*, **8** (6), e65420.
104. Arai, H. *et al.* (1990) Cloning and expression of a cDNA encoding an endothelin receptor. *Nature*, **348** (6303), 730–732.
105. Daniel, E.E. (1991) *Neuropeptide Function in the Gastrointestinal Tract*, CRC Press, Boca Raton, FL, 510 p.
106. Petho, G. and Reeh, P.W. (2012) Sensory and signaling mechanisms of bradykinin, eicosanoids, platelet-activating factor, and nitric oxide in peripheral nociceptors. *Physiol. Rev.*, **92** (4), 1699–1775.
107. Vander, A.J., Sherman, J.H., and Luciano, D.S. (1994) *Human Physiology: The Mechanisms of Body Function*, 6th edn, McGraw-Hill, New York.
108. von Euler, U. (1935) Über die spezifische blutdrucksenkende Substanz des menschlichen Prostata-und Samenblasensekretes. *J. Mol. Med.*, **14** (33), 1182–1183.
109. Woodward, D.F., Jones, R.L., and Narumiya, S. (2011) International Union of Basic and Clinical Pharmacology. LXXXIII: classification of prostanoid receptors, updating 15 years of progress. *Pharmacol. Rev.*, **63** (3), 471–538.
110. Devane, W.A. *et al.* (1992) Isolation and structure of a brain constituent that binds to the cannabinoid receptor. *Science*, **258** (5090), 1946–1949.
111. Di Marzo, V. and De Petrocellis, L. (2012) Why do cannabinoid receptors have more than one endogenous ligand? *Philos. Trans. R. Soc. London, Ser. B*, **367** (1607), 3216–3228.
112. Arai, K.-I. *et al.* (1990) Peptide growth factors and their receptors, in *Handbook of Experimental Pharmacology*, Springer-Verlag, Berlin, New York.

113. Mangelsdorf, D.J. *et al.* (1995) The nuclear receptor superfamily: the second decade. *Cell*, **83** (6), 835–839.
114. McCarthy, M.M. (2008) Estradiol and the developing brain. *Physiol. Rev.*, **88** (1), 91–124.
115. Hadley, M.E. (1996) *Endocrinology*, 4th edn, Prentice Hall, Englewood Cliffs, NJ, xxi, 518 p.
116. Phoenix, C.H. *et al.* (1959) Organizing action of prenatally administered testosterone propionate on the tissues mediating mating behavior in the female guinea pig. *Endocrinology*, **65**, 369–382.
117. Wolf, G. (2004) The discovery of vitamin D: the contribution of Adolf Windaus. *J. Nutr.*, **134** (6), 1299–1302.
118. Rang, H.P. (2006) The receptor concept: pharmacology's big idea. *Br. J. Pharmacol.*, **147** (Suppl. 1), S9–S16.
119. Ehrlich, P. (1913) Address in pathology, on chemiotherapy: delivered before the seventeenth international congress of medicine. *Br. Med. J.*, **2** (2746), 353–359.
120. Kenakin, T.P. (2006) *A Pharmacology Primer: Theory, Applications, and Methods*, 2nd edn, Academic Press, Amsterdam, Boston, MA, xviii, 299 p.
121. Ebbing, D.D. and Gammon, S.D. (2010) *General Chemistry*, 9th edn, Brooks Cole, Belmont, CA.
122. Hill, A.V. (1909) The mode of action of nicotine and curari, determined by the form of the contraction curve and the method of temperature coefficients. *J. Physiol.*, **39** (5), 361–373.
123. Clark, A.J. (1926) The reaction between acetyl choline and muscle cells. *J. Physiol.*, **61** (4), 530–546.
124. Ariens, E.J. (1983) Receptors: the materialization of a concept. *Pharm. Weekbl. Sci.*, **5** (4), 121–128.
125. Furchgott, R.F. (1966) The use of beta-haloalkylamines in the differentiation of receptors and in the determination of dissociation constants of receptor-agonist complexes. *Adv. Drug Res.*, **3**, 21–55.
126. Stephenson, R.P. (1956) A modification of receptor theory. *Br. J. Pharmacol. Chemother.*, **11** (4), 379–393.
127. Limbird, L.E. (2005) *Cell Surface Receptors: A Short Course on Theory and Methods*, 3rd edn, Springer Science+Business Media, New York, xii, 219 p.
128. Del Castillo, J. and Katz, B. (1957) Interaction at end-plate receptors between different choline derivatives. *Proc. R. Soc. Lond. Ser. B: Biol. Sci.*, **146** (924), 369–381.
129. De Lean, A., Stadel, J.M., and Lefkowitz, R.J. (1980) A ternary complex model explains the agonist-specific binding properties of the adenylate cyclase-coupled beta-adrenergic receptor. *J. Biol. Chem.*, **255** (15), 7108–7117.
130. Costa, T. and Herz, A. (1989) Antagonists with negative intrinsic activity at delta opioid receptors coupled to GTP-binding proteins. *Proc. Natl. Acad. Sci.* U.S.A., **86** (19), 7321–7325.
131. Samama, P. *et al.* (1993) A mutation-induced activated state of the beta 2-adrenergic receptor. Extending the ternary complex model. *J. Biol. Chem.*, **268** (7), 4625–4636.
132. Kenakin, T. (2001) Inverse, protean, and ligand-selective agonism: matters of receptor conformation. *FASEB J.*, **15** (3), 598–611.
133. Gbahou, F. *et al.* (2003) Protean agonism at histamine H3 receptors in vitro and in vivo. *Proc. Natl. Acad. Sci.* U.S.A., **100** (19), 11086–11091.
134. Neubig, R.R. (2007) Missing links: mechanisms of protean agonism. *Mol. Pharmacol.*, **71** (5), 1200–1202.
135. Weiss, J.M. *et al.* (1996) The cubic ternary complex receptor-occupancy model. III. resurrecting efficacy. *J. Theor. Biol.*, **181** (4), 381–397.
136. Brink, C.B. *et al.* (2004) Recent advances in drug action and therapeutics: relevance of novel concepts in G-protein-coupled receptor and signal transduction pharmacology. *Br. J. Clin. Pharmacol.*, **57** (4), 373–387.
137. Kenakin, T. (2007) Functional selectivity through protean and biased agonism: who steers the ship? *Mol. Pharmacol.*, **72** (6), 1393–1401.
138. Schulster, D. and Levitzki, A. (1980) *Cellular Receptors for Hormones and Neurotransmitters*, John Wiley & Sons, Ltd, Chichester, New York, xx, 412 p.
139. Paton, W.D. and Rang, H.P. (1965) The uptake of atropine and related drugs by intestinal smooth muscle of the guinea-pig in relation to acetylcholine receptors. *Proc. R. Soc. Lond. B Biol. Sci.*, **163**, 1–44.
140. Kahn, C.R. (1976) Membrane receptors for hormones and neurotransmitters. *J. Cell Biol.*, **70**(2, Pt. 1), 261–286.
141. Motulsky, H., Spannard, P., and Neubig, R. (1994) *Graphpad Prism (version 1.0)*, GraphPad Prism Software Inc, San Diego, CA.
142. Roets, E., Burvenich, C., and Roberts, M. (1992) Muscarinic receptor subtypes, beta-adrenoceptors and cAMP in the tracheal smooth muscle of conventional and double-muscled calves. *Vet. Res. Commun.*, **16** (6), 465–476.
143. Cheng, Y. and Prusoff, W.H. (1973) Relationship between the inhibition constant (K1) and the concentration of inhibitor which causes 50 per cent inhibition (I50) of an enzymatic reaction. *Biochem. Pharmacol.*, **22** (23), 3099–3108.
144. Munson, P.J. and Rodbard, D. (1988) An exact correction to the "Cheng-Prusoff" correction. *J. Recept. Res.*, **8** (1–4), 533–546.
145. Ruoho, A.E. *et al.* (1973) The mechanism of photoaffinity labeling. *Proc. Natl. Acad. Sci.* U.S.A., **70** (9), 2567–2571.
146. Tomizawa, M. *et al.* (2007) Defining nicotinic agonist binding surfaces through photoaffinity labeling. *Biochemistry*, **46** (30), 8798–8806.
147. Kabbani, N. (2008) Proteomics of membrane receptors and signaling. *Proteomics*, **8** (19), 4146–4155.
148. Roberts, M.F., Chilgren, J.D., and Zygmunt, A.C. (1989) Effect of temperature on alpha-adrenoceptor affinity and contractility of rabbit ear blood vessels. *Blood Vessels*, **26** (4), 185–196.
149. Roberts, M. *et al.* (2002) Adrenoceptor and local modulator control of cutaneous blood flow in thermal stress. *Comp. Biochem. Physiol. A Mol. Integr. Physiol.*, **131** (3), 485–496.
150. Goldstein, A., Aronow, L., and Kalman, S.M. (1973) *Principles of Drug Action: The Basis of Pharmacology*, A Wiley Biomedical-Health Publication, 2d edn, John Wiley & Sons, Inc, New York, xv, 854 p.
151. Funder, J.W. (1987) Receptors, hummingbirds, and refrigerators. *News Physiol. Sci.*, **2**, 231–232.
152. Bernard, C. (1974) *Lectures on the Phenomena of Life Common to Animals and Plants*, vol. **1**, Thomas.
153. Plato and Jowett, B. (1937) *The Works of Plato*, Tudor Publishing Co, New York.

154. Hille, B. (1989) The Sharpey-Schafer Lecture. Ionic channels: evolutionary origins and modern roles. *Q. J. Exp. Physiol.*, **74** (6), 785–804.
155. Chen, T.Y. and Hwang, T.C. (2008) CLC-0 and CFTR: chloride channels evolved from transporters. *Physiol. Rev.*, **88** (2), 351–387.
156. Steinbacher, S. *et al.* (2007) in *Current Topics in Membranes* (ed P.H. Owen), Academic Press, pp. 1–24.
157. Martinac, B., Saimi, Y., and Kung, C. (2008) Ion channels in microbes. *Physiol. Rev.*, **88** (4), 1449–1490.
158. Ward, J.M., Maser, P., and Schroeder, J.I. (2009) Plant ion channels: gene families, physiology, and functional genomics analyses. *Annu. Rev. Physiol.*, **71**, 59–82.
159. Verkhratsky, A., Krishtal, O.A., and Petersen, O.H. (2006) From Galvani to patch clamp: the development of electrophysiology. *Pflugers Arch.*, **453** (3), 233–247.
160. Pearce, J.M. (2001) Emil Heinrich Du Bois-Reymond (1818–1896). *J. Neurol. Neurosurg. Psychiatry*, **71** (5), 620.
161. Hille, B. (2001) *Ion Channels of Excitable Membranes*, 3rd edn, Sinauer, Sunderland, MA, xviii, 814 p.
162. Kamada, T. (1934) Some observations on potential differences across the ectoplasm membrane of Paramecium. *J. Exp. Biol.*, **11** (1), 94–102.
163. Young, J.Z. (1936) Structure of nerve fibres and synapses in some invertebrates, in *Cold Spring Harbor Symposia on Quantitative Biology*, Cold Spring Harbor Laboratory Press.
164. Katz, B. (1966) *Nerve, Muscle, and Synapse*, McGraw-Hill Series in the New Biology, McGraw-Hill, New York, ix, 193 p.
165. Hodgkin, A.L. and Huxley, A.F. (1939) Action potentials recorded from inside a nerve fibre. *Nature*, **144** (3651), 710–711.
166. Hodgkin, A.L. and Huxley, A.F. (1952) A quantitative description of membrane current and its application to conduction and excitation in nerve. *J. Physiol.*, **117** (4), 500–544.
167. Neher, E. and Sakmann, B. (1976) Single-channel currents recorded from membrane of denervated frog muscle fibres. *Nature*, **260** (5554), 799–802.
168. Neher, E. (1992) Nobel lecture. Ion channels for communication between and within cells. *EMBO J.*, **11** (5), 1672–1679.
169. Salkoff, L.W., Wei, A.D., Baban, B., Butler, A., Fawcett, G., Ferreira, G., and Santi, C.M. (2005) Potassium Channels in C. Elegans, http://www.wormbook.org/chapters/www_function/function.html (accessed 06 July 2015).
170. Ahern, C.A. and Kobertz, W.R. (2009) Chemical tools for K(+) channel biology. *Biochemistry*, **48** (3), 517–526.
171. MacKinnon, R. (2004) Potassium channels and the atomic basis of selective ion conduction (Nobel Lecture). *Angew. Chem. Int. Ed.*, **43** (33), 4265–4277.
172. Gouaux, E. and Mackinnon, R. (2005) Principles of selective ion transport in channels and pumps. *Science*, **310** (5753), 1461–1465.
173. Nishida, M. and MacKinnon, R. (2002) Structural basis of inward rectification: cytoplasmic pore of the G protein-gated inward rectifier GIRK1 at 1.8 A resolution. *Cell*, **111** (7), 957–965.
174. Payandeh, J. *et al.* (2011) The crystal structure of a voltage-gated sodium channel. *Nature*, **475** (7356), 353–358.
175. Alexander, S.P., Mathie, A., and Peters, J.A. (2011) Guide to receptors and channels (GRAC). *Br. J. Pharmacol.*, **164** (s1), S1–S2.
176. Catterall, W.A. (2011) Voltage-gated calcium channels. *Cold Spring Harbor Perspect. Biol.*, **3** (8), a003947.
177. Dolphin, A.C. (2012) Calcium channel auxiliary $\alpha_2\delta$ and β subunits: trafficking and one step beyond. *Nat. Rev. Neurosci.*, **13**, 542–555.
178. Gees, M., Colsoul, B., and Nilius, B. (2010) The role of transient receptor potential cation channels in Ca2+ signaling. *Cold Spring Harb. Perspect. Biol.*, **2** (10), a003962.
179. Nilius, B. and Owsianik, G. (2011) The transient receptor potential family of ion channels. *Genome Biol.*, **12** (3), 218.
180. Liu, D. and Liman, E.R. (2003) Intracellular Ca2 and the phospholipid PIP2 regulate the taste transduction ion channel TRPM5. *Proc. Natl. Acad. Sci. U.S.A*, **100**, 15160–15165.
181. Boudker, O. *et al.* (2007) PDB ID 2NWW: coupling substrate and ion binding to extracellular gate of a sodium-dependent aspartate transporter. *Nature*, **445** (7126), 387–393.
182. Caveney, S. *et al.* (2006) Ancestry of neuronal monoamine transporters in the Metazoa. *J. Exp. Biol.*, **209**(Pt. 24), 4858–4868.
183. Skou, J.C. (1998) The identification of the sodium–potassium pump (Nobel lecture). *Angew. Chem. Int. Ed. Engl.*, **37** (17), 2320–2328.
184. Locher, K.P., Lee, A.T., and Rees, D.C. (2002) PDB ID 1L7V: the E. coli BtuCD structure: a framework for ABC transporter architecture and mechanism. *Science*, **296** (5570), 1091–1098.
185. Lee, J.Y. *et al.* (2014) PDB ID 4MRN: structural basis for heavy metal detoxification by an Atm1-type ABC exporter. *Science*, **343** (6175), 1133–1136.
186. Lee, J.K., Finer-Moore, J.S., and Stroud, R.M. (unpublished work) PDB ID 3NE2: Archaeoglobus fulgidus aquaporin.
187. King, L.S. and Agre, P. (1996) Pathophysiology of the aquaporin water channels. *Annu. Rev. Physiol.*, **58**, 619–648.
188. Wang, Y., Schulten, K., and Tajkhorshid, E. (2005) What makes an aquaporin a glycerol channel? A comparative study of AqpZ and GlpF. *Structure*, **13** (8), 1107–1118.
189. Tasneem, A. *et al.* (2005) Identification of the prokaryotic ligand-gated ion channels and their implications for the mechanisms and origins of animal Cys-loop ion channels. *Genome Biol.*, **6** (1), R4.
190. Hilf, R.J. and Dutzler, R. (2009) Structure of a potentially open state of a proton-activated pentameric ligand-gated ion channel. *Nature*, **457** (7225), 115–118.
191. Voss, R.S. and Jansa, S.A. (2012) Snake-venom resistance as a mammalian trophic adaptation: lessons from didelphid marsupials. *Biol. Rev. Camb. Philos. Soc.*, **87** (4), 822–837.
192. Lewis, R.J. *et al.* (2012) Conus venom peptide pharmacology. *Pharmacol. Rev.*, **64** (2), 259–298.
193. Wang, S.Y. *et al.* (2006) How batrachotoxin modifies the sodium channel permeation pathway: computer

modeling and site-directed mutagenesis. *Mol. Pharmacol.*, **69** (3), 788–795.

194. Spurny, R. *et al.* (2012) PDB ID 2YOE: pentameric ligand-gated ion channel ELIC is activated by GABA and modulated by benzodiazepines. *Proc. Natl. Acad. Sci.* U.S.A., **109** (44), E3028–E3034.
195. Doble, A. and Martin, I.L. (1992) Multiple benzodiazepine receptors: no reason for anxiety. *Trends Pharmacol. Sci.*, **13** (2), 76–81.
196. Brenner, G.M. (2000) *Pharmacology*, Saunders Text and Review Series, 1st edn, W.B. Saunders, Philadelphia, PA, ix, 509 p.
197. Hucho, F. (1993) *Neurotransmitter Receptors. New Comprehensive Biochemistry*, Elsevier, Amsterdam, New York, xvi, 369 p.
198. Fleck, M.W., Thomson, J.L., and Hough, L.B. (2012) Histamine-gated ion channels in mammals? *Biochem. Pharmacol.*, **83** (9), 1127–1135.
199. Ringstad, N., Abe, N., and Horvitz, H.R. (2009) Ligand-gated chloride channels are receptors for biogenic amines in C. elegans. *Science*, **325** (5936), 96–100.
200. Foreman, J.C. and Johansen, T. (eds) (2003) *Textbook of Receptor Pharmacology*, 2nd edn, CRC Press, Boca Raton, FL, 289 p.
201. North, R.A. (2002) Molecular physiology of P2X receptors. *Physiol. Rev.*, **82** (4), 1013–1067.
202. Burnstock, G. (2008) Purinergic signalling and disorders of the central nervous system. *Nat. Rev. Drug Discov.*, **7** (7), 575–590.
203. Fountain, S.J. *et al.* (2008) Permeation properties of a P2X receptor in the green algae Ostreococcus tauri. *J. Biol. Chem.*, **283** (22), 15122–15126.
204. Damodaran (2010) Musings on Markets, http://aswathdamodaran.blogspot.com/2010/11/are-complex-models-answer.html (accessed 9 February 2014).
205. Nihoul, J. (1994) Do not use a simple model when a complex one will do. *J. Mar. Syst.*, **5** (6), 401–406.
206. Fredriksson, R. and Schioth, H.B. (2005) The repertoire of G-protein-coupled receptors in fully sequenced genomes. *Mol. Pharmacol.*, **67** (5), 1414–1425.
207. Gurevich, V.V. and Gurevich, E.V. (2008) How and why do GPCRs dimerize? *Trends Pharmacol. Sci.*, **29** (5), 234–240.
208. Hall, Z.W. (1992) *An Introduction to Molecular Neurobiology*, Sinauer Associates, Sunderland, MA, 555 p.
209. Moepps, B. and Fagni, L. (2003) Mont Sainte-Odile: a sanctuary for GPCRs. Confidence on signal transduction of G-protein-couple receptors. *EMBO Rep.*, **4** (3), 237–243.
210. Gehring, W.J. (2004) Historical perspective on the development and evolution of eyes and photoreceptors. *Int. J. Dev. Biol.*, **48** (8-9), 707–717.
211. Maggio, R., Vogel, Z., and Wess, J. (1993) Coexpression studies with mutant muscarinic/adrenergic receptors provide evidence for intermolecular "cross-talk" between G-protein-linked receptors. *Proc. Natl. Acad. Sci.* U.S.A., **90** (7), 3103–3107.
212. Terrillon, S. and Bouvier, M. (2004) Roles of G-protein-coupled receptor dimerization. *EMBO Rep.*, **5** (1), 30–34.
213. Nordstrom, K.J. *et al.* (2011) Independent HHsearch, Needleman--Wunsch-based, and motif analyses reveal the overall hierarchy for most of the G protein-coupled receptor families. *Mol. Biol. Evol.*, **28** (9), 2471–2480.
214. Yarmolinsky, D.A., Zuker, C.S., and Ryba, N.J. (2009) Common sense about taste: from mammals to insects. *Cell*, **139** (2), 234–244.
215. Axel, R. (1995) The molecular logic of smell. *Sci. Am.*, **273** (4), 154–159.
216. Davis, P.J. *et al.* (2010) Mini-review: cell surface receptor for thyroid hormone and nongenomic regulation of ion fluxes in excitable cells. *Physiol. Behav.*, **99** (2), 237–239.
217. Sutherland, E.W. and Cori, C.F. (1951) Effect of hyperglycemic-glycogenolytic factor and epinephrine on liver phosphorylase. *J. Biol. Chem.*, **188** (2), 531–543.
218. Berthet, J., Rall, T.W., and Sutherland, E.W. (1957) The relationship of epinephrine and glucagon to liver phosphorylase. IV. Effect of epinephrine and glucagon on the reactivation of phosphorylase in liver homogenates. *J. Biol. Chem.*, **224** (1), 463–475.
219. Krebs, E.G. and Fischer, E.H. (1956) The phosphorylase b to a converting enzyme of rabbit skeletal muscle. *Biochim. Biophys. Acta*, **20** (1), 150–157.
220. Rodbell, M. (1980) The role of hormone receptors and GTP-regulatory proteins in membrane transduction. *Nature*, **284** (5751), 17–22.
221. Cuatrecasas, P. (1974) Membrane receptors. *Annu. Rev. Biochem.*, **43**, 169–214.
222. Cassel, D. and Selinger, Z. (1977) Mechanism of adenylate cyclase activation by cholera toxin: inhibition of GTP hydrolysis at the regulatory site. *Proc. Natl. Acad. Sci.* U.S.A., **74** (8), 3307–3311.
223. Northup, J.K. *et al.* (1980) Purification of the regulatory component of adenylate cyclase. *Proc. Natl. Acad. Sci.* U.S.A., **77** (11), 6516–6520.
224. Milligan, G. and Kostenis, E. (2006) Heterotrimeric G-proteins: a short history. *Br. J. Pharmacol.*, **147** (Suppl. 1), S46–S55.
225. Hurley, J.B. *et al.* (1984) Homologies between signal transducing G proteins and ras gene products. *Science*, **226** (4676), 860–862.
226. Holinstat, M., Oldham, W.M., and Hamm, H.E. (2006) G-protein-coupled receptors: evolving views on physiological signalling: symposium on G-protein-coupled receptors: evolving concepts and new techniques. *EMBO Rep.*, **7** (9), 866–869.
227. Gilman, A.G. (1987) G proteins: transducers of receptor-generated signals. *Annu. Rev. Biochem.*, **56**, 615–649.
228. Lodowski, D.T. *et al.* (2005) The role of G beta gamma and domain interfaces in the activation of G protein-coupled receptor kinase 2. *Biochemistry*, **44** (18), 6958–6970.
229. Logothetis, D.E. *et al.* (1987) The beta gamma subunits of GTP-binding proteins activate the muscarinic K+ channel in heart. *Nature*, **325** (6102), 321–326.
230. Wang, P., Perfect, J.R., and Heitman, J. (2000) The G-protein beta subunit GPB1 is required for mating and haploid fruiting in Cryptococcus neoformans. *Mol. Cell. Biol.*, **20** (1), 352–362.
231. Rebecchi, M.J. and Pentyala, S.N. (2000) Structure, function, and control of phosphoinositide-specific phospholipase C. *Physiol. Rev.*, **80** (4), 1291–1335.

232. Lott, J.S. *et al.* (1999) The ordered visual transduction complex of the squid photoreceptor membrane. *Mol. Neurobiol.*, **20** (1), 61–80.

233. Takefuji, M. *et al.* (2013) RhoGEF12 controls cardiac remodeling by integrating G protein- and integrin-dependent signaling cascades. *J. Exp. Med.*, **210** (4), 665–673.

234. Horinouchi, T. *et al.* (2013) Endothelin receptor signaling: new insight into its regulatory mechanisms. *J. Pharmacol. Sci.*, **123** (2), 85–101.

235. Bos, J.L., Rehmann, H., and Wittinghofer, A. (2007) GEFs and GAPs: critical elements in the control of small G proteins. *Cell*, **129** (5), 865–877.

236. Dohlman, H.G. and Thorner, J.W. (2001) Regulation of G protein-initiated signal transduction in yeast: paradigms and principles. *Annu. Rev. Biochem.*, **70**, 703–754.

237. Lefkowitz, R.J. (2013) A brief history of G-protein coupled receptors (Nobel Lecture). *Angew. Chem. Int. Ed.*, **52** (25), 6366–6378.

238. Kobilka, B.K. *et al.* (1988) Chimeric alpha 2-,beta 2-adrenergic receptors: delineation of domains involved in effector coupling and ligand binding specificity. *Science*, **240** (4857), 1310–1316.

239. Langley, J.N. (1901) Observations on the physiological action of extracts of the supra-renal bodies. *J. Physiol.*, **27** (3), 237–256.

240. Donckier, J.E. *et al.* (2001) Cardiovascular effects of beta 3-adrenoceptor stimulation in perinephritic hypertension. *Eur. J. Clin. Invest.*, **31** (8), 681–689.

241. von Homeyer, P. and Schwinn, D.A. (2011) Pharmacogenomics of beta-adrenergic receptor physiology and response to beta-blockade. *Anesth. Analg.*, **113** (6), 1305–1318.

242. Civantos Calzada, B. and Aleixandre de Artinano, A. (2001) Alpha-adrenoceptor subtypes. *Pharmacol. Res.*, **44** (3), 195–208.

243. Flavahan, N.A. *et al.* (1985) Cooling and alpha 1- and alpha 2-adrenergic responses in cutaneous veins: role of receptor reserve. *Am. J. Physiol.*, **249** (5, Pt. 2), H950–H955.

244. Nott, J., Byse, M., Fulgencio, R., and Roberts, M. (2002) How alpha-adrenoceptors affect blood vessel constriction in thermoregulation. *Proc. Oreg. Acad. Sci.*, **38**, 5–6.

245. Dale, H.H. (1914) *The Action of Certain Esters and Ethers of Choline and Their Relation to Muscarine*, Wellcome Physiological Research Laboratories.

246. Hussain, A. *et al.* (2013) High-affinity olfactory receptor for the death-associated odor cadaverine. *Proc. Natl. Acad. Sci.* U.S.A., **110** (48), 19579–19584.

247. Drury, A.N. and Szent-Gyorgyi, A. (1929) The physiological activity of adenine compounds with especial reference to their action upon the mammalian heart. *J. Physiol.*, **68** (3), 213–237.

248. Burnstock, G. (1976) Purinergic receptors. *J. Theor. Biol.*, **62** (2), 491–503.

249. Burnstock, G. (2007) Purine and pyrimidine receptors. *Cell. Mol. Life Sci.*, **64** (12), 1471–1483.

250. Burnstock, G. (2013) Introduction to purinergic signalling in the brain. *Adv. Exp. Med. Biol.*, **986**, 1–12.

251. Ricciotti, E. and FitzGerald, G.A. (2011) Prostaglandins and inflammation. *Arterioscler. Thromb. Vasc. Biol.*, **31** (5), 986–1000.

252. Castillo, P.E. *et al.* (2012) Endocannabinoid signaling and synaptic function. *Neuron*, **76** (1), 70–81.

253. Pert, C.B. and Snyder, S.H. (1973) Opiate receptor: demonstration in nervous tissue. *Science*, **179** (4077), 1011–1014.

254. Simon, E.J., Hiller, J.M., and Edelman, I. (1973) Stereospecific binding of the potent narcotic analgesic (3H) Etorphine to rat-brain homogenate. *Proc. Natl. Acad. Sci.* U.S.A., **70** (7), 1947–1949.

255. Hughes, J. *et al.* (1975) Identification of two related pentapeptides from the brain with potent opiate agonist activity. *Nature*, **258** (5536), 577–580.

256. Satoh, M. and Minami, M. (1995) Molecular pharmacology of the opioid receptors. *Pharmacol. Ther.*, **68** (3), 343–364.

257. Fry, B.G. *et al.* (2009) A central role for venom in predation by Varanus komodoensis (Komodo Dragon) and the extinct giant Varanus (Megalania) priscus. *Proc. Natl. Acad. Sci.* U.S.A., **106** (22), 8969–8974.

258. Fry, B.G. *et al.* (2010) Novel venom proteins produced by differential domain-expression strategies in beaded lizards and gila monsters (genus Heloderma). *Mol. Biol. Evol.*, **27** (2), 395–407.

259. Kratz, E., Dugas, J.C., and Ngai, J. (2002) Odorant receptor gene regulation: implications from genomic organization. *Trends Genet.*, **18** (1), 29–34.

260. Mosinger, B. *et al.* (2013) Genetic loss or pharmacological blockade of testes-expressed taste genes causes male sterility. *Proc. Natl. Acad. Sci.* U.S.A., **110** (30), 12319–12324.

261. Fischer, J.A., Muff, R., and Born, W. (2002) Functional relevance of G-protein-coupled-receptor-associated proteins, exemplified by receptor-activity-modifying proteins (RAMPs). *Biochem. Soc. Trans.*, **30** (4), 455–460.

262. Galvez, T. and Pin, J.P. (2003) How do G-protein-coupled receptors work? The case of metabotropic glutamate and GABA receptors. *Med. Sci. (Paris)*, **19** (5), 559–565.

263. Fredriksson, R. *et al.* (2003) The G-protein-coupled receptors in the human genome form five main families. Phylogenetic analysis, paralogon groups, and fingerprints. *Mol. Pharmacol.*, **63** (6), 1256–1272.

264. Troemel, E.R. *et al.* (1995) Divergent seven transmembrane receptors are candidate chemosensory receptors in C. elegans. *Cell*, **83** (2), 207–218.

265. Ciesla, J., Fraczyk, T., and Rode, W. (2011) Phosphorylation of basic amino acid residues in proteins: important but easily missed. *Acta Biochim. Pol.*, **58** (2), 137–148.

266. Cohen, P. (2002) The origins of protein phosphorylation. *Nat. Cell Biol.*, **4** (5), E127–E130.

267. Hubbard, S.R. and Miller, W.T. (2007) Receptor tyrosine kinases: mechanisms of activation and signaling. *Curr. Opin. Cell Biol.*, **19** (2), 117–123.

268. Segarini, P.R. *et al.* (1987) Membrane binding characteristics of two forms of transforming growth factor-beta. *J. Biol. Chem.*, **262** (30), 14655–14662.

269. Massague, J. and Gomis, R.R. (2006) The logic of TGF-beta signaling. *FEBS Lett.*, **580** (12), 2811–2820.

270. Roitt, I.M., Brostoff, J., and Male, D.K. (2001) *Immunology*, 6th edn, Mosby, Edinburgh, New York, 480 p.

271. Barclay, A.N. (2003) Membrane proteins with immunoglobulin-like domains--a master superfamily of interaction molecules. *Semin. Immunol.*, **15** (4), 215–223.
272. Bossemeyer, D. (1995) Protein kinases--structure and function. *FEBS Lett.*, **369** (1), 57–61.
273. Lemmon, M.A. and Schlessinger, J. (2010) Cell signaling by receptor tyrosine kinases. *Cell*, **141** (7), 1117–1134.
274. Hubbard, S.R. (2009) The juxtamembrane region of EGFR takes center stage. *Cell*, **137** (7), 1181–1183.
275. Taniguchi, C.M. *et al.* (2006) Divergent regulation of hepatic glucose and lipid metabolism by phosphoinositide 3-kinase via Akt and PKClambda/zeta. *Cell Metab.*, **3** (5), 343–353.
276. Heldin, C.H. (2013) Targeting the PDGF signaling pathway in tumor treatment. *Cell Commun. Signal.*, **11**, 97.
277. Shim, A.H. *et al.* (2010) Structures of a platelet-derived growth factor/propeptide complex and a platelet-derived growth factor/receptor complex. *Proc. Natl. Acad. Sci.* U.S.A., **107** (25), 11307–11312.
278. Molloy, N.H., Read, D.E., and Gorman, A.M. (2011) Nerve growth factor in cancer cell death and survival. *Cancers (Basel)*, **3** (1), 510–530.
279. Hinck, A.P. (2012) Structural studies of the TGF-βs and their receptors – insights into evolution of the TGF-β superfamily. *FEBS Lett.*, **586** (14), 1860–1870.
280. Pandey, K.N. (2011) Guanylyl cyclase/atrial natriuretic peptide receptor-A: role in the pathophysiology of cardiovascular regulation. *Can. J. Physiol. Pharmacol.*, **89** (8), 557–573.
281. Kuhn, M. (2012) Endothelial actions of atrial and B-type natriuretic peptides. *Br. J. Pharmacol.*, **166** (2), 522–531.
282. Dautry-Varsat, A. and Lodish, H.F. (1984) How receptors bring proteins and particles into cells. *Sci. Am.*, **250** (5), 52–58.
283. Brown, M.S. and Goldstein, J.L. (1984) How LDL receptors influence cholesterol and atherosclerosis. *Sci. Am.*, **251** (5), 58–66.
284. Dieckmann, M., Dietrich, M.F., and Herz, J. (2010) Lipoprotein receptors--an evolutionarily ancient multifunctional receptor family. *Biol. Chem.*, **391** (11), 1341–1363.
285. Avila, J.L. and Kissil, J.L. (2013) Notch signaling in pancreatic cancer: oncogene or tumor suppressor? *Trends Mol. Med.*, **19** (5), 320–327.
286. Shaw, S. (1988) Immunology: a short course by Eli Benjamini and Sidney Leskowitz, Alan R. Liss, Inc.,1988. pound15.75, $22.95 (xv + 390 pages) ISBN 0 8451 4228 3. *Immunol. Today*, **9** (6), 182–183.
287. Jerne, N.K. (1955) The natural-selection theory of antibody formation. *Proc. Natl. Acad. Sci.* U.S.A., **41** (11), 849–857.
288. Talmage, D.W. (1957) Allergy and immunology. *Annu. Rev. Med.*, **8**, 239–256.
289. Burnet, F.M. (1964) Immunological recognition of self. Nobel lecture (december 12, 1960), in *Nobel Lectures in Physiology or Medicine 1942–1962*, Elsevier, Amsterdam.
290. Raloff, J. and Raloff, J. (1995) Beyond estrogens. *Sci. News*, **148**, 44.
291. Marrack, P. and Kappler, J. (1986) The T cell and its receptor. *Sci. Am.*, **254** (2), 36–45.
292. Weiss, R.A. (1996) HIV receptors and the pathogenesis of AIDS. *Science*, **272** (5270), 1885–1886.
293. Tsai, M.J. and O'Malley, B.W. (1994) Molecular mechanisms of action of steroid/thyroid receptor superfamily members. *Annu. Rev. Biochem.*, **63**, 451–486.
294. Sladek, F.M. (2011) What are nuclear receptor ligands? *Mol. Cell. Endocrinol.*, **334** (1-2), 3–13.
295. Thornton, J.W., Need, E., and Crews, D. (2003) Resurrecting the ancestral steroid receptor: ancient origin of estrogen signaling. *Science*, **301** (5640), 1714–1717.
296. Clever, U. and Karlson, P. (1960) Induction of puff changes in the salivary gland chromosomes of Chironomus tentans by ecdysone. *Exp. Cell Res.*, **20**, 623–626.
297. Jensen, E.V. (1962) On the mechanism of estrogen action. *Perspect. Biol. Med.*, **6**, 47–59.
298. Karlson, P. (1963) New concepts on the mode of action of hormones. *Perspect. Biol. Med.*, **6** (2), 203–214.
299. Jacob, F. and Monod, J. (1961) Genetic regulatory mechanisms in the synthesis of proteins. *J. Mol. Biol.*, **3**, 318–356.
300. Toft, D. and Gorski, J. (1966) A receptor molecule for estrogens: isolation from the rat uterus and preliminary characterization. *Proc. Natl. Acad. Sci.* U.S.A., **55** (6), 1574–1581.
301. O'Malley, B.W. and McGuire, W.L. (1968) Studies on the mechanism of estrogen-mediated tissue differentiation: regulation of nuclear transcription and induction of new RNA species. *Proc. Natl. Acad. Sci.* U.S.A., **60** (4), 1527–1534.
302. Jensen, E.V. *et al.* (1968) A two-step mechanism for the interaction of estradiol with rat uterus. *Proc. Natl. Acad. Sci.* U.S.A., **59** (2), 632–638.
303. Ptashne, M. (1967) Specific binding of the lambda phage repressor to lambda DNA. *Nature*, **214** (5085), 232–234.
304. Beato, M. and Klug, J. (2000) Steroid hormone receptors: an update. *Hum. Reprod. Update*, **6** (3), 225–236.
305. Majewska, M.D. (2007) Steroids and ion channels in evolution: from bacteria to synapses and mind. Evolutionary role of steroid regulation of GABA(A) receptors. *Acta Neurobiol. Exp. (Wars)*, **67** (3), 219–233.
306. Kutschera, U. and Wang, Z.Y. (2012) Brassinosteroid action in flowering plants: a Darwinian perspective. *J. Exp. Bot.*, **63** (10), 3511–3522.
307. Devarakonda, S. *et al.* (2003) Structure of the heterodimeric ecdysone receptor DNA-binding complex. *EMBO J.*, **22** (21), 5827–5840.
308. Beato, M. (1989) Gene regulation by steroid hormones. *Cell*, **56** (3), 335–344.
309. Alberts, B. (2008) *Molecular Biology of the Cell*, 5th edn, Garland Science, New York.
310. Denis, M. and Gustafsson, J.A. (1989) The Mr approximately 90,000 heat shock protein: an important modulator of ligand and DNA-binding properties of the glucocorticoid receptor. *Cancer Res.*, **49** (8 Suppl), 2275s–2281s.
311. Klinge, C.M. (2000) Estrogen receptor interaction with co-activators and co-repressors. *Steroids*, **65** (5), 227–251.

312. Centenera, M.M. *et al.* (2008) The contribution of different androgen receptor domains to receptor dimerization and signaling. *Mol. Endocrinol.*, **22** (11), 2373–2382.
313. Liu, Y.Y. and Brent, G.A. (2010) Thyroid hormone crosstalk with nuclear receptor signaling in metabolic regulation. *Trends Endocrinol. Metab.*, **21** (3), 166–173.
314. Laudet, V. (2011) The origins and evolution of vertebrate metamorphosis. *Curr. Biol.*, **21** (18), R726–R737.
315. Fleet, J.C. *et al.* (2012) Vitamin D and cancer: a review of molecular mechanisms. *Biochem. J.*, **441** (1), 61–76.
316. Hayward, D.C. *et al.* (1999) The sequence of Locusta RXR, homologous to Drosophila Ultraspiracle, and its evolutionary implications. *Dev. Genes Evol.*, **209** (9), 564–571.
317. Bain, J. (2007) The many faces of testosterone. *Clin. Interv. Aging*, **2** (4), 567–576.
318. McEwen, B.S. (1991) Non-genomic and genomic effects of steroids on neural activity. *Trends Pharmacol. Sci.*, **12** (4), 141–147.
319. Uotinen, N. *et al.* (1999) Distribution of progesterone receptor in female mouse tissues. *Gen. Comp. Endocrinol.*, **115** (3), 429–441.
320. Krug, A.W. *et al.* (2011) Cell membrane-associated mineralocorticoid receptors? New evidence. *Hypertension*, **57** (6), 1019–1025.
321. Necela, B.M. and Cidlowski, J.A. (2004) Mechanisms of glucocorticoid receptor action in noninflammatory and inflammatory cells. *Proc. Am. Thorac. Soc.*, **1** (3), 239–246.
322. Villena, J.A. *et al.* (2004) Desnutrin, an adipocyte gene encoding a novel patatin domain-containing protein, is induced by fasting and glucocorticoids: ectopic expression of desnutrin increases triglyceride hydrolysis. *J. Biol. Chem.*, **279** (45), 47066–47075.
323. Freudzon, L. *et al.* (2005) Regulation of meiotic prophase arrest in mouse oocytes by GPR3, a constitutive activator of the Gs G protein. *J. Cell Biol.*, **171** (2), 255–265.
324. Williams, J.S. and Williams, G.H. (2003) 50th anniversary of aldosterone. *J. Clin. Endocrinol. Metab.*, **88** (6), 2364–2372.
325. McDonnell, D.P. and Wardell, S.E. (2010) The molecular mechanisms underlying the pharmacological actions of ER modulators: implications for new drug discovery in breast cancer. *Curr. Opin. Pharmacol.*, **10** (6), 620–628.
326. Moss, R.L., Gu, Q., and Wong, M. (1997) Estrogen: nontranscriptional signaling pathway. *Recent Prog. Horm. Res.*, **52**, 33–68; discussion 68-69.
327. Revelli, A. *et al.* (1994) Steroid receptors in human spermatozoa. *Hum. Reprod.*, **9** (5), 760–766.
328. Wehling, M. (1996) Non-genomic steroid action--take a closer look, it's not rare! An invited commentary. *Eur. J. Endocrinol.*, **135** (3), 287–288.
329. Walker, W.H. (2010) Non-classical actions of testosterone and spermatogenesis. *Philos. Trans. R. Soc. Lond. B Biol. Sci.*, **365** (1546), 1557–1569.
330. Levin, E.R. (2009) Plasma membrane estrogen receptors. *Trends Endocrinol. Metab.*, **20** (10), 477–482.
331. Paul, S.M. and Purdy, R.H. (1992) Neuroactive steroids. *FASEB J.*, **6** (6), 2311–2322.
332. Rodrigo, G.J. (2006) Rapid effects of inhaled corticosteroids in acute asthma: an evidence-based evaluation. *Chest*, **130** (5), 1301–1311.
333. Lindemann, L. *et al.* (2005) Trace amine-associated receptors form structurally and functionally distinct subfamilies of novel G protein-coupled receptors. *Genomics*, **85** (3), 372–385.
334. O'Malley, B.W. (2005) A life-long search for the molecular pathways of steroid hormone action. *Mol. Endocrinol.*, **19** (6), 1402–1411.
335. Denner, L.A. *et al.* (1990) Regulation of progesterone receptor-mediated transcription by phosphorylation. *Science*, **250** (4988), 1740–1743.
336. Sladek, R. *et al.* (1997) Chromosomal mapping of the human and murine orphan receptors ERRalpha (ESRRA) and ERRbeta (ESRRB) and identification of a novel human ERRalpha-related pseudogene. *Genomics*, **45** (2), 320–326.
337. Torday, J.S. (2013) Evolution and cell physiology. 1. Cell signaling is all of biology. *Am. J. Physiol. Cell Physiol.*, **305** (7), C682–C689.
338. Ryan, T.J. and Grant, S.G. (2009) The origin and evolution of synapses. *Nat. Rev. Neurosci.*, **10** (10), 701–712.
339. Matschinsky, F.M. (2002) Regulation of pancreatic beta-cell glucokinase: from basics to therapeutics. *Diabetes*, **51** (Suppl. 3), S394–S404.
340. Turner, R.W., Anderson, D., and Zamponi, G.W. (2011) Signaling complexes of voltage-gated calcium channels. *Channels (Austin)*, **5** (5), 440–448.
341. Csaba, G. (1984) The present state in the phylogeny and ontogeny of hormone receptors. *Horm. Metab. Res.*, **16** (7), 329–335.
342. Kolenbrander, P.E. *et al.* (1993) Coaggregation: specific adherence among human oral plaque bacteria. *FASEB J.*, **7** (5), 406–413.
343. Wilson, H. (1907) On some phenomena of coalescence and regeneration in sponges. *J. Exp. Zool.*, **5** (2), 245–258.
344. Moscona, A.A. (1963) Studies on cell aggregation: demonstration of materials with selective cell-binding activity. *Proc. Natl. Acad. Sci.* U.S.A., **49** (5), 742–747.
345. Humphreys, T. (1963) Chemical dissolution and in vitro reconstruction of sponge cell adhesions. I. Isolation and functional demonstration of the components involved. *Dev. Biol.*, **8**, 27–47.
346. Gilbert, S.F. (2006) *Developmental Biology*, 8th edn, Sinauer Associates, Sunderland, MA, xviii, 817 p.
347. Williams, A.F. and Barclay, A.N. (1988) The immunoglobulin superfamily--domains for cell surface recognition. *Annu. Rev. Immunol.*, **6**, 381–405.
348. Desgrosellier, J.S. and Cheresh, D.A. (2010) Integrins in cancer: biological implications and therapeutic opportunities. *Nat. Rev. Cancer*, **10** (1), 9–22.
349. Liu, P. *et al.* (2009) Targeting the phosphoinositide 3-kinase pathway in cancer. *Nat. Rev. Drug Discov.*, **8** (8), 627–644.
350. Nelson, J. (2008) *Structure and Function in Cell Signaling*, John Wiey & Sons, Ltd, Chichester.
351. Massoner, P. *et al.* (2010) Insulin-like growth factors and insulin control a multifunctional signalling network of significant importance in cancer. *Br. J. Cancer*, **103** (10), 1479–1484.

352. Ryter, S.W. and Choi, A.M. (2013) Carbon monoxide: present and future indications for a medical gas. *Korean J. Intern. Med.*, **28** (2), 123–140.

353. Gilbert, S.F. (2008) All I really needed to know I learned during gastrulation. *CBE Life Sci. Educ.*, **7** (1), 12–13.

354. Conti, V. *et al.* (2013) Adrenoreceptors and nitric oxide in the cardiovascular system. *Front. Physiol.*, **4**, 321.

355. Guo, W., Cheng, Z.Y., and Zhu, Y.Z. (2013) Hydrogen sulfide and translational medicine. *Acta Pharmacol. Sin.*, **34** (10), 1284–1291.

356. Patten, B.M. and Carlson, B.M. (1981) *Patten's Foundations of Embryology*, McGraw-Hill Series in Organismic Biology, 4th edn, McGraw-Hill, New York, xv, 672 p.

357. Kirk, D.L. (2005) A twelve-step program for evolving multicellularity and a division of labor. *Bioessays*, **27** (3), 299–310.

358. King, N. (2004) The unicellular ancestry of animal development. *Dev. Cell*, **7** (3), 313–325.

359. Le Bail, A. *et al.* (2010) Auxin metabolism and function in the multicellular brown alga Ectocarpus siliculosus. *Plant Physiol.*, **153** (1), 128–144.

360. Knoll, A.H. (2011) The multiple origins of complex multicellularity. *Annu. Rev. Earth Planet. Sci.*, **39**, 217–239.

361. Hallmann, A. and Kirk, D.L. (2000) The developmentally regulated ECM glycoprotein ISG plays an essential role in organizing the ECM and orienting the cells of Volvox. *J. Cell Sci.*, **113** (Pt. 24), 4605–4617.

362. King, N. *et al.* (2008) The genome of the choanoflagellate Monosiga brevicollis and the origin of metazoans. *Nature*, **451** (7180), 783–788.

363. Reinke, C.A. and Carthew, R.W. (2008) BMP signaling goes posttranscriptional in a microRNA sort of way. *Dev. Cell*, **15** (2), 174–175.

364. Adamska, M. *et al.* (2007) Wnt and TGF-beta expression in the sponge Amphimedon queenslandica and the origin of metazoan embryonic patterning. *PLoS One*, **2** (10), e1031.

365. Leys, S.P. and Degnan, B.M. (2001) Cytological basis of photoresponsive behavior in a sponge larva. *Biol. Bull.*, **201** (3), 323–338.

366. Leys, S.P., Mackie, G.O., and Meech, R.W. (1999) Impulse conduction in a sponge. *J. Exp. Biol.*, **202** (Pt. 9), 1139–1150.

367. Elliott, G.R. and Leys, S.P. (2010) Evidence for glutamate, GABA and NO in coordinating behaviour in the sponge, Ephydatia muelleri (Demospongiae, Spongillidae). *J. Exp. Biol.*, **213** (13), 2310–2321.

368. Degnan, B.M. (2010) The rise of genomics sheds light on the dawn of animals. *Evol. Dev.*, **12** (5), 425–427.

369. Levin, M. (2005) Left-right asymmetry in embryonic development: a comprehensive review. *Mech. Dev.*, **122** (1), 3–25.

370. Baguna, J. *et al.* (2008) Back in time: a new systematic proposal for the Bilateria. *Philos. Trans. R. Soc. London Ser. B: Biol. Sci.*, **363** (1496), 1481–1491.

371. De Robertis, E.M. (2009) Spemann's organizer and the self-regulation of embryonic fields. *Mech. Dev.*, **126** (11-12), 925–941.

372. Samuel, G., Miller, D., and Saint, R. (2001) Conservation of a DPP/BMP signaling pathway in the nonbilateral cnidarian Acropora millepora. *Evol. Dev.*, **3** (4), 241–250.

373. Neill, A.T. and Vacquier, V.D. (2004) Ligands and receptors mediating signal transduction in sea urchin spermatozoa. *Reproduction*, **127** (2), 141–149.

374. Epel, D. (1977) The program of fertilization. *Sci. Am.*, **237** (5), 128–138.

375. Moy, G.W. *et al.* (1996) The sea urchin sperm receptor for egg jelly is a modular protein with extensive homology to the human polycystic kidney disease protein, PKD1. *J. Cell Biol.*, **133** (4), 809–817.

376. Ohlendieck, K. and Lennarz, W.J. (1995) Role of the sea urchin egg receptor for sperm in gamete interactions. *Trends Biochem. Sci.*, **20** (1), 29–33.

377. Zigler, K.S. (2008) The evolution of sea urchin sperm bindin. *Int. J. Dev. Biol.*, **52** (5-6), 791–796.

378. Kamei, N. and Glabe, C.G. (2003) The species-specific egg receptor for sea urchin sperm adhesion is EBR1,a novel ADAMTS protein. *Genes Dev.*, **17** (20), 2502–2507.

379. Abassi, Y.A. and Foltz, K.R. (1994) Tyrosine phosphorylation of the egg receptor for sperm at fertilization. *Dev. Biol.*, **164** (2), 430–443.

380. Wassarman, P.M. and Litscher, E.S. (1995) Sperm--egg recognition mechanisms in mammals. *Curr. Top. Dev. Biol.*, **30**, 1–19.

381. Clark, G.F. (2011) The molecular basis of mouse sperm-zona pellucida binding: a still unresolved issue in developmental biology. *Reproduction*, **142** (3), 377–381.

382. Schultz, R.M. and Kopf, G.S. (1995) Molecular basis of mammalian egg activation. *Curr. Top. Dev. Biol.*, **30**, 21–62.

383. Spemann, H. and Mangold, H. (1924) Über induktion von Embryonalanlagen durch Implantation artfremder Organisatoren. *Dev. Genes Evol.*, **100** (3), 599–638.

384. Crick, F. (1970) Diffusion in embryogenesis. *Nature*, **225** (5231), 420–422.

385. McMahon, D. (1974) Chemical messengers in development: a hypothesis. *Science*, **185** (4156), 1012–1021.

386. de Larco, J.E. and Todaro, G.J. (1978) Growth factors from murine sarcoma virus-transformed cells. *Proc. Natl. Acad. Sci.* U.S.A., **75** (8), 4001–4005.

387. Erwin, D.H. and Davidson, E.H. (2002) The last common bilaterian ancestor. *Development*, **129** (13), 3021–3032.

388. Holland, L.Z. (2000) Body-plan evolution in the Bilateria: early antero-posterior patterning and the deuterostome-protostome dichotomy. *Curr. Opin. Genet. Dev.*, **10** (4), 434–442.

389. Weinstein, D.C. and Hemmati-Brivanlou, A. (1999) Neural induction. *Annu. Rev. Cell Dev. Biol.*, **15**, 411–433.

390. Coutu, D.L. and Galipeau, J. (2011) Roles of FGF signaling in stem cell self-renewal, senescence and aging. *Aging (Albany)*, **3** (10), 920–933.

391. Wolpert, L. (2007) *Principles of Development*, 3rd edn, Oxford University Press, Oxford, New York, xxiii, 551 p.

392. Moses, K. (1991) The role of transcription factors in the developing Drosophila eye. *Trends Genet.*, **7** (8), 250–255.

393. Muha, V. and Muller, H.A. (2013) Functions and mechanisms of Fibroblast Growth Factor (FGF) signalling in drosophila melanogaster. *Int. J. Mol. Sci.*, **14** (3), 5920–5937.

394. Ganfornina, M.D., Sánchez, D., and Bastiani, M.J. (1996) The role of the cell surface in neuronal pathfinding. *Bioscience*, **46** (5), 344–354.
395. Sperry, R.W. (1963) Chemoaffinity in the orderly growth of nerve fiber patterns and connections. *Proc. Natl. Acad. Sci.* U.S.A., **50**, 703–710.
396. Raff, M.C. (1992) Social controls on cell survival and cell death. *Nature*, **356** (6368), 397–400.
397. Koonin, E.V. and Aravind, L. (2002) Origin and evolution of eukaryotic apoptosis: the bacterial connection. *Cell Death Differ.*, **9** (4), 394–404.
398. Clarke, P.G.H. (1985) Neuronal death in the development of the vertebrate nervous system. *Trends Neurosci.*, **8**, 345–349.
399. Micheau, O. and Tschopp, J. (2003) Induction of TNF receptor I-mediated apoptosis via two sequential signaling complexes. *Cell*, **114** (2), 181–190.
400. Abbey, E. (1991) *The Journey Home: Some Words in Defense of the American West*, Plume.
401. Maugham, R. (1978) *Conversations with Willie: Recollections of W. Somerset Maugham*, Allen, London, 184 p., 8 leaves of plates.
402. Griffiths, A.J.F. (2008) *Introduction to Genetic Analysis*, 9th edn, W.H. Freeman and Co, New York, xxiii, 838 p.
403. Rebay, I., Fehon, R.G., and Artavanis-Tsakonas, S. (1993) Specific truncations of Drosophila Notch define dominant activated and dominant negative forms of the receptor. *Cell*, **74** (2), 319–329.
404. Gorska, A.E. *et al.* (1998) Dominant-negative interference of the transforming growth factor beta type II receptor in mammary gland epithelium results in alveolar hyperplasia and differentiation in virgin mice. *Cell Growth Differ.*, **9** (3), 229–238.
405. Biebermann, H. *et al.* (1997) Congenital hypothyroidism caused by mutations in the thyrotropin-receptor gene. *N. Engl. J. Med.*, **336** (19), 1390–1391.
406. Cain, S.M. and Snutch, T.P. (2011) Voltage-gated calcium channels and disease. *Biofactors*, **37** (3), 197–205.
407. George, A.L. Jr., (2005) Inherited disorders of voltage-gated sodium channels. *J. Clin. Invest.*, **115** (8), 1990–1999.
408. Conti-Fine, B.M., Milani, M., and Kaminski, H.J. (2006) Myasthenia gravis: past, present, and future. *J. Clin. Invest.*, **116** (11), 2843–2854.
409. Li, C. *et al.* (2007) Spatiotemporal coupling of cAMP transporter to CFTR chloride channel function in the gut epithelia. *Cell*, **131** (5), 940–951.
410. Welsh, M.J. *et al.* (1994) in *Current Topics in Membranes*, Chapter 7 (ed B.G. William), Academic Press, pp. 153–171.
411. Cassel, D. and Pfeuffer, T. (1978) Mechanism of cholera toxin action: covalent modification of the guanyl nucleotide-binding protein of the adenylate cyclase system. *Proc. Natl. Acad. Sci.* U.S.A., **75** (6), 2669–2673.
412. McLachlan, S.M. *et al.* (2007) The link between Graves' disease and Hashimoto's thyroiditis: a role for regulatory T cells. *Endocrinology*, **148** (12), 5724–5733.
413. Effraimidis, G. and Wiersinga, W.M. (2014) Mechanisms in endocrinology: autoimmune thyroid disease: old and new players. *Eur. J. Endocrinol.*, **170** (6), R241–R252.
414. Pinterova, M., Kunes, J., and Zicha, J. (2011) Altered neural and vascular mechanisms in hypertension. *Physiol. Res.*, **60** (3), 381–402.
415. Orlov, S.N., Tremblay, J., and Hamet, P. (1996) cAMP signaling inhibits dihydropyridine-sensitive Ca2+ influx in vascular smooth muscle cells. *Hypertension*, **27**(3, Pt. 2), 774–780.
416. Sitzler, G. *et al.* (1998) Vascular beta-adrenergic receptor adenylyl cyclase system from renin-transgenic hypertensive rats. *Hypertension*, **31** (5), 1157–1165.
417. Homcy, C.J., Vatner, S.F., and Vatner, D.E. (1991) Beta-adrenergic receptor regulation in the heart in pathophysiologic states: abnormal adrenergic responsiveness in cardiac disease. *Annu. Rev. Physiol.*, **53**, 137–159.
418. Weigle, D.S. and Kuijper, J.L. (1996) Obesity genes and the regulation of body fat content. *Bioessays*, **18** (11), 867–874.
419. Schwartz, M.W. *et al.* (1996) Specificity of leptin action on elevated blood glucose levels and hypothalamic neuropeptide Y gene expression in ob/ob mice. *Diabetes*, **45** (4), 531–535.
420. Banks, W.A. *et al.* (2004) Triglycerides induce leptin resistance at the blood-brain barrier. *Diabetes*, **53** (5), 1253–1260.
421. Parks, E.J. (2002) Dietary carbohydrate's effects on lipogenesis and the relationship of lipogenesis to blood insulin and glucose concentrations. *Br. J. Nutr.*, **87** (Suppl. 2), S247–S253.
422. Duman, R.S., Heninger, G.R., and Nestler, E.J. (1997) A molecular and cellular theory of depression. *Arch. Gen. Psychiatry*, **54** (7), 597–606.
423. Chatterjee, S.S. *et al.* (1998) Hyperforin as a possible antidepressant component of hypericum extracts. *Life Sci.*, **63** (6), 499–510.
424. Malberg, J.E. *et al.* (2000) Chronic antidepressant treatment increases neurogenesis in adult rat hippocampus. *J. Neurosci.*, **20** (24), 9104–9110.
425. Shen, L.H., Liao, M.H., and Tseng, Y.C. (2012) Recent advances in imaging of dopaminergic neurons for evaluation of neuropsychiatric disorders. *J. Biomed. Biotechnol.*, **2012**, 259349.
426. Dreyer, M. and Rudiger, H.W. (1988) Genetic defects of human receptor function. *Trends Pharmacol. Sci.*, **9** (3), 98–102.
427. Lee, P.A. and Witchel, S.F. (1997) The influence of estrogen on growth. *Curr. Opin. Pediatr.*, **9** (4), 431–436.
428. Mani, S.K., Blaustein, J.D., and O'Malley, B.W. (1997) Progesterone receptor function from a behavioral perspective. *Horm. Behav.*, **31** (3), 244–255.
429. Schneider, J.S. *et al.* (2003) Progesterone receptors mediate male aggression toward infants. *Proc. Natl. Acad. Sci.* U.S.A., **100** (5), 2951–2956.
430. Lamberts, S.W. (2001) Hereditary glucocorticoid resistance. *Ann. Endocrinol. (Paris)*, **62** (2), 164–167.
431. Raloff, J. (1994) The gender benders. *Sci. News*, **145** (2), 24.
432. Levitt, P.M. and Guralnick, E.S. (1979) *The Cancer Reference Book: Direct and Clear Answers to Everyone's Questions*, Paddington Press, New York; distributed by Grosset & Dunlap, 271 p.

433. Rubin, L.L. and de Sauvage, F.J. (2006) Targeting the Hedgehog pathway in cancer. *Nat. Rev. Drug Discov.*, **5** (12), 1026–1033.

434. Bishop, J.M. (1983) Cellular oncogenes and retroviruses. *Annu. Rev. Biochem.*, **52**, 301–354.

435. Owen, A.J., Pantazis, P., and Antoniades, H.N. (1984) Simian sarcoma virus--transformed cells secrete a mitogen identical to platelet-derived growth factor. *Science*, **225** (4657), 54–56.

436. Lemoine, N.R. (1994) Molecular biology of breast cancer. *Ann. Oncol.*, **5** (Suppl. 4), 31–37.

437. Cheng, L. *et al.* (2012) Molecular pathology of lung cancer: key to personalized medicine. *Mod. Pathol.*, **25** (3), 347–369.

438. Chia, C.Y., Kumari, U., and Casey, P.J. (2014) Breast cancer cell invasion mediated by Galpha12 signaling involves expression of interleukins-6 and -8, and matrix metalloproteinase-2. *J. Mol. Signaling*, **9**, 6.

439. Litzenburger, B.C. and Brown, P.H. (2014) Advances in preventive therapy for estrogen-receptor-negative breast cancer. *Curr. Breast Cancer Rep.*, **6**, 96–109.

440. May, F.E. (2014) Novel drugs that target the estrogen-related receptor alpha: their therapeutic potential in breast cancer. *Cancer Manage. Res.*, **6**, 225–252.

441. Savage-Rumbaugh, E.S. and Lewin, R. (1994) *Kanzi: The Ape at the Brink of the Human Mind*, Wiley.

442. Quinn, W.G. and Gould, J.L. (1979) Nerves and genes. *Nature*, **278** (5699), 23.

443. Jekely, G. (2011) Origin and early evolution of neural circuits for the control of ciliary locomotion. *Proc. Biol. Sci.*, **278** (1707), 914–922.

444. Kandel, E.R. (2012) The molecular biology of memory: cAMP, PKA, CRE, CREB-1, CREB-2, and CPEB. *Mol. Brain*, **5**, 14.

445. Koshland, D.E. Jr., (1977) A response regulator model in a simple sensory system. *Science*, **196** (4294), 1055–1063.

446. Vladimirov, N. and Sourjik, V. (2009) Chemotaxis: how bacteria use memory. *Biol. Chem.*, **390** (11), 1097–1104.

447. Tagkopoulos, I., Liu, Y.C., and Tavazoie, S. (2008) Predictive behavior within microbial genetic networks. *Science*, **320** (5881), 1313–1317.

448. Kandel, E.R. (2013) *Principles of Neural Science*, Chapter 1, 5th edn, McGraw-Hill, New York, 1709 p.

449. Hejnol, A. and Martindale, M.Q. (2008) Acoel development supports a simple planula-like urbilaterian. *Philos. Trans. R. Soc. London Ser. B: Biol. Sci.*, **363** (1496), 1493–1501.

450. Kotikova, E.A. and Raiikova, O.I. (2008) Architectonics of the central nervous system in Acoela, Plathelminthes, and Rotifera. *Zh. Evol. Biokhim. Fiziol.*, **44** (1), 83–93.

451. Cajal, S. (1967) *Nobel Lectures, Physiology or Medicine 1901–1921*, Elsevier Pubishing Company, Amsterdam.

452. Hokfelt, T. *et al.* (1980) Peptidergic neurones. *Nature*, **284** (5756), 515–521.

453. Arbas, E.A., Meinertzhagen, I.A., and Shaw, S.R. (1991) Evolution in nervous systems. *Annu. Rev. Neurosci.*, **14**, 9–38.

454. Brown, A. (2003) Axonal transport of membranous and nonmembranous cargoes: a unified perspective. *J. Cell Biol.*, **160** (6), 817–821.

455. Bliss, T.V. and Lomo, T. (1973) Long-lasting potentiation of synaptic transmission in the dentate area of the anaesthetized rabbit following stimulation of the perforant path. *J. Physiol.*, **232** (2), 331–356.

456. Kandel, E.R. and Schwartz, J.H. (1982) Molecular biology of learning: modulation of transmitter release. *Science*, **218** (4571), 433–443.

457. Kandel, E.R. (2003) in *Nobel Lectures, Physiology or Medicine 1996–2000* (ed H. Jornvall), World Scientific Publishing Co, Singapore.

458. Zakharenko, S.S. *et al.* (2003) Presynaptic BDNF required for a presynaptic but not postsynaptic component of LTP at hippocampal CA1-CA3 synapses. *Neuron*, **39** (6), 975–990.

459. Snyder, S.H. (1977) Opiate receptors and internal opiates. *Sci. Am.*, **236** (3), 44–56.

460. Al-Hasani, R. and Bruchas, M.R. (2011) Molecular mechanisms of opioid receptor-dependent signaling and behavior. *Anesthesiology*, **115** (6), 1363–1381.

461. Cuatrecasas, P. (1971) Insulin--receptor interactions in adipose tissue cells: direct measurement and properties. *Proc. Natl. Acad. Sci.* U.S.A., **68** (6), 1264–1268.

462. Stefano, G.B. *et al.* (2012) Endogenous morphine: up-to-date review 2011. *Folia Biol. (Praha)*, **58** (2), 49–56.

463. Way, B.M. and Lieberman, M.D. (2010) Is there a genetic contribution to cultural differences? Collectivism, individualism and genetic markers of social sensitivity. *Soc. Cogn. Affect. Neurosci.*, **5** (2-3), 203–211.

464. Csaba, G. (1986) Why do hormone receptors arise? *Experientia*, **42** (7), 715–718.

465. Doolittle, R.F. (1995) The multiplicity of domains in proteins. *Annu. Rev. Biochem.*, **64**, 287–314.

466. Trams, E.G. (1981) On the evolution of neurochemical transmission. *Differentiation*, **19** (3), 125–133.

467. Burnstock, G. (1996) Purinoceptors: ontogeny and phylogeny. *Drug Dev. Res.*, **39** (3-4), 204–242.

468. Jukes, T. and Cantor, C. (1969) in *Mammalian Protein Metabolism* (ed H.N. Munro), Academic Press, New York, pp. 21–132.

469. Li, W.-H. and Graur, D. (1991) *Fundamentals of Molecular Evolution*, Sinauer Associates, Sunderland, MA, xv, 284 p.

470. Ohno, S. (1970) *Evolution by Gene Duplication*, Springer-Verlag, London, New York, xv, 160 p.

471. Kasahara, M. (2007) The 2R hypothesis: an update. *Curr. Opin. Immunol.*, **19** (5), 547–552.

472. Hittinger, C.T. and Carroll, S.B. (2007) Gene duplication and the adaptive evolution of a classic genetic switch. *Nature*, **449** (7163), 677–681.

473. Li, H. *et al.* (1996) Emergence of preferred structures in a simple model of protein folding. *Science*, **273** (5275), 666–669.

474. Force, A. *et al.* (1999) Preservation of duplicate genes by complementary, degenerative mutations. *Genetics*, **151** (4), 1531–1545.

475. Thornton, J.W. (2001) Evolution of vertebrate steroid receptors from an ancestral estrogen receptor by ligand exploitation and serial genome expansions. *Proc. Natl. Acad. Sci.* U.S.A., **98** (10), 5671–5676.

476. Freeman, M.E. *et al.* (2000) Prolactin: structure, function, and regulation of secretion. *Physiol. Rev.*, **80** (4), 1523–1631.
477. Fekete, E.M. and Zorrilla, E.P. (2007) Physiology, pharmacology, and therapeutic relevance of urocortins in mammals: ancient CRF paralogs. *Front. Neuroendocrinol.*, **28** (1), 1–27.
478. Bilek, R., Bicikova, M., and Safarik, L. (2011) TRH-like peptides. *Physiol. Res.*, **60** (2), 207–215.
479. Kastin, A.J. (2013) *Handbook of Biologically Active Peptides*, 2nd edn, Elsevier, Amsterdam, liii, 1942 p.
480. Pertseva, M. (1991) The evolution of hormonal signalling systems. *Comp. Biochem. Physiol. A Comp. Physiol.*, **100** (4), 775–787.
481. Christopher, G.K. and Sundermann, C.A. (1995) Isolation and partial characterization of the insulin binding sites of Tetrahymena pyriformis. *Biochem. Biophys. Res. Commun.*, **212** (2), 515–523.
482. Bathgate, R.A. *et al.* (2013) Relaxin family peptides and their receptors. *Physiol. Rev.*, **93** (1), 405–480.
483. Ebendal, T. (1992) Function and evolution in the NGF family and its receptors. *J. Neurosci. Res.*, **32** (4), 461–470.
484. Huang, X. *et al.* (2009) Discovery of a novel prolactin in non-mammalian vertebrates: evolutionary perspectives and its involvement in teleost retina development. *PLoS One*, **4** (7), e6163.
485. Anderson, P.A. and Greenberg, R.M. (2001) Phylogeny of ion channels: clues to structure and function. *Comp. Biochem. Physiol. B: Biochem. Mol. Biol.*, **129** (1), 17–28.
486. Lefkowitz, R.J. *et al.* (1985) Reconstitution of the beta-adrenergic receptor. *J. Membr. Biol.*, **87** (1), 1–12.
487. Strotmann, R. *et al.* (2011) Evolution of GPCR: change and continuity. *Mol. Cell. Endocrinol.*, **331** (2), 170–178.
488. Oguma, T., Asano, K., and Ishizaka, A. (2008) Role of prostaglandin D(2) and its receptors in the pathophysiology of asthma. *Allergol. Int.*, **57** (4), 307–312.
489. Stevens, C.W. (2009) The evolution of vertebrate opioid receptors. *Front. Biosci. (Landmark Edition)*, **14**, 1247–1269.
490. Nordstrom, K.J. *et al.* (2009) The Secretin GPCRs descended from the family of Adhesion GPCRs. *Mol. Biol. Evol.*, **26** (1), 71–84.
491. Cao, J. *et al.* (2009) Evolution of the class C GPCR Venus flytrap modules involved positive selected functional divergence. *BMC Evol. Biol.*, **9**, 67.
492. Schulte, G. (2010) International union of basic and clinical pharmacology. LXXX. The class Frizzled receptors. *Pharmacol. Rev.*, **62** (4), 632–667.
493. Yokoyama, S. and Starmer, W.T. (1992) Phylogeny and evolutionary rates of G protein alpha subunit genes. *J. Mol. Evol.*, **35** (3), 230–238.
494. Oka, Y. *et al.* (2009) The fifth class of Galpha proteins. *Proc. Natl. Acad. Sci.* U.S.A., **106** (5), 1484–1489.
495. Scheeff, E.D. and Bourne, P.E. (2005) Structural evolution of the protein kinase-like superfamily. *PLoS Comput. Biol.*, **1** (5), e49.
496. King, N. and Carroll, S.B. (2001) A receptor tyrosine kinase from choanoflagellates: molecular insights into early animal evolution. *Proc. Natl. Acad. Sci.* U.S.A., **98** (26), 15032–15037.
497. Wouters, M.A. *et al.* (2005) Evolution of distinct EGF domains with specific functions. *Protein Sci.*, **14** (4), 1091–1103.
498. Robinson, D.R., Wu, Y.M., and Lin, S.F. (2000) The protein tyrosine kinase family of the human genome. *Oncogene*, **19** (49), 5548–5557.
499. Bridgham, J.T. *et al.* (2010) Protein evolution by molecular tinkering: diversification of the nuclear receptor superfamily from a ligand-dependent ancestor. *PLoS Biol.*, **8** (10), pii: e1000497. doi: 10.1371/journal.pbio.1000497
500. Bridgham, J.T., Carroll, S.M., and Thornton, J.W. (2006) Evolution of hormone-receptor complexity by molecular exploitation. *Science*, **312** (5770), 97–101.
501. Baker, M.E. (2003) Evolution of adrenal and sex steroid action in vertebrates: a ligand-based mechanism for complexity. *Bioessays*, **25** (4), 396–400.
502. Ollmann, M.M. *et al.* (1998) Interaction of Agouti protein with the melanocortin 1 receptor in vitro and in vivo. *Genes Dev.*, **12** (3), 316–330.

Index